'Raising awareness and knowledge around and creating safe and challenging space t need to do. This challenging, and, by desi that – frameworks, tools, techniques that the eco-phase cycle and find their way in c
Magdalena Nowicka Mook, C

'There are many important books on the climate crisis. Coaches have begun to recognise they have a role but also to begin for many the journey of moving from concern to action. What we have been missing is a guide to that journey and a framework for conceptualising the challenge we face. Ecological and Climate Conscious Coaching provides that guide and that framework. It takes us through the process of understanding how we might recognise and respond to the crisis confronting us all. Providing both concepts and tools for reflection and action it invites to rethink how we approach our role as coaches and our contribution as both professionals and citizens. I have worried about, puzzled over and pursued concerns for the health of the environment for more than fifty years. At times I have been despairing of our lack of action. This book makes me more hopeful that our profession can make a difference if we accept the authors invitation. Some books are important. This one is essential. I implore all coaches to both read and engage with the journey it offers. The world and our future demands no less.'
Professor David A Lane, *Professional Development Foundation*

'In our work, some of the richest material are the things not being talked about. And chief among those is climate change. How can coaches and mentors be concerned for the wellbeing of their clients if they avoid discussing the very environment in which the work takes place? This fascinating dialogue explores ways to open up a vital conversation that underpins the flourishing of us all. By turns philosophical, practical, spiritual and realistic, this fascinating and necessary read offers productive ways out of denial and into action.'
Margaret Heffernan, *CEO and Author,* Wilful Blindness *and* Uncharted

'In the face of the climate emergency and ecological breakdown, every profession is called to reinvent itself. It's wonderful to see the coaching profession taking on the challenge is a profound and wise manner.'
Frederic Laloux, *author of* Reinventing Organizations

'This book is a work of co-creative love. It is enriched by the comradeship between the authors, animated by their long and committed coaching experience and eco-systemic knowledge, and fuelled by their enduring love for the Earth and all who share this miraculous planet. The reader is invited to share a journey with them and the many others who people the pages through seven days of stories, information, poetry, provocations and helpful practices. While the authors do not flinch from laying out the facts this is anything but a book of doom. I found it a deeply absorbing and, dare I say it, enjoyable, gripping read as I was guided through lively conversations rich with images and personal adventure. This book is a vital offering in these times of devastation. It will support us to take our next steps and be our good companion on the long and perilous journey.'

Hetty Einzig, *leadership coach, consultant, author; the Association for Coaching Director of Publications Strategy & Executive Editor of Coaching Perspectives*

'Through following the 7 days of this book's journey you will discover how to commit to spiritual action to create an empowering emerging future. May this book inspire readers so that if our collective grandchildren ask us, 'what were you doing when the earth was being plundered?' you will have an answer.'

Ram S Ramanathan, *Coacharya*

'A brilliant and thought-provoking book that captures the spirit of coaching. I enjoyed the narrative that weaves together multiple perspectives to raise awareness and deepen reflection. This book will definitely help leaders, coaches, OD practitioners, and individuals make a broader systemic impact through a climate-conscious mindset.'

Anjali Nair, *Founder-The HR Studio; Member, Board of Governors – International Association of Coaching; OD Consultant; leadership coach*

'A very timely and forward-looking collective masterpiece reminding us to redefine the role of our profession when it comes to focus on what needs to be sustained. The book inspires us all about how the world could be a better place when, as humans, we start living in harmony with ourselves and with the universe surrounding us. So a great companion in our journey from ego-centric to an eco-centric way of life … '

Dr Riza Kadilar, *EMCC Global President*

'Stop feeling isolated in grief about our climate challenge by joining global co-conveners who offer a poignant and practical companion guide for you to break-through inertia. Learn from direct experience shared by practitioners and become part of a community with a mutual commitment to feel deeply through the stages of curiosity, awareness, information, engagement and ultimately the ecological literacy essential to sustain useful eco-action. In the words of Jane Goodall, beyond passion and fantasy, real hope demands action and engagement. This guide is a timeless resource for every person to generate real hope for each other.'

Janet M. Harvey, *CEO inviteCHANGE*

This book is timely and essential. Drawing the reader in with captivating stories from contributors across the world, the authors approach climate and ecological crisis in a fresh and thoughtful way. Coaching is on a path to reinvention, and this book will quickly become a staple of the new curriculum. A must read!

Dr. Marshall Goldsmith, *Thinkers50 #1 Executive Coach and New York Times bestselling author of* Triggers, Mojo, *and* What Got You Here Won't Get You There

'This is a beautiful, inspiring and empowering treasure of a book, packed not only with resources and tips, but love, joy and hope.'

Liz Hall, *leadership coach and editor of* Coaching at Work *magazine*

'This is a book of insightful journeys over seven days, reflecting on the ecological and climate crisis and the important role coaching can play. Through the creative development of eco-engaged coaching, supervision and training, the authors provide a strategic vision and direction for the future. This exciting book will inform both trainee and experienced coaching practitioners with what they need to know about eco-engaged practice.'

Stephen Palmer *PhD, Faculty of Climate Change and Coaching Psychology, ISCP International Centre for Coaching Psychology Research*

'Many of us wonder what work we can do as coaches and human beings, to address the challenges of climate change. These authors don't wonder, they do something that is quite remarkable and consequential! This brilliant book offers us practical ways to become eco-informed by articulating a framework for regenerative coaching which begins with our self-work, as the key to engaging differently with our clients be they individuals, teams, or broader organizations. Chapter by chapter the authors create an experiential workshop for use, offering provocative questions for our thoughtful consideration while enriching each chapter by sharing their own, to evoke our deepest reflections.'

Pam McLean, *Ph.D., Co-Founder & Chief Knowledge Officer, Hudson Institute of Coaching*

'In an ever-busier world, leaders have to fight harder and harder to escape from short-term thinking. Paradoxically, attending to the farther horizon of climate change and the wider eco-system enhances decision making in the now. This book provides an essential guide for future-focused leadership.'

David Clutterbuck, *Special Ambassador, EMCC*

'This companion guide to evolving coaching practice respectfully invites the reader on a journey around the eco-cycle with the understanding that this journey in non-linear and does not reach an end. I suspect that it is not only the eco-cycle that we will keep coming back to, but also this guide. On the journey, we receive guidance on how to introduce the ecology into our coaching work respectfully without being evangelical or unethical. The guide addresses supervision, eco-engaged coach development and training and ways of resourcing ourselves and our work – so we work from and are supported by 'Source'. The book is written in a way as to invite people to work with others as they journey around the eco-cycle. Eve, Josie, Peter and Ali – as I read, I could hear your voices and feel your vulnerability as you spoke from the heart. You are right it is not a "comfortable" book but a vital resource to urgently have us move individually and collectively from ego-centric to eco-centric. I urge all of us to form hubs and to work through this guide and to action what emerges so that we can confidently answer our great grandchildren when they ask us what it is that we did.'

Colleen Qvist, *CQ Associates (Pty) Ltd, COMENSA Master Coach, COMENSA Master Mentor, National Vice President Coaches and Mentors of South Africa (COMENSA), COMENSA representative In the Group of Signatories to the Joint Global Statement on Climate Change*

'This is literally an invaluable book. That as a species we need to change our behaviour is indisputable. Coaching is a profoundly effective approach to enabling that change.'

Myles Downey, *author of* Enabling Genius, Effective Modern Coaching *and* The Enabling Manager

'There is no doubt in my mind that this book is a labour of love, perseverance, and hard work. While incorporating experiential frameworks and activities, it also includes in-depth questions, scientific information, tools, techniques and reflections, making it a transformational book and agent of change. It impels us to embark on a journey that is both caring for mother earth and food for our souls.'

Dr. Anne Dolly Kuzhimadathil, *Founder & CEO, SALT & Immediate Past President, APAC*

'Ecological and Climate Conscious Coaching is a unique work: creative, spiritual, practical and thought-provoking. The authors highlight the idea that the ecological crisis is an epistemological crisis. In so doing, they demonstrate the value of coaching and the role it plays in transforming hearts and minds: by connecting with what matters, and challenging unhelpful thought patterns. The book is very much a journey of inquiry; one which makes a sincere, enlightened and focussed attempt to save our world. It comes at the right time.'

Dr Caroline Horner, *i-coach academy*

'It is said that we don't just inherit the earth from our ancestors but borrow it from our collective children. This book addresses the most important issue facing our and many other beings, the climate and ecological crisis. It does so by drawing us in through the diverse and enticing stories from contributors worldwide, and through the passion, dialogue and compassion in its pages, taking us on a journey with trusted companions. If there is one book you buy that will touch your very being, I would recommend this.'

Damian Goldvarg, *Leadership Development expert and author. Past President, International Coaching Federation*

'This very timely anthology captures deep and original insights from a collective of global voices to address the future of humanity and our planet. The authors pose a vital question: how can coaching, mentoring, supervising and training make the most beneficial contribution to the great challenges of our time? It is a call for all of us: how can climate and ecology be integrated into our practice? Based on a seven-day 'eco-journey' with four convenors, over 60 contributors ask and answer challenging questions about what it means to be a practising coach today – and in what way coaching needs to transform. What, in fact, are the wider responsibilities of a coach to society? And do practitioners enable those economic, political, and ecological systems that perpetuate themselves, thus dangerously harming the planet? David Lane points out, coaches work with decision makers in industries that have a big impact on the environment. Sir John Whitmore defined the purpose of coaching as raising awareness so that people would take responsibility and ownership for their choices. He apparently did so with full awareness of our impending climate and ecological crisis. If coaching is good at helping people take a larger view of their reality, then what can coach practitioners and supervisors do to help clients to transform their businesses?The book identifies a role for coaches to support leaders in transforming the rules of the game itself. But to shift our perspectives as coaches, we need to look at the 'regeneration' of coach development and training and develop the courage to act. This book asks difficult questions that as practitioners we need to begin to address urgently. As Peter Hawkins says, the coaching profession needs to move from being individual and ego centric to being eco centric.'

Dr Sunny Stout-Rostron, *University of Stellenbosch School of Business*

'This book is so much more than a book. In addition to its deeply informed systemic analysis and rich storytelling, Ecological and Climate-conscious Coaching offers a grounded guide for a revolution in consciousness, rich in conversations and camaraderie, packed with diverse perspectives and practices. This book is not only a must read for all coaching professionals and business and community leaders, it is also a template of transformation for anyone seeking understanding, support and agency in response to the ecological and climate crises. No matter who you are, find yourself a few companions and embark on this inspired interactive journey guided by the wise minds and loving hearts of its authors.'

Sally Gillespie, *PhD, author of* Climate Crisis and Consciousness: Re-imagining our world and ourselves

Ecological and Climate-Conscious Coaching

This book takes you on a seven-day journey with your guides: 66 coaches and thought-leaders from around the world. Through storytelling, poetry and other creative approaches, readers can follow this programme alone or with others and take a practical and empowering look at the impact of the climate emergency on their practice and how they might respond.

Ecological and Climate-Conscious Coaching: A Companion Guide to Evolving Coaching Practice describes methods for adapting your practice while making a livelihood, reframing your work with urgency and action through exploration of the five-stage 'Eco-phase' cycle, moving from 'Eco Curious', 'Eco-Informed', 'Eco-Aware', to 'Eco-Engaged' and 'Eco Active'. Designed to encourage discussion, raise awareness, and increase confidence about stepping into a leadership role, the book explores the difference that coaching can make in the world as a result of greater eco-awareness and systemic understanding.

Featuring powerful stories from around the world, and with a treasure trove of resources and practical tools and methods, supported by reflective and practical exercises, this book will be an inspiring read not only for those involved in coaching, supervision, mentoring and leadership development, but also for leaders.

Alison Whybrow was a globally renowned Coaching Psychologist, Supervisor, Facilitator, and Climate Coaching Alliance (CCA) co-founder. Campaigner, community-builder, and transformational leader, she inspired people to reach their higher purpose in service of a living earth.

Eve Turner is a renowned coach and supervisor globally. Immediate past Chair of professional body, the Association of Professional Executive Coaching and Supervision, she is highly active volunteering, including co-founding the CCA and founding the Global Supervisors' Network.

Josie McLean was an early pioneer of professional coaching in Australia and the President of the International Coaching Federation Australasia. CCA co-founder, she works with organizations evolving adaptive cultures to shape sustainable organizations, communities, and planet.

Peter Hawkins is a global thought leader in Systemic Coaching, Executive Teams and Board Development. He's a leading international consultant, coach, supervisor, researcher, and best-selling author including team coaching, leadership, and culture change.

Ecological and Climate-Conscious Coaching

A Companion Guide to Evolving
Coaching Practice

Edited by
Alison Whybrow, Eve Turner and Josie
McLean, with Peter Hawkins

Routledge
Taylor & Francis Group

LONDON AND NEW YORK

Cover image: © Adobe Stock Images

First published 2023
by Routledge
4 Park Square, Milton Park, Abingdon, Oxon OX14 4RN

and by Routledge
605 Third Avenue, New York, NY 10158

Routledge is an imprint of the Taylor & Francis Group, an informa business

British Library Cataloguing-in-Publication Data
A catalogue record for this book is available from the British Library

Library of Congress Cataloging-in-Publication Data
Names: Whybrow, Alison, 1968- editor.
Title: Ecological and climate-conscious coaching : a companion guide to evolving coaching practice / edited by Alison Whybrow, Eve Turner and Josie McLean, with Peter Hawkins.
Description: Abingdon, Oxon ; New York, NY : Routledge, 2023. |
Includes bibliographical references and index. |
Identifiers: LCCN 2022028693 |
Subjects: LCSH: Personal coaching. | Environmental psychology. |
Climatic changes--Psychological aspects.
Classification: LCC BF637.P36 E296 2023 | DDC 158.3--dc23/eng/20220824
LC record available at https://lccn.loc.gov/2022028693

ISBN: 978-0-367-72198-5 (hbk)
ISBN: 978-0-367-72200-5 (pbk)
ISBN: 978-1-003-15382-5 (ebk)

DOI: 10.4324/9781003153825

Typeset in Times New Roman
by Taylor & Francis Books

Dedicated to future generations with apologies for what we have done to our collective home. And in the hope that together we can make a difference for the future.

and

to Alison, our friend and colleague, and to her supportive and loving family, Gordon, Georgie, and Millie. Alison, your sudden and unexpected death after a short illness, in February 2022, is a loss beyond measure. We hope that your energy and inspiration, your courage and passion, and most of all your love, will encourage others to follow in your big footsteps. You have left a huge imprint, and we hope, in finishing the book without you, that we have served your legacy. You will always be with us, in our hearts and minds.

and

to Stop Ecocide International to support your important global work, and which will receive the author royalties, matched by the publisher, for this book.

Contents

List of illustrations		xv
Poem credits		xvii
Preface from the authors		xviii
Acknowledgements		xix
List of contributors		xx

1	Welcome letter	1
2	An introduction to the ecological and climate crisis	10
3	Stories of Eco-Awakening: Day One Morning	16
4	Awakening in the Coaching Profession: Day One Afternoon	35
5	Eco-Informed: Day Two Morning: 'Listen to the Science'	51
6	Eco-Informed: Day Two Afternoon: Listening to the Earth	69
7	Eco-Aware – Processing our Emotional Responses: Day Three Morning	90
8	Eco-Aware – Shifting our Thinking: Day Three Afternoon	111
9	Eco-Aware – Spiritually Connecting: Day Three Evening	120
10	Eco-Engaged Coaching: Day Four Morning	134
11	Eco-Engaged Coaching: Day Four Afternoon: Stories of transforming our coaching	152
12	Eco-Engaged Supervision: Day Five Morning	170
13	Eco-Engaged Coach Development and Training: Day Five Afternoon	187

14   Eco-Active: Day Six Morning: Impacting the wider world and
      developing the coaching profession   205

15   Eco-Active: Day Six Afternoon: Eco-Active – developing our vision
      and values, and transforming our own business   226

16   Integrating our Learning: Final Day Morning   245

17   Resourcing Ourselves for the Future Journey and Farewell: Final
      Day Afternoon   256

      Glossary   264

      Appendix: Climate Coaching Alliance   269
      References   281
      Index   299

Illustrations

Figures

1.1	Eco-Phase Cycle	4
4.1	Use of coaching skills from a sociological perspective	42
5.1	Planetary boundaries	59
5.2	Responsibility for excess emissions	61
5.3	The world's richest 10 percent are responsible for half of global consumption emissions	62
5.4	Human-Made Mass v Living Biomass, 1900–2025	64
5.5	Percentage of mammals	65
6.1	Deep Time Walk timeframe	76
6.2	Coaching and Supervision for Social Change	81
7.1	Ecological Awareness Cycle	96
8.1	Holarchy illustrating the interdependence of us all	112
10.1	Neil's Wheel (2022)	142
10.2	Examples of Neil's Wheel	144
12.1	Neil's Wheel	177
12.2	The Seven-eyed model	180
14.1	Nine domains of a climate conscious coach	208
15.1	Vision as a cradle for experiments to learn your way forward (McLean, 2020: 72)	231
15.2	Segments representing Hawken's questions	243

Tables

12.1	Changing supervisory questions	176
13.1	Current training	191
13.2	Future training	193
13.3	Awareness of joint professional body statement	193
13.4	Discussing joint professional statement with students	194
15.1	Vision and values audit	233
15.2	Beneficial social and ecological value assessment	242

17.1 Activities designing our way forward   260

Boxes

The seven narratives of denial   97
Leaving his cells behind him   114
He Karakia Timatanga (To open a meeting)   199
Karakia Whakamutunga Tawhito (To close a meeting)   199
Sustainable Development Goals   236
The four sacred gifts   250

Poem Credits

Preface from the authors

This book has been a labour of love. It has also been a challenge to decide what to include and what to leave out. We do hope that you find what is included as inspiring, challenging, enlightening, and joyful to you in your learning, as it has been to ours, as we have researched, practiced, and written it. Six of us started on this journey: Alison, David, Eve, Josie, Peter, and Zoe, but for various reasons we each had to take a back step at times. And while three of us got it over the line – Eve, Josie, and Peter – without Alison's inspiration and dedication to it, before her untimely death, and David and Zoe's input, this would not have happened. Alison's passion is her lasting legacy, shining throughout, while David's understanding and passion for story-telling and Zoe's passion for the changes that are needed to support Earth were central to the book.

As you read this book, and particularly as you read about the setting up of the Climate Coaching Alliance (CCA) – you will notice our passion for living systems and non-conventional structures. This has made an order of authors very difficult, given the conventions of the authors going in order of the amount they have written. Instead, we have unconventionally gone for alphabetical order by first name. Each of the four of us have brought a large and significant but different contribution due to life unfolding differently from how we expected, as we wrote.

Alison stood up to take the lead in the middle of the project, when we needed to make more progress shaping much of the material we already had and bringing her deep spirit and commitment to indigenous spiritual wisdom. Eve led on launching the book project, working with the publisher and doing much of the painstaking background work, as well as providing her significant deep coaching and supervision work in this field and her amazing capacity to network and include voices from around the world. Josie brought her deep experience, developing her own approach to systemic organizational coaching as she worked with sustainability in complex systems and culturally mixed communities – she added her truth-telling to our rich mix. Peter, who has supported the CCA's co-founders since before its inception, brought his long experience of working with these issues and provided the book structure and the ways of making it an experiential workshop and journey, alongside his extensive writing. His voice can be heard throughout the book.

Thank you for joining us on this journey. We are so glad to have you with us.

Acknowledgements

Our thanks go to Janet Harvey, Catherine Carr, and Sam Wells for their generosity in providing critiques and suggestions on the draft manuscript, which have improved the narrative.

To the team at Coaches Rising for their support with the research on coaches and their relationship with climate change and the ecological crisis. This included mailing the survey to coaches subscribed to their conference and providing an additional workshop facility to share experiences. Particular thanks to Laurens van Aarle, Joel Monk, and Ellen Waenink.

To Dr Christiaan Röell, Postdoctoral Teaching Fellow at Roehampton University, UK, for his outstanding work analyzing the survey data and providing suggestions related to the research.

To The Renewal Foundation for its generosity in funding the research and manuscript preparation.

To Fiona Benton for her sterling work getting the book ready for the publishers.

To Alexis O'Brien, our editor at Routledge, for her support through much of the process and Katie Randall, the Editorial Assistant who became our editor getting this into print, answering questions quickly and patiently, and to Susannah Frearson and Heather Evans at Routledge, whose initial support and enthusiasm for the project was much appreciated.

To all our colleagues across the world for their generous contributions both to the book and to our collective and individual learning, including our many CCA friends and volunteers.

To our families for supporting and tolerating us through the many hours of writing and editing!

Contributors

Dr Alison Whybrow was always curious, circling ever outwards and more deeply to understand how the world works. Her early work provided insight into how systems work and how change really happens. Alison's professional foundation was as a chartered occupational psychologist, before diving into counselling and coaching. Here, she worked with colleagues to bring coaching and psychology together. She worked across industries and local and global communities. Co-founder of the CCA, Alison described herself as a mother, sister, author, and gardener. With a strong spiritual practice she lived as simply as possible.

Eve Turner has had three careers. Starting in music, she learned the power of collaboration, the whole being more than the sum of the parts and the inter-relationship between parts, conducting an orchestra. She was a journalist and senior leader at the BBC (British Broadcasting Corporation) before becoming an internal, then external coach and supervisor. Eve has won many awards for her research, writing, coaching, and supervision and alongside a busy practice is hugely active as a writer and volunteer, co-founding the CCA, founding and leading the Global Supervisors' Network, and the immediate past Chair of the coaching and supervision professional body, the Association of Professional Executive Coaching and Supervision.

After a career as a financial analyst and in corporate strategic planning, Dr **Josie McLean** became a coach. Living in Australia, she was recognized for her contributions towards the development of the International Coaching Federation (ICF) Australasia through the ICF President's Award in 2009. Josie completed doctoral studies in organizational sustainability in 2017. Her talents are shared with people in organizations and sustainability practitioners. In addition, she shares systemic practices with coaches. She is an author, speaker, teacher, coach, and catalyst for systemic change. Josie is a co-founder of the CCA. She works for the future of her/our three adult children.

Professor Peter Hawkins has been a life-long spiritual seeker, learning from many traditions, trainings, and from the wider world of nature. His work and life have taken many forms, including being a psychotherapist, researcher, organizational development consultant to a great variety of organizations, teacher, trainer, entrepreneur, professor of leadership, author of many books, coach, supervisor, systemic team coach, spiritual guide, gardener, woodlander, lover, parent, and grandparent. He continues to be a life-long learner.

Dr David Drake is CEO of The Moment Institute (MI; www.themomentinstitute. com) and the founder of Narrative Coaching. His new work, *Integrative Development*, is transforming how we help others to learn and grow. In the MI LAB, graduates are using this work to address the issues of our time. David is a thought leader for the Institute of Coaching at Harvard, and he gave the opening keynote for the first International Coaching Psychology Congress in London. He has authored some 60 publications, including as co-editor of *SAGE Handbook of Coaching* (2017) and as author of *Narrative Coaching: The Definitive Guide to Bringing New Stories to Life* (2018).

Zoe Cohen is a highly experienced accredited coach and supervisor who has worked with senior leaders across almost every sector in the economy. Zoe has been passionate about nature, social justice, and sustainability all her life. She now dedicates most of her energies to activism, including non-violent civil disobedience. She has developed a significant 'voice' and following on LinkedIn for her forthright posting on all aspects of the systemic drivers and consequences of the climate and ecological emergency. Zoe has also been 'catalyzing' her own profession in this regard, in the UK and internationally for the past three years. Zoe initiated and co-wrote an open letter to the coaching profession and professional bodies in July 2019. She also speaks and writes on related topics.

Abena Amponsaa Baafi is a result-oriented project management professional with over eight years' varied experience. She has managed projects on climate change, gender and entrepreneurship in Ghana. She holds an executive master's in development policies and Practices, Master of Science in Environmental Science, Policy and Management and Bachelor's degree in Geography.

Adrian G. Tsukamoto's areas of focus are organizational change management and coaching for leaders who need to stay relevant in an increasingly globalized environment. Adrian specializes in developing strategies to instil and realize behavioural change within organisations. Adrian's professional mission is to support leaders who can represent and articulate their perspectives globally.

Akua Amoa Okyere-Nyako is a gender and climate change expert with over 10 years' experience. She has worked on various climate change projects. She holds a Master of Science degree in Environmental and Energy Management and a Bachelor's degree in Land Economy.

Alice Howard-Vyse is a strategic designer, facilitator and collaboration coach focussed on social-environmental innovation. Through her business, Humanise

This, Alice supports pioneering leaders and their teams to stay the course for making positive change. In early 2020, in response to Australia's bushfire crisis, Alice organized a series of community talking circles to help people who were feeling hopeless and helpless in the face of the disaster to re-find a sense of hope, community, and resilience.

Andra Morosi is a seasoned executive coach, team coach, mentor and supervisor, and founder of International Milestones. She is an active member of the CCA and the founder of the French-speaking CCA community. Her work with leaders, teams, and organizations is aimed at enabling sustainable systemic organizational, cultural, and behavioural evolutions.

Andy Miller. Inspired by Mowgli from *The Jungle Book*, Andy's childhood was spent living in trees to study nature close-up. He wrote and published a bird book that won him the regional science fair at age 11. He was Lead Scientist and Indigenous Liaison with major environmental non-governmental organizations (NGOs) for 20 years. Andy has founded: Indigenous Work Force: youth business coaching; People Climate Earth: family climate coaching; and Climate Earth Consulting: corporate climate coaching.

Anita Sanchez, PhD, Nawat (Aztec) and Mexican American, is a transformational leadership consultant, speaker, coach, and recipient of the 2019 International Latino Book Award for *The Four Sacred Gifts: Indigenous Wisdom for Modern Times*. She bridges indigenous teachings with the latest science to inspire and equip women and men to enjoy meaningful, empowered lives and careers. She is a board member of Bioneers, the Evolutionary Business Council, and the Pachamama Alliance.

Aya Usui, leadership coach, founder of Lucky Gakuen LLC, PCC, is trilingual and internationally educated. Her mission is to accompany leaders to discover their potential and guide their thinking in new directions. Her recent focus is to create "Sustainable Leaders" who are self-sustaining, highly conscious, and carry out action to mitigate or reverse climate change.

Catherine Gorham is a senior practitioner coach and supervisor accredited by the European Mentoring and Coaching Council. Trained in ecotherapy, she specializes in nature as dynamic co-partner and psychological safety, equipping practitioners to integrate Nature as mirror to the inner landscape. She has authored a chapter in *Coaching Supervision Groups: Resourcing Practitioners* (ed. Jo Birch, 2021).

Charmaine Roche, currently engaged in PhD research at Leeds Beckett University, is an executive coach and Coaching Supervision Academy (CSA)-accredited coach supervisor, as well as director of Lifeflowbalance Coaching and Consulting Ltd. She is a compassionate disruptor working with decoloniality as a systemic lens as a social justice change agent and equity and belonging consultant.

Clover Hogan is a climate activist, researcher on eco-anxiety, and the founding executive director of Force of Nature – a youth non-profit mobilizing mindsets for climate action. Clover has worked alongside the world's leading authorities on sustainability, consulted within the boardrooms of Fortune 500 companies, and supported students in over 50 countries to realize their power as change-makers. Her podcast can be found at www.forceofnature.xyz.

Cory McGowan is a leadership and transformational coach based in Minakami, Japan. An American by birth, global citizen, and resident in Japan for over 20 years, his passion is partnering with people to create new possibility in their lives, in particular by combining transformational leadership coaching with outdoor adventures. Learn more at www.adventure-partner.net.

Professor David Lane is a psychologist and coach with a long-standing interest in ethics and environmental health. He has been involved across a number of professional bodies in projects to develop ethical standards of practice and accreditation. His work explores the complexity of global issues and inter-cultural sensitivity and relational ethics.

Denis Opio is an advocate, mentor and coach, and founder of Inspiring star International in Uganda. He is a global youth power leader and member of the CCA and Climate and Biodiversity Coaching Community. Denis enjoys using his skills to empower and inspire youth to become change-makers. He is a graduate from Makerere University with a Bachelor's Degree in Development Studies.

Dr Diana Collett is a consultant/coach dedicated to creating a better global future that works for all. Her interest and expertise lie in how to use power and rank to include diversity, and improving work/life balance, happiness, and productivity by purposefully recrafting screen dependence to achieve significant goals.

Diana Tedoldi, an MSc in Philosophy, holds a degree in Plants Intelligence (University of Florence, Italy), PCC, and is founder of The Nature Coaching Academy, where she teaches how to coach in partnership with nature. She has been working in the corporate environment for over 20 years, focusing on leadership and teamwork development, community-building, and personal and ecological sustainability and well-being. www.naturecoaching.net

Dominique Barbes, MCC, is an executive coach and consultant working one to one and with teams in both French and English. She is also a coach trainer with International Mozaik and a trained supervisor with the CSA. Dominique works and lives in Canada (Québec) and France and is a member of CCA global and its French-speaking community. She is active with coaches, company leaders, and field collectives.

Elsa Valdivielso Martínez is an accredited coach, facilitator, and well-being practitioner. Her work focuses on leadership, career and personal development. She also helps her clients deal with the ethical dilemmas arising from the current ecological crisis. Elsa uses a nature-centred approach and holds her sessions outdoors.

Giles Hutchins is a leadership coach and regenerative business practitioner, drawing on over 25 years' organizational change experience and specializing in learning and organizational development inspired by nature. He is author and co-author of several books and articles, including *Leading by Nature* (2022), *Regenerative Leadership* (2019), and *Future Fit* (2016). His work can be found at www.gileshutchins.com.

Gillian Walter is a coach, coaching supervisor, mentor coach, artist, owner of Inside-Out Coaching, and author of *Choir of Brave Voices: Creative Reflection for a Seasonal Journey of Self-Discovery.* Accredited by the ICF, EMCC, EASC, and CSA, she works with creative, narrative, and somatic methodologies. British-born, she now lives and works in Switzerland with her family and Schnauzer.

Gosia Henderson is a sustainability lead at TrueSustainability Consulting, specializing in human-centred organizational transformation, and a co-founder of Ecollective – a coaching collective with a mission to catalyze our human journey to a regenerative future. She works with leaders and organizations that aim to be a force for good for nature and society and engage every individual and team in their sustainability transformation.

Heather Monro spent her 20s and 30s amassing medals at international level in wilderness running and orienteering. On retirement from full-time competition, she brought her passion for understanding human consciousness and becoming more of ourselves into the coaching realm. As a coach, she sees herself as a ruthlessly compassionate thinking partner for aspirational leaders. Her work is explicitly transformative and informed by vertical development theory. www.brightspacethinking.co.uk

Heather-Jane Gray, RN, FCPHR, ICF MCC, is the chief executive of Synergy Global (www.synergy-global.com). With a strong background in health and wellbeing, Heather-Jane has been an advocate for sustainability for many years. As a mother/grandmother/member of the Climate Reality Leadership Corps, she volunteers in her spare time to support a variety of climate-related organizations and facilitates a range of sustainability webinars and programmes.

Jaime Blakeley-Glover is a coach, team coach, and facilitator who combines his coaching with the benefits of working in nature and core belief in the importance of human development in creating a more sustainable world. Jaime helps to shape the CCA global community as well as being an active member of his local CCA Community and the Global Sustainability pod. He is co-founder of Orientate, a nature-based leadership and team development partner and Ecollective, a coaching collective.

Jan Brause is a professional coach and coach supervisor who is committed to enabling senior leaders, business owners, and individuals to excel in their roles and to inspire those around them. She is curious about engaging her clients in conversations about what it means if we treat the planet as a stakeholder.

Janet Mrenica is certified at the ACC level by the International Coaching Federation, and is an Integral Professional, Climate Change and Trauma Informed Coach. She's also a Fellow Chartered Professional Accountant/Certified Management Accountant (Canada) and Fellow Chartered Public Finance Accountant (UK). As a former Canadian federal government executive, she now coaches and holds space for community conversations grounded in relations systems thinking (i.e. Indigenous ways and Theory U), Circle Way, The Work that Reconnects, and the Art of Hosting. Janet is also co-convenor of the CCA-Canada monthly Circle.

Jaya Bhateja, MCC, is a student of life dedicated to creating tangible results for organizations, teams, and coaches through coaching. Jaya runs a boutique leadership coaching and coach education firm, Abhyudaya, which is a Sanskrit word (deriving from Abhay = Sun; and daya = rise). Jaya has been recognized globally on several occasions, one of them being the Young Leader Award from the ICF in 2019. www.abhyudayacoach.com

Jeanine Bailey is co-founder and director of Empower World, a coaching and professional coach-training organization. Jeanine is an International Coaching Federation MCC Coach, coach mentor, and certified coach supervisor working globally and specifically in the Middle East, Australasia, and the UK. Jeanine passionately supports organizations and individuals to achieve their aspirations and unlock their powerful potential.

Jen Horn plants seeds for reconnection and regeneration as a certified transformational coach with Haraya Coaching, a learning facilitator for sustainability at Ateneo de Manila University, writer, host, and founder of Muni. She completed her MSc in Sustainable Development at the University of Surrey, UK, on understanding the personal motivations of sustainability leaders in the Philippines. More about her work and writing can be found at https://linktr.ee/jenhorn

Jennifer Uchendu is an ecofeminist, sustainability communicator, and founder of sustyvibes, a youth-led organization making sustainability actionable for young people in Nigeria. Jennifer's work supports community-led climate adaptation and advocacy projects that involve women and youths and focuses on exploring eco-emotions in young people through The Eco-anxiety Africa Project. Jennifer holds a Master's degree in Development Studies (Institute of Development Studies, UK) under the prestigious Chevening Scholarship. A 2018 Mandela Washington Fellow, Bill and Melinda Gates GoalKeeper, Jennifer is co-author of the e-book: *A Guide to Business Sustainability in Nigeria.*

Joel DiGirolamo is Vice-President of Research and Data Science at the International Coaching Federation. He is the author of two books and several research papers. Joel holds a Master's degree in industrial and organizational psychology, an MBA, a Bachelor's degree in electrical engineering, and is an associate editor for *Consulting Psychology Journal.*

Jeremy Lewis coaches aspiring leaders in local government, housing, and education. He believes in the power of coaching to help leaders to tackle the climate crisis. In his supervision practice, he transforms how coaches think and feel about themselves, so that they get out of their own way and become the coach that they always imagined they would be. https://growthecoach.com.

Dr John Wood, founding director of Leadership Solutions Global, is a psychologist specializing in leadership development, executive coaching and team-building, and the application of stages of adult development and mindful practices to corporate settings. He is the author of several leadership articles and practices in Australia and overseas and is a faculty member for two international programs.

Jojo Mehta co-founded Stop Ecocide in 2017, alongside barrister and legal pioneer the late Polly Higgins, to support the establishment of ecocide as a crime at the International Criminal Court. As key spokesperson and Executive Director of Stop Ecocide International, Jojo has overseen its growth with teams in 15 countries and websites in nine languages and works with grassroots campaigns, international lawyers, politicians, diplomats, and influencers.

Kanishka Sikri is a writer, theorist, and strategic adviser thinking about violability: the practice that marks certain lives, bodies, and worlds to the possibility of violence. She is currently a PhD Candidate at York University in the UK. Her work can be found at kanishkasikri.com..

Katerina Kanelidou, MCC, is a Leadership and Team Coach based in Greece, founder of Six Steps Ahead, and co-founder of Team Coaching Global Alliance. She is also co-leader of the ICF Team and Group Coaching Community of Practice.

Kenza Khomsi, PhD, is an ICF-ACC credentialed coach, certified mentor, and trainer, with a special interest in children's and young adults' environmental education, aiming to raise awareness about climate change and its various effects on eco-systems. Kenza is an expert in Climate and Air Quality, having membership of the World Meteorology Organisation Study Group for Integrated Health Services and the Study Group on Integrated Urban Services. She is also a co-ordinating lead author of the African integrated assessment of air pollution and climate change led by the Climate and Clean Air Coalition and has a PhD in Climatology and Climate Change.

Kevin Snorf is a coach specializing in leadership, men's work, change initiatives, and embodiment. He draws on backgrounds in mixed martial arts, warrior yoga, acrobatics, ecological design, change management, and adult development to pioneer customized approaches for change leaders to be more influential, impactful, and resourceful in their work.

Lilith Joanna Flanagan, MCC, is the author of *Environmental coaching methodology* and founder of the Evolution Coaching Academy. She integrated her ICF Master Certified Coach credential with MSc Holistic Science and MBA. Lilith has coached individuals and teams in global organizations and NGOs and served in leadership roles in multinational corporations and ICF Chapters.

Lily Seto, MA, PCC, ESIA, EIA, Diploma in Coaching SuperVision, acknowledges that she lives and works on the traditional territories of the W̱SÁNEĆ people, also known as North Saanich BC in Canada. She is honoured to be invited into service and partnership with various Indigenous communities and organizations in British Columbia, where the learning and the connections that she has made have had a profound effect on her.

Liz Hall is a leadership coach and editor of *Coaching at Work* magazine and the author of publications including *Mindful coaching, Coach your team*, and *Coaching in times of crisis and transformation*. A vegan, she lives in southern Spain with various loved ones including a host of rescue animals.

Mark McMordie is CEO of the Conscious Leader and Co-Founder of Caerus Change. He works with CEOs and leaders to build psychologically safe, innovative organizations that leave the world better than they found it and helps leaders to develop their inner and outer capacities for more inclusive, inquiry-based leadership and organization transformation.

Michelle Degroot, MSc, BA, is from the Secwepemc Nation and a member of the Tk'emlups First Nation. She currently resides on the ancestral unceded territory of the Tsleil-Waututh Nation. Michelle works for the First Nation Health Authority and has been in the field of Indigenous health and wellness for two decades. Her interest in coaching is to support others to find their solutions regardless of their field, sector, or area of interest. Her approach to lifelong learning supports her inquisitive nature and her appreciation when others share their stories and experiences with her.

Musa Nxumalo is a cross-pollinator whose 2020 book reflects that: *I am a Cross-Pollinator: Afrikologist Thought Experiments for Leaders Navigating from Ego-Ism to Eco-Ism*. His reason for being, as a coach and a facilitator, is to help leaders to become more fruitful in what they do using biomimicry principles.

Neil Scotton supports people, organizations, and communities, taking on the challenge of real, positive, systemic change. He coaches, writes, speaks, walks, grows vegetables, plays the guitar, does tai chi, hugs his family, and gets a healthy slice of humble pie as a citizen supporting a charity helping disadvantaged people.

Paula Downey is a partner at Downey Youell Associates and co-creator of CultureWork®, a living-systems approach to organization, culture, and change. She has a Masters in Responsibility and Business Practice from the University of Bath and writes and speaks on the realities and opportunities of whole-system change. www.culturework.ie

Rashmi Shetty, PCC, is an international award-winner for scripting and narrating radio documentaries. Her approach of coaching for social impact places the climate as an important stakeholder. She primarily coaches women leaders transitioning to senior leadership roles. She co-leads CCA Asia and also has her own podcast.

Reshma Aziz Khan is Kenyan and grew up in Nairobi, the 'Green City in the Sun'. Reshma set up K'enso Consulting to support social impact leaders in their conscious leadership, through her leadership coaching, team effectiveness, and inclusion facilitation work. Reshma lives a low-waste life in Nairobi with her husband and three dogs, making all her own body products and growing much of her own food, in her commitment to doing better for the planet.

Rita Symons is a coach, mentor, supervisor, and leadership development specialist, combining work in these areas with her role as a global citizen, playing a small part in positive action. She does most of her work in the public sector, having over twenty years of previous experience in the UK's National Health Service, ten of which were at board level.

Rob Hopkins is the founder of the Transition movement, author of several books, including *The Transition Handbook, The Power of Just Doing Stuff*, and most recently *From What Is to What If*. He presents the *From What If to What Next* podcast and is a director of the UK's first 100% community-owned brewery. He lectures and writes widely on community-led change and imagination. www.robhopkins.net.

Roselyne Lécuyer is a member of ICF France and the Climate Coaching Alliance Francophone. Her background is in human resources management and development. Working with leading international companies, Roselyne has contributed to the development of talents and the transformation of skills related to business, retail, and luxury service for internal audiences and distributors. Roselyne has worked across all continents and lived in Asia and Africa, where she also worked for an NGO in agroforestry and research for sustainable development.

Sally Gillespie, PhD, writes, lectures, and facilitates workshops on climate psychology and ecopsychology. Her book *Climate Crisis and Consciousness: Reimagining Our World and Ourselves* explores the psychological challenges and developmental processes of climate engagement for individuals and societies.

Samantha Suppiah is a South East Asian design strategist for sustainability and regeneration, weaving Global South regenerative dialogue, leadership, and design. With a background in mechanical engineering, building physics, architecture, and urbanism in Europe, Samantha has a broad global perspective that qualifies her as a systems navigator, and creative disruptor.

Stephan Ulrich works for the United Nations in Vietnam. He uses coaching for his team and colleagues in his role as manager as well as outside of work to support individuals in the environmental space to enhance their impact.

Tabitha Jayne is a Professional Certified Coach with the ICF and the founder and director of Earthself, which helps leaders and teams to create resilient life-sustaining organizations. Tabitha has trained more than 100 coaches to work with nature and planet Earth and is the author of *The Nature Process* and *Nature Embodied*.

Tensei Yoshida, MCC, is the founder of Mindfulness Based Coach Camp and an organizational change coach. He specializes in coaching in leadership development and practice for managers in major corporations and entrepreneurs, further improving communication in organizations. Tensei also focuses on designing and training Sustainable Development Goals in corporations, local government, and schools.

Tina Joanes, PhD, is an environmental psychologist and coach. She researches and teaches at Justus Liebig University Giessen, Germany, and works mostly with students in her coaching practice. Her research focuses on explaining and changing environmentally relevant behaviour.

Tyson Yunkaporta is an author, academic, educator, indigenous thinker, maker (traditional wood carving), arts critic, researcher, and poet. His book *Sand Talk – How Indigenous Thinking Can Save the World* looks at global systems from an indigenous perspective. From the Apalech clan in north Queensland, Tyson now lives in Melbourne and has adoptive and community/cultural ties across Australia, from western New South Wales to Perth, Western Australia.

Vaishnavi Viswanathan is an entrepreneur who owns two firms: one working with organizations and the other with and for nature. She has been a leadership skills facilitator for over 20 years and an expressive arts therapy practitioner and life coach for over a decade. Vaishnavi is also a certified forest therapy practitioner and founded Nature Connections India in 2018 with the aim of enabling people and organizations to get closer to nature.

Winfred Nelson has over 20 years' experience in development planning and climate change. He has pioneered a mentoring and coaching programme for climate change in Ghana. Winfred holds an Master's degree in Environmental Studies and a Bachelor's degree in Social Sciences. He works for the National Development Planning Commission in Ghana as Chief Analyst.

Chapter 1

Welcome letter

Thank you for being one of the many participants joining us on this very important collective journey. We: Josie, Peter, Eve, and Alison, are looking forward to sharing it and our collective past journeys with you. As you might know, we have each been exploring this broad and deep topic for a long time. We have also been exploring it collectively within the Climate Coaching Alliance since 2019. This journey includes over 60 different contributors. We are bringing them together here as a clarion call for transformation to our lives and work - for a better future for humanity and the more than human world.

Humanity is in peril, and we need to change our ways. Specifically, this book, this journey, seeks to help us to reveal to ourselves why, and in what ways, we need to transform our lives and our professional coaching, supervision, mentoring, and leadership development practices.

The journey is in part a workshop, where we will hear from coaches and thought leaders around the world, about their experiences and practices in moving from ego-centric to eco-centric coaching. Together we invite you to engage in many experiential and reflective exercises and experiments. Throughout the Days and Chapters, you will find *Exercise*, which is where the talking ends and an experiential activity starts. Some of these you can do alone, and some are ideally done with others. If possible, you might wish to think about convening a small group of colleagues or perhaps a group of friends, neighbours, or others in your community, so they can undertake this journey and work through parts of this workshop with you. In this way, you will enhance your learning, which occurs most easily in a social setting, and invite others into the transformational change that is now required. If you have access to reliable technology, you could set up a group using a video or audio system. It really does not matter if everyone is a coach, a mentor, a supervisor. This book is for everyone, and anyone can join in and benefit from the work.

We include many resources such as practical frameworks, tools, techniques, and reflective exercises to support ourselves to work with the climate and ecology in our practice, our businesses, and our communities. Some of these resources were already available before the writing began and have become more prominent in the current context. Many were discovered through joyful connections as we have

DOI: 10.4324/9781003153825-1

explored our collective crisis with many colleagues. Others have evolved in the process of writing together.

And one word of explanation. We use the word 'convener', conveying the sense of calling us together, when one of the four of us – Alison, Eve, Josie, or Peter – is offering explorations, exercises or connections between contributions. We identify ourselves only when we write specific elements.

The journey is also a collective and collaborative inquiry for none of us alone has the answers to how we address the enormous challenges of climate change, ecological crisis, social inequality, and loss of bio-diversity and how we heal the relationship between the human and the 'more than human world'. The collaborative inquiry will require all of us to walk together, with an open mind, open heart and open will (Scharmer and Kaufer, 2013), to the learning edge to discover collectively what is required.

We will all need to engage in unlearning, as well as reframing and catalyzing thinking and action. Whatever our starting point, we aim to take each other to a place of greater understanding, confidence, and agency to discover, create, and shape how we respond in our practice, as a citizen and across every system we engage with.

The depth and scale of what we are facing requires us to look fully at the situation and disrupt our normal ways of seeing and doing and therefore being. We hope that this is, at times, an uncomfortable journey and read. If it is not, then we have failed in the disruption that is required. We also hope that this book is joyful, loving, and inspiring. If it is not, we have failed to bring the depth of abundance and love that will carry us forward.

The journey is also a pilgrimage in and through nature, where we will be recognizing that we are the youngest child of creation and exploring how we can learn from the many connected aspects of the 'more than human' world, about what is required of us and what they can teach us.

What does it mean to show up on this journey together as pilgrims? We can all learn from the history of pilgrimages across different cultures.

The first requirement is to take off our privilege. This starts with our expensive clothes, designer brand suits and dresses, jewellery: 'make-up' that attempts to mark us out as 'important' or 'successful' and wear ordinary clothes that show we are the same as others. Whatever your status, those who go on the Hajj to Mecca or become a Sanyasi in India, leave their old clothes behind, along with their roles and titles and become humble everyday members of humanity.

It is not just our adornments we need to leave behind but our past successes, titles and status.

Then we need to let go of attachment to beliefs and habits that keep us imprisoned in our old ways of being. Just remember all the above will be waiting for you when you get back and you will have the choice of which ones you want to put back on or step into, and which ones you now want to give away or compost.

The second requirement is that we need to rely on others. In the 1960s Satish Kumar, when he went on his pilgrimage for peace from India to the four capital cities whose countries had nuclear weapons at the time, took no food or money, so he had to rely on what he was given each day by the people of the countries he walked through.

The third requirement is to bring all of ourselves. Not just our intellectual curiosity, our notebook waiting to be filled, but the fullness of our being – our feelings, hopes, dreams, failures, grief, shame, traumas, hearts, and bodies.

The final requirement is to come as a seeker. 'Seeking for what?' you might ask. Well not a pot of gold at the end of the rainbow. Nor for easily packaged answers that you can turn into a training programme which you can sell. Nor for simple competencies and certificates you can add to your long CV.

The seeking is for something much more and much less than this. Seeking responses to questions of the heart such as:

- What can I best contribute that the world of tomorrow needs?
- What would make my life truly worthwhile?
- What is the best legacy I can leave for those who come after me?
- What do future generations need from me and others right now?
- How can I and my work give more back to our wider ecology than we take?
- How can I make a positive difference and who can I partner with?

Remember your seeking is not just for you alone; you carry with you the needs and questions of your family, those you coach, the organizations you work for, your communities and the 'more than human' world.

We welcome all seekers.

We would love you to use this work as a trusted companion, a friend, a place where you can, at any time, find inspiration, insight, and practical ideas to enable you to respond to yours and your clients' own inner and outer worlds, as we step forward together into a planet whose life support systems are deeply damaged. In coming together, in recognizing this is 'all of us', there is a sense of hope.

We invite you to join us on this journey, building from the words of Chris Johnston (2021), to explore how we live a life that we have a heart in. And to look closely at the story that happens through us.

A guide for your journey and through this book

This book is a guide in our process of unfolding to what it means to be a person living in the first half of what is being defined as a pivotal decade, the early 2020s. What does it mean to be a coach practicing now? Learning is a constant and consistent feature of all living organisms (Bateson, 1972; Maturana and Varela, 1987). As the environment within which a living organism is embedded changes, so the organism learns from its environment and changes with reference to its identity (or DNA). The organism must be able to pay attention and listen.

You are developing the skills to bring the topic appropriately into every aspect of your work

You are taking responsibility to influence and create change through your wider profession and all your stakeholders

Environmental Social Economic Responsibility

You are emotionally working through your various reactions when facing the impact of human actions, as well as beginning to shift your mindsets and beliefs.

You are open to discovering more about the environmental, social and economic crisis and your own part within it

You are looking at the data and science and listening to Nature, to find out and understand what is happening and what is needed.

Figure 1.1 Eco-Phase Cycle

For those of us brought up with distracted and disconnected western, consumeristic lifestyles, we have found many ways not to pay attention. This book offers, perhaps, a bridge or path back to what we have collectively lost and forgotten.

The journey and workshop, as well as the book structure, will follow the Eco-Phase Cycle, initially developed by Peter Hawkins and then further co-developed by him with the Climate Coaching Alliance membership, in preparation for the first 24-hour global conversation in March 2020. That conversation helped us all to realize the power of this simple model to enable the climate and ecological crisis conversation to become one of inclusion rather than polarization. It also helped us to realize that we are all constantly moving in a cycle that might be a spiral of constant movement, not unlike the living systems that we exist within and are. It also emphasizes how the ecological crisis is a symptom, not a cause, and is intrinsically connected to issues of social inequity, economics, culture, and conflict.

The Eco-Phase Cycle provides a map, of sorts, for the journey and the book. The book is broken into seven days, each mapping onto one of the Eco-Phase Cycles, with two days for the Eco-Engaged phase which addresses how we transform our work as coaches, supervisors, and trainers.

Day One: Eco-Awakening - the journey begins

We will begin Day One by arriving and coming into community. We will hear from each of the workshop leaders as well as several of the participants, their personal story of awakening to the ecological challenges and teachings in which all our work is embedded. Everyone will be invited to create and share their story of how they awakened to the wider 'more than human' world and the ecological crisis in this precious planet we share with so many other sentient beings.

In the afternoon of Day One we will step beyond our own awakening to how coaching itself is awakening to the wider social and ecological challenges of our shared Earth.

We will consider how we overcome our weddedness to the current systems, and what deeper awareness we might need to craft and hone, to support coaching to awaken its potential increased contribution more fully to the challenges of our time. We will also look at the places and practices where coaching is already transforming and the difference this is creating.

Day Two: Eco-Informed

Our inquiry on Day Two starts with exploring what we individually and collectively need to know and hold in mind to be appropriately ecologically informed. This is the requisite knowledge we need to have a strong foundation, when the external winds of social media, political posturing, company greenwashing, and client denial blow around and through us. We start by following Greta Thunberg's injunction to listen to the science. We will then consider how we join the dots of information and data, and how we might find a balance between those who try and convince us that the ecological crisis is not as bad as we think, or that human ingenuity will solve it, or, from the other direction, it is all too late, and we might as well give up doing anything to remedy the situation.

But science and data are only two ways of being informed and of knowing. In the afternoon of Day Two we will explore becoming Eco-Informed in experiential and embodied ways: Learning from indigenous traditions, learning from attuning to nature, learning from re-walking the 4.6 billion years of this Earth's life.

In this process we re-ground, discover insights and wisdom cognitively, somatically and in our souls. This section is about a new knowing that cannot be unknown, and it is designed to be felt. We hope to deepen the knowing so that we grow a new way of attending and engaging with the crisis we face and the abundance of life.

Day Three: Eco-Aware

During Day Three we will explore what we individually and collectively need to work through and transform in and between us at three levels: (i) the emotional reactions; (ii) cognitive mindsets through which we experience the world; and (iii) the spiritual ways in which we experience connection.

Emotionally working through the various reactions, when facing what we humans have done, is deep work. We are joining the dots of our own story with the wider global story. We hold this unfolding with love and compassion because we are each in it and moving through it. Each of us might travel through worry and fear, for the lives of our children and grandchildren; we might experience deep grief for the loss of so many species we shared this planet with; and guilt and shame that we have been part of the ecocide; anger and rage at those with even more power than us for being wilfully blind and ego-centric; our own overwhelm and denial. We are dismantling denial in and between ourselves as we unfold to what is possible when we stand in truth and interdependence.

Gregory Bateson (1972) posited that the ecological crisis was fundamentally an epistemological crisis, rooted in the ways of thinking originating from Newtonian atomistic thinking and the white western industrial revolution (see Glossary pages 264–268 for an exploration of these terms). As we move through this day, we will start to explore the nature of the transformation required in our cultural paradigms and how we move from atomistic objectifying ways of splitting the world through systems thinking, to systemic awareness, eco-systemic and participatory consciousness, realizing a sense of interbeing. We will develop the awareness of our relationship with the wider Earth and the guidance and alignment we might receive about what is needed.

Day Four: Eco-Engaged coaching

On Day Four we will focus on developing the skills and methods to attend to the climate and ecological crisis in our coaching and mentoring, and the alignment towards a flourishing Earth, society and community in every coaching relationship. Many, particularly the pragmatist learners, might be asking: 'this is all very well to learn the science and explore our feelings about it, but how can we bring this into our coaching?'. Those of you who identify with this will be pleased to hear that Day Four is about becoming practical and our key inquiry question for this day is: **'What we can do as coaches and in our coaching that will make the best beneficial contribution to the great challenges of our time.'**

Day Five: Eco-Engaged supervision and training

On Day Five we explore how supervision can be eco-centric and a place for developing the eco-systemic capacity of each of us and our work.

In the afternoon we will inquire together how an eco-centric approach can be embedded into coach training at all levels and offer some examples from those who have already been doing this.

Day Six: Eco-Active

On this sixth and penultimate day we will look at what each of us can do beyond our coaching work, in every aspect of our lives. We will look at what we are each directly responsible for and can change by ourselves. Secondly, we will explore what we are indirectly contributing to, and how we can help change this in partnership with others. What we are participating in can influence others to change. Here, we really engage with the work of rethreading our lives, examining how the climate and ecological crisis is present in every one of our behaviours and habits. Rethreading takes time and attention. Together we can catalyze and accelerate this process with ideas and possibilities, courage and confidence we receive from each other, that would otherwise elude us.

Day Seven: Our final day

In the previous six days as we have journeyed together, we have been right round the Eco-Phase Cycle, but we have not come to a place of completion, as this is a cycle not a linear journey. We will all need to continue to cycle many times around and through all these stages, in the hope that each time we will be able to do so in a deeper and more inclusive way. Learning is not linear, and each phase informs each other phase in a beautiful emergent and complex way. Even before you arrive for Day One, you each will have travelled many of these paths, individually, collectively, retracing, looping back, looking again and each time, seeing something different, learning something more richly and shifting more fully. We are still on these pathways, now alongside and in recognition of each other. We do not, and cannot, do this journey alone, and on our final day we will explore how we can resource ourselves, ask for help from fellow travellers and from the 'more than human' world.

We hope that this book becomes a well-used and dog-eared companion that continues to give generously over the years to come.

Before you join the journey on Day One, we invite you to pause and answer a few reflective questions.

Exercise

- What do you really hope for as a result of this journey and reading some or all of this book?
- What questions are you bringing with you?
- What do you believe you need to leave behind to go on this journey?
- What is the most important gift you will bring to the community that you share this journey with?

We also invite you to write a letter to future generations or our collective grandchildren about your hopes and fears for the future and what you did to make a difference.

*

As we complete our pre-work, and connect with the programme ahead, we share a piece by Drew Dellinger, the American poet and activist, with thanks to him for his generosity in allowing us to use it. Drew is internationally renowned and a powerful catalyst in shifting global consciousness. We recommend a short video made of this poem by Bioneers to music and pictures, narrated by the poet and available on YouTube (Dellinger, 2022) to capture its full beauty.

hieroglyphic stairway

it's 3:23 in the morning
and I'm awake
because my great great grandchildren
won't let me sleep
my great great grandchildren
ask me in dreams
what did you do while the planet was plundered?
what did you do when the earth was unraveling?

surely you did something
when the seasons started failing?

as the mammals, reptiles, birds were all dying?

did you fill the streets with protest
when democracy was stolen?

what did you do
once
you
knew?

I'm riding home on the Colma train
I've got the voice of the milky way in my dreams

I have teams of scientists
feeding me data daily
and pleading I immediately
turn it into poetry

I want just this consciousness reached
by people in range of secret frequencies
contained in my speech

I am the desirous earth
equidistant to the underworld
and the flesh of stars

I am everything already lost

the moment the universe turns transparent
and all the light shoots through the cosmos

I use words to instigate silence

I'm a hieroglyphic stairway
in a buried Mayan city

suddenly exposed by a hurricane

a satellite circling earth
finding dinosaur bones
in the Gobi desert
I am telescopes that see back in time

I am the precession of the equinoxes,
the magnetism of the spiralling sea

I'm riding home on the Colma train
with the voice of the milky way in my dreams

I am myths where violets blossom from blood
like dying and rising gods

I'm the boundary of time
soul encountering soul
and tongues of fire

it's 3:23 in the morning
and I can't sleep
because my great great grandchildren
ask me in dreams
what did you do while the earth was unraveling?

I want just this consciousness reached
by people in range of secret frequencies
contained in my speech

(Drew Dellinger, 2003)

Thank you for joining. We are glad to have you with us,

Eve, Peter, Josie, and Alison.
P.S. We enclose some pre-reading to set the scene for Day One, which is: *An introduction to the ecological and climate crisis and the role coaching needs to play.*
There is a glossary of key terms on page 264.

Chapter 2

An introduction to the ecological and climate crisis

A provocation

There is an extraordinary plaque at Borgarfjordur, in western Iceland that commemorates Okjökull, the first of the country's glaciers to be completely lost to climate change. It was declared 'dead' in 2014, when it was no longer thick enough to flow across the landscape, as it had done for centuries. Framed as 'A letter to the future', the plaque reads (in Icelandic and English):

> In the next 200 years all our glaciers are expected to follow the same path. This monument is to acknowledge that we know what is happening and what needs to be done. Only you know if we did it.

Along with the date, the memorial carries the words '415ppm CO_2'. In January 2022 atmospheric carbon dioxide was measured at 418 parts per million, higher than at any point since humans have lived on Earth (Global Monitoring Laboratory, 2022).

What does it mean when parts of the body that enable life, shut down and die?

The living system of Earth operates in a time frame that is beyond our lived human consciousness. That does not mean it is not alive. Many of us have forgotten or diminished the importance of Earth as a body that sustains the life of all who know this planet as home. Many of us have forgotten what it means to live in a truly sustainable way. We have forgotten we are co-evolving with all living beings and living systems with which we share this precious Earth.

Forgetting started to happen many hundreds, if not thousands, of years ago. That sounds like a really long time. In Earth time, though, it is merely the blink of an eye. We could be considered merely an experiment, a test and a pilot study. Are humans the Earth's own artificial intelligence experiment?

If so, we might be an experiment that is inviting Earth to close us down, creating imbalances in Earth's living cycles that will lead to a violent period of rebalancing.

Now we are seeing Earth change in a timeframe that makes sense in the lived experience of the human lifespan. When things move this quickly, we are called to

DOI: 10.4324/9781003153825-2

pay attention. Earth, as we have known her for millennia, is now calling for the emergency team.

This journey is about remembering, awakening and acting with this new awareness in our lives and our practice.

A line from *The Overstory* (Powers, 2018) gives words to the felt sense of what we might experience when we look through an individual lens '…all trying to bail out the ocean of capitalism with an acorn cap…' (p. 300). And yet, what if this were a collective story of a great metanoia or turning point? When we come together, combining our resourcefulness, creativity, love, and courage, what might then emerge? As Sally Gillespie notes: 'We can only step up to collective crisis through collective means and practices' (Gillespie, 2020: XX).

The current state

We are stealing the future, selling it to the present and calling it GDP (Hawken, 2017).

We have been stealing the future for a long time and we are seeing the consequences of that in significant and irretrievable ways. As we write, our world news has been full of items about floods in China, Germany, and Japan; heatwaves, cyclones, and temperatures previously unrecorded in North America and Siberian permafrost melting to name but a few headlines, in amongst daily updates about the Covid-19 pandemic and war.

If we pause, and pay attention, it is very quickly and easily overwhelming.

Amid the noise, we need to stop and discover the patterns sitting behind this apocalyptic emergency and explore how we might respond and consider what is ours to do.

To understand what is happening in the world on a global scale, we will, on Day Two of the journey, learn from the hard science, and review the International Panel on Climate Change and the COP26 reports. We will look at what this means for me, you, us and for our one planet. However, the abstract nature of science can objectify the message, and to bring it alive we need stories. As Gregory Bateson (1972) pointed out, the only language that can reflect the subtly and complexity of the interconnection of living relationships is the language of story.

We are in the age of the Anthropocene, the sixth mass extinction our planet has witnessed and the first to be brought about by just one species, so called *Homo sapiens*. Just like all other great extinctions, the dynamic relationships of living systems means that we will not have the final say in what happens next. But what we choose to do next shapes the critical context of the next act.

We are at the moment where the *Great Turning* (Macy and Johnstone, 2012) is possible. However, we are yet to create the required turning in human consciousness, ways of being and ways of living that are required, and it is not guaranteed that we will avoid ecological and social collapse.

Why coaches and why coaching?

When Sir John Whitmore defined the purpose of coaching as raising awareness so that people would take responsibility and ownership for their choices, he did so with his own full awareness of our impending ecological crisis, wondering whether 'coaching has emerged to be a midwife for a new era?' (Whitmore, 2017: 22). The call for coaching to transform comes from others too: 'Coaching needs to become part of …. The shift in collective human consciousness that our collective ecology is requiring from us as a species' (Hawkins, 2015: 45–46). Palmer (2015) draws attention to the importance of our natural environment for our well-being, shaping eco-psychology approaches within coaching practice. Stansfield (in Einzig, 2017: 4) writes: 'The role of coaches in the 21st century is changing significantly; coaches need to be part activists, part counsellors and part change agents'. Chris Johnstone asks: 'What would it be like if the story of the great turning happened through [coaching] as a profession?' (Johnstone, 2021).

The way of life many of us have built, is creating most of the pain, but at the same time protects those of us with privilege from feeling the pain or seeing the reality of the impact of the choices we make. This dream world provides a sense of safety, as we continue the habits of being and doing that wreck life on Earth. The father of British venture capitalism, Sir Ronald Cohen (2020), stated that our economic system is responsible for the environmental degradation we see around us. It could not be clearer. It is a truth our neo liberal and neo-conservative systems are not designed to address.

Coaching is good at helping people take a larger view of their reality and what we are here for. Without alignment to a flourishing community, society, and ecology, we might be helping people to be highly effective, but in the wrong direction. Coaches could help themselves and others to revisit some of the decisions that might be leading us into this place of collapse. Coaches can help clients and their sponsors find their part to play.

How coaches can step up to the global challenges

There has been a growing call for coaching and coaching psychology to raise their game and there is acknowledgement of the role we might play (Brick and Van der Linden, 2018; Hayhoe, 2019; Cohen, Aspey and Whybrow, 2019; Hawkins and Turner, 2020). The profession is beginning to address the challenge, with, for example, a revised *Global Code of Ethics* including social and environmental impact within the realm of coaching focus (2021), underpinned by a *Joint Global Statement* signed by 11 professional coaching, mentoring, psychology, and supervision bodies (Joint Global Statement on Climate Change, 2020).

How are we equipping ourselves to reach into this new landscape for coaching? How are we meeting the possibly unspoken needs of our clients and their businesses, communities and employees? How are we meeting the needs of our own communities and bringing coaching home to support transformation where we are?

The self-inflicted wound is much deeper and the required response more fundamental than a simple shift to low carbon choices. It requires a fundamental rethink of our way of living, and a move away from the addiction to more and more consumption where each year we use more resources than the Earth can replenish in the same period (in 2022 we had used a year's worth of resources by 28 July). This is not within easy reach especially when we as coaches are stuck playing the same game. It builds on. Einzig's work: 'One of the questions I sought to explore with this book is the potential for coaching to support leaders in transforming the rules of the game itself' (2017: 2).

Shifting our lives and our practice in a way that responds effectively to our existential emergency is neither easy, nor simple. We need to clearly see the water we are swimming in and that requires new lenses providing varifocal vision (Hawkins and Turner, 2020, p. 23), new instruments, a different level of attention and an expansion of our emotional, cognitive, and spiritual capacities.

As Scharmer and Kaufer (2013) note, the system needs to see and sense itself in order for change to happen. Covid-19 has shown us how we are all vulnerable and inextricably inter-connected. As coaches, we need to hold the door of that sensing and awareness open long enough for ourselves and our coachees to step through. Dialogue, action through dialogue, and compassion are healing. Perhaps coaching now needs to step fully into that space of healing, to allow the system to sense itself without turning away.

So, the possibility of coaching and coaches to play an important role is here. The need is here, and the intention has been signalled across the profession.

Now individually and collectively we need to build pathways from where we are today to transform towards restorative approaches that align with a life-giving planet.

Taking stock of where coaching is currently

If coaching holds part of the answer, where are we now as a profession, as practitioners?

We carried out two online global surveys between October 2020 and January 2021. Among those signed up to the Coaches Rising 2020 summit, 351 out of the 353 coaches and aligned professionals that completed our coaches' surveys reported concern about the state of the environment (Hawkins, Turner, and Roell, 2022). In both surveys, around 90 percent of respondents reported being always or often concerned. People shared insight into this ark of concern, with a consistent or near consistent presence of concern, a love of nature, a minute noticing, a macro awareness, and in the fear and despair, and occasional hope.

> I am generally concerned about our climate and the environment and can also get sucked into life in general and lose sight of what is important.

The planet will exit us because we have destroyed the conditions for our own survival. I think that`s sad. On a nearer-term scale, it`s the cause of enormous suffering.

I am very aware of the effects of climate change. At the same time, I believe in Nature's regenerative capabilities. Yes, we as humans influence the planet's climate and need to do all we can to reduce that effect. At the same time let's not underestimate what Nature herself is capable of.

Within this concern, there are answers. There are people who have been exploring this issue for much, if not all, of their adult lives. Whilst some respondents felt highly confident bringing in climate and environmental issues into their practice with grace and skill, most did not. When the coachee is not directly mentioning issues of climate and ecology, coaches are less confident that they can hold a wider frame of awareness. For many there was a sense of stuckness about what it looks like to bring the ecology and climate to the fore as the following quote illustrates:

My biggest concern with leading with environmental issues is alienating my clientele. Most people have their head in the sand when it comes to climate change and the impact they personally are having on the planet. It feels very confrontational to bring it front and center. I'd feel more confident if I could develop a framework that was inclusive of these issues without beating someone over the head with it.

Some have taken direct action to change who they work with because of their values around the climate and ecological crisis:

The hardest and the easiest was to 'fire' a client who insisted on travel and in-person sessions when we could create an effective online environment. What made it hard, was letting go of a group of participants who were greatly helped and appreciative of the work we were doing together.

Interestingly, most respondents have changed many aspects of their personal lives as a result of climate and environmental concerns. It is the practice changes that can leave us stuck.

Like us, many are waking up to the possibility and joy that they might be able to make a shift in their practice that will better contribute to a flourishing Earth:

I guess I haven't opened my eyes as to how to make shifts in my coaching practice in relation to climate change and it feels exciting to think about it.

Asking around purpose in the coming decade, the themes of moving from ego to eco in practice and ways of being, raising individual and collective consciousness, developing a sense of purpose and collective well-being, were the themes that emerged.

We asked what support was needed, what learning was required or useful to share, and the greatest concerns that coaches had.

We have used the responses and more to inform this book.

Chapter 3

Stories of Eco-Awakening

Day One Morning

> In the twentieth century the glory of the human has become the desolation of the Earth. And now the desolation of the Earth is becoming the destiny of the human. From here on the primary judgement of all human institutions, professions, programs and activities will be determined by the extent to which they inhabit, ignore or foster a mutually enhancing human-Earth relationship.
>
> Thomas Berry (2006a)

Convener

The Convener rises and looks around, catching our eyes. Then in a resonant voice speaks these words of the great poet Mevlana Rumi:

> Come, come whomever you are
> Worshipper, Lover, Cynic or Friend
> Ours is not a caravan of despair
> Even if you have broken your vows a thousand times
> Welcome, welcome, welcome once again.

It is the beginning of the journey. We gather in this large tent, open to the winds in all four directions. We are meeting each other and here to discover the journey and the vision quest we each need to go on, both together and individually.

Each of us is invited to tell our story – what brought us here and now? How did we awaken to the wider 'more than human' world and the ecological crisis in this precious planet we share with so many other sentient beings?

Exercise

What is your story of awakening? You can either share it now and write or record it or pass and wait till you have heard the stories from your fellow travellers. But

DOI: 10.4324/9781003153825-3

do not wait too long. We each need to tell our story to start the next phase of our journey together.

*

The first to rise is Sally Gillespie:

A pivotal moment in my own experience of developing an ecological sense of self came through a dream that felt more like a vision. In 2008 I organised a panel about depth psychology and climate change. Not long after, I had a vivid dream of global climate catastrophe. In this dream, I swung on a rope high above the Earth as oceans swallowed land masses. Beneath me, millions of people in the oceans desperately clung to fast disappearing shores. I knew I had to join them. Letting go of the rope, I dropped into the midst of this catastrophe and became one of many grasping on to heaving shores. A desperate poodle swam into my arms, who I cared for as best I could, while feeling the futility of everyone's struggle to survive.

This dream catapulted me out of my habitual life and worldview. I shook for the vulnerability of all beings as my consciousness opened up to the realities of collective fate. While I did not believe my dream was precognitive or prophetic in any literal sense, I did feel that my world, the one I knew as a reality, was crumbling. My dream crashed through denials and rationalisations, rupturing foundational beliefs about individual autonomy and independence. I felt in my gut that I had to acknowledge the full seriousness of global warming, and that, in one way or another, I would spend the rest of my life acting in response.

Over the next years, I had more dreams connected to ecological themes, all set in or around water. In each, I was approached by an animal who looked for care and connection from me. These dreams navigated me through a sea of feelings. Waves of despair, guilt, judgement, grief, and confusion made way over time for currents of tenderness, connection, delight, wonder and love. No longer panicked, guilty, or inattentive, I was able to become more focused on responding to climate crisis from a place of love and connection with the living world. Taken together, my dreams felt like a kind of initiation bringing me face to face with my human self, my animal relations, and my place in the world. They also inspired me to do PhD research into the psychological experience of ongoing climate engagement which led me to facilitating reflective climate conversations in groups.

We can only step up to collective crisis through collective means and practices. Reflective group conversations in response to ecological crises enable people to articulate and listen to a diversity of experiences and understandings, while grappling with the grief, despairs, and fears of living in times of profound changes in world and worldview. Only through sitting and listening to one another can we ask fundamental questions about what matters most in life and death, while sharing and exploring painful emotions, deep-seated assumptions, values, conflicts, images, dreams, and meanings. Safe, respectful, and open dialogues engender warmth, tolerance, validation and a diversity of understandings on how to engage with ecological crises. They also fuel activism and prevent burnout.

When I woke from my despairing dream of drowning in a disintegrating world twelve years ago, all I could think was 'how do I respond to this?' Now the question I hold is 'how do we respond to this?' – whether 'this' is climate emergency, biodiversity collapse or a pandemic. Only by engaging in open dialogues where we can share our thoughts, feelings and imaginings together, can we become fully present, laying the foundation for communities capable of bearing uncertainties and loss, honouring differences, experimenting with change, cultivating compassion and respecting the ecological foundations of life.

The next to rise to her feet is Liz Hall:

Thank you Sally. My story is about awakening to the need to act. We each need to ask 'if not now, when, and if not me, who?'

As a leadership coach, mindfulness teacher and writer/journalist, I'm committed to supporting healing and flourishing, and sparking transformation in myself, and others.

I seek to take a systemic stance in my coaching, and interconnectedness is at the heart of the beautiful spiritual teachings I follow and seek to embody, including those of Zen master Thich Nhat Hanh, who talked about 'inter-being'. I've rescued animals, I've been a vegan for several years, I love walking in the mountains or along the coast, and I've been pretty good at recycling for some time. I've given lots of space to the topic of the climate crisis in the pages of Coaching at Work magazine.

And yet, when it came to the climate emergency, I was strangely disconnected somehow. Yes, I'd feel sad when I read stories about entire species disappearing, or oil spills killing creatures, and I'd feel a little panicky when I thought about the future and my children. But it all felt pretty distant. Over there. Not here. Not yet.

Then I woke up. I can't quite pinpoint a single moment in time when this happened. But it was in 2019. Extinction Rebellion's activities in the UK and Spain, where I live, played a part: the beauty and chilling and haunting quality of the Red Brigade, the elders who'd never taken to the streets before, Greta Thunberg. I connected deeply to my own pain, to my grief, to my anger. I had a strong sense, just as I now have around diversity and being an ally around race equity, that I could no longer pretty much turn a blind eye. If not now, when on Earth? And if not me, then who? As Greta so powerfully said, 'How dare you?' How dare I?

What did, and what do, I do? I joined Extinction Rebellion (XR) locally, taking part in various protests in Madrid and Alicante (as the Blue Brigade). I've stepped back a little from actively demonstrating although I now plan to offer coaching to XR activists. I joined the excellent Climate Coaching Alliance (CCA) and signaled the support of Coaching at Work for the organisation. And I launched Climate Coaching Action Day, a global initiative that first took place on 5 March 2020. People from all over the world take part. I've participated in all of the CCA's global conversations and festivals around climate, with the first marking Climate Coaching Action Day. As part of the latter, I also kicked off the day with a joint mindfulness-based intention setting gathering with Mark McMordie, delivered a webinar with climate coach Charly Cox on climate coaching, and offered pro-bono climate coaching. And in November 2020, Coaching at Work launched its first Climate Coaching Award, honouring both Neil Scotton, and the CCA and its co-founders Eve, Josie and Alison for their contributions.

Key learnings for me include:

- Turning towards our understandable 'difficult' emotions such as grief and anger is necessary- not doing so keeps us in paralyzing pain and in the illusion of separateness from fellow beings and the rest of nature. Turning towards and being with these emotions is also the beginning of a much-needed healing process, which we are going to explore on Day Three.
- Coaching, mindfulness, compassion and somatics (working with the body) can help in building our capacity to be with these emotions, to identify and align with our true values and purpose, and feel into our interconnectedness- which can be inspiring and a source of strength.
- Getting together in small 'pods' with others to explore what's coming up for us & supporting each other in relation to the climate emergency can be very healing and resourcing. Community is very important.
- Communing with nature is paramount: hugging trees, walking barefoot in the sand, watching clouds, planting seeds.
- The climate emergency IS our agenda: it's the agenda of our coaching clients, the leaders we work with, our friends and colleagues, our teams, our employees. But I believe the best approach is to invite others into the space, gently and compassionately.

Forgiveness of ourself and others is key – no blaming or shaming. We are where we are, it's what next that matters.

As Liz sits down, Josie McLean rises to her feet:

Thank you Liz.

In 2003 I was talking with John Whitmore during the ICF Australia (ICFA) second regional conference in Sydney. I had invited John to Australia to offer a keynote and I was more than a little awed to meet one of the pioneers of professional coaching within organisations. Me – a new coach in an emerging profession on the other side of the world. I was in a state of high excitement because I had chaired the development of this conference and here it was. Here we were – just talking.

As I recall it now, I am watching myself in conversation with John. The conference is buzzing around us as we talk. We are sitting at a small room lounge table with comfy chairs on either side of the table. I recall the smile that comes easily to his face. Tears are welling in my eyes now as I recall this. John was disarmingly open and humble. No *primo uomo* here.

Maybe I asked John what he was currently working on – maybe our conversation just drifted towards this topic. Either way, he is telling me about his work with a company in the Middle East that is engaged in expanding water sources using desalination plants. We start talking about the possibility of future wars being fought over water supplies as Earth warms.

I am now feeling as though a sound of these words, the topic, is reverberating inside me. It is as though the words themselves are clashing inside me – there is a sound in my ears as they clash. It is a momentary experience, in 'real time', only seconds. But it feels like long minutes to me where my world has stopped, and I am unable to think or speak. During this time, I am aware that John and all the other people around me may have noticed that I am no longer there. Maybe the outside has come inside me. I can't explain it or understand it, but my body is vibrating internally with the experience. I don't say anything about this experience for quite some time - certainly not at the time to John. (I now wish I had because he would have helped me understand it.) I had no words to explain it and, in the moment, I was unable to digest it.

What I was left with was a clear sense of the trajectory of humanity towards an abyss. An image of a car travelling at great speed towards the edge of a cliff, placing the foot on the accelerator instead of the brake, or turning the steering wheel. I expect this metaphor came later – first was the feeling or sense of alarm and impending doom.

I became aware of my part in helping companies exploit people and planet. I vowed not to return until I had an alternative to offer. I commenced a quest to learn how to live and work sustainably. I don't pretend I am a saint in my own practices, but gradually over time, I aligned my business niche with my values and significantly reduced my carbon footprint by developing a principle for air travel. I decided to fly only if I had something unique to offer or specific to learn that could only be achieved by travelling.

My journey was a largely lonely one in Australia but in 2009, working with Kim Lisson and David Drake, I chaired the ICFA regional conference centred on coaching and sustainability. It was the first time an indigenous healer presented within our conference.

Then doctoral research followed, exploring how to shape a 'sustaining organisation' which was an important part of my quest. As a part of this study, I travelled internationally for a while to learn with leading systems thinkers. That learning is still unfolding as I sink deeper into an awareness of emergent change and living systems as a way of engaging systemically.

CONVENER: Thanks Josie, I am just noticing here in Liz, Josie's and Sally's stories an awareness of the impending danger to people and planet developing at different rates. But the awakening they 'felt' was a physical visceral shift that happened that moved them from an unawareness and distant knowing, to an intimate knowing – almost as if the Earth was standing and screaming 'No, LOOK AT ME – don't you dare turn away'. We have got used to looking without seeing and we need someone to get hold of our chins and turn our face towards Earth and demand that we really look. A heart shift is required to move us from knowing but not knowing. And we have to be open to allowing that insistence to penetrate.

Exercise

What are you noticing as you hear and read these? Just like art, we all see and notice different patterns and threads.

*

Paula Downey, can you now share your story about learning to say 'no'.

PAULA: The first 'no' expressed itself in my body.

As I signed in at the lobby, clipped on my visitor's badge and became an OD consultant, an uneasiness would descend. I could feel my stomach tighten. Working in these corporate environments was wildly different from the creative spaces where I'd spent the first part of my career, collaborating on scripts and storyboards or directing location shoots or putting it all together in editing suites and sound studios where bringing your whole self to the team and the work is the beating heart of the creative process.

Inside 'corporate' worlds, people, for the most part, seem to me committed to a stage-managed way of operating. The things that matter most are largely hidden or silenced.

Something didn't feel right but back in the mid-nineties I had neither the concepts nor the language to name it. I could just feel it. Few were talking about our broken climate system or the collapse of our planet's biodiversity or the radical inequality and destruction upon which our lives are built and certainly not inside the organisations that are the cause of these problems. Instead, the chimera of 'Corporate Social Responsibility' was drawing a veil across deeper truths.

I realised that we were implicated too. After all, we'd won awards for work that had helped maintain the disturbing status quo. It's easy to unwittingly collude when you're ignorant of the facts of our lives. As I slowly awoke to what we're doing to our planet home the gap between my increasing awareness and my professional life widened to a chasm that could no longer be crossed.

The next 'no' was a conscious rejection.

There was a specific moment when I was required to stay silent about something that deserved to be brought into the open and discussed: the impacts of the client's technology on the human and natural world. Legitimate questions, asked by their own people, that needed to be acknowledged.

On the outside, nothing much happened. I stayed silent. Said nothing. On the inside, I completely lost it. That was it. I couldn't do this any more.

When I joined the cutting-edge Masters in Responsibility and Business Practice at the University of Bath in England - twenty years ago - I knew I was swimming in life-giving water. It felt like the right place to be. I'd said 'no' to a corporate offer to fund my studies. I wanted to ask big questions about our entire system and find answers that might be inconvenient. My research explored the largely dysfunctional relationship between business, politics and the media and I discovered firsthand the depth of our dilemma - how we are complicit in a destructive system that few fully understand and which most of us, including the so-called powerful, feel powerless to change.

I also discovered that this is sacred knowledge. For decades, even centuries, voices from the margins who've tried to raise the alarm about the deadly system we've created have been drowned-out by the cacophony of the majority, who choose to inhabit the safer, shallower waters of the mainstream. Who look away from uncomfortable truths or believe in fairy tales that claim it's possible to get to a new place via the worn-out roads of business-as-usual. Or that transformation can happen without sacrifice; that by some miracle we can undertake the mountainous trek ahead without breaking sweat or jettisoning whole territories of a way of life that has brought about the sixth mass extinction of life on Earth. That we can arrive somewhere new, unruffled and undisturbed.

Graduating from the University of Bath wasn't the end of the journey, just the end of the beginning. Everything had changed. My eyes had been opened to the intelligence of nature and living systems and my heart knew this was the key to any future worth having. I believe passionately that only by reintegrating life's intelligence into our professional practices and using our work and workplaces to create conditions conducive to life will we have any hope of rescuing ourselves or saving whatever can be saved now.

We are nature. But we lost our grip on this reality so long ago that to many people the implications of working the way nature works seem impractical. Even today, the appetite for radically new ways of seeing and working remains limited.

That 'no' was to become a long dark night of the soul. A test of the spirit. Refusing to go back, at one point I had ninety-six cents in my bank account, and we had a debt the size of a modest house. Saying 'no' for me meant being willing to go broke. We lived frugally and kept writing and speaking about new possibilities and new ways of working. Shining a light into dark corners.

Slowly people began to show up. Most were women, mostly leaders of nonprofits confronting the worst effects of our current economic model or heading up local government or education or faith-based communities. These early adopters were trying to address complex issues for which there are no easy answers. They knew full well that the certainties of hierarchy and control are well-dressed illusions and were excited to discover a better way to see, to think and to work. They were willing to trust something else and try something new: participation, collaboration, community, democracy. We learned together. We created new spaces together. And through this work I grew into a new kind of practitioner.

I learned that I must say 'no' to create space for the new 'Yes'. At first, the way forward may not be obvious. My 'no' took me to new frontiers of learning, none of which was easy. But at least now I was contributing to purposes worthy of my life, developing new skills, finding my voice.

Saying 'no' to what doesn't give life, is a sacred act. It creates new space for me, for you, for life itself. Saying 'no' is the day to day defining of our relationship with ourselves and the world around us. Like all important relationships, we're asked to commit and recommit, over and over, as life confronts us with choices, inviting us to say 'no', again and again, to the seductions of comfort and false certainty.

The moment you say 'no', the curse of ambivalence begins to lift. What follows is the rarely easy but always necessary journey towards integrity and ever-increasing wholeness.

CONVENER: Thank you, Paula. I wonder what it is that we each need to say 'no' to as we wake up to the ecological crisis? This takes us beyond awakening to the underbelly that perhaps we are trying not to see. The story charts the challenges and flourishing as we let go, reset and realign.

So let me invite one of the people that supported us in the early stages of our planning, David Drake, to tell his story.

DAVID: One of the joys of growing up in a pre-Silicon Valley northern California was being able to roam freely across the hills and through the orchards. Before all the fences came, you could move about wherever your heart and friends would take you. Sure, there were dogs and grumpy neighbours to avoid in some places as well as the inevitable call for dinner that interrupted your adventures. What stands out for me now is the deep appreciation I gained for how everything was connected and that I was merely a part of a much larger system.

Part of this system was the suburban Protestant church through which I learned the visceral power and bonds of community and a deep respect for all that had been created. I experienced the power of ritual to remind us of what is sacred. My repertoire of rituals has grown over time, largely through practices I've developed to guide people across thresholds as part of passages of life and death. With Blake, they remind me that we can 'Hold Infinity in the palm of our hand. And Eternity in an hour.' (Blake, 1950)

My awakening about the Earth deepened over time through many sojourns into the wilderness, immersive spiritual practices, years as a master gardener, and more. In the first half of my life, this often led me to engage in social action and facilitate dialogues about difficult social issues. In the process I came to see that there are many ways to contribute, and our role often is to invite people to take the next step up for them. I came to see that if enough molecules of water find the same course, even if taking different paths to get there, they can form a river which brings new life to a person, a team, a community.

In developing narrative coaching in the early 2000s, I wrote extensively about the need for a more systemic approach to evidence and coaching. The transformations our clients were seeking required attention to their personal and collective stories. I created 'Integrative Development' to bring them together in one practice and stop sending changed people back into unchanged environments. I used this approach to design projects and programs in which people changed themselves and their environments at the same time. We are now launching labs to teach others how to work this way in service of issues that matter to them.

When I look into the span of history, I am a pessimist for whom nothing matters. This frees me to be more creative and less attached as I work. When I look into the eyes of another human, I am an optimist for whom everything matters. This frees me to be more compassionate and less presumptive as I work. When I investigate whatever is in front of me, I am a pragmatist for whom the opening matters. This frees me to be more attentive to what this person needs most right now and less worried about what the crowds are doing.

Moving between these views, I travel more lightly these days and on this journey we share together, as I live into the questions, 'Who are you?'; 'Who am I to be?'; and 'What is mine to do?' The more I awaken, the more I can stand at thresholds with and for others to help them do the same.

Eve Turner rises to her feet and with a big smile says:

Thank you David you have posed some important questions for all of us. My story starts at around six when I first got on the stage at Peter Pan's Playground in Brighton on the English south coast where 'Uncle Jack' invited children to entertain the audience. I was probably enticed by the lollipop each child received! I remember him asking 'what do you want to do when you grow up?' I was completely clear then and I said determinedly, 'I'm going to be a vet.' I always loved animals so that seemed natural. Unfortunately, over the next few years I discovered the reality of training as a vet (dissections in college and euthanasia as a necessary part of practice) and reluctantly changed my mind. But as I read more, I discovered the harsh reality of factory farming and animal testing, the cruelty done to our fellow sentient beings, and in my teens, I refused to eat meat and began a lifelong practice of only using natural beauty products that also guaranteed they were not tested on animals. I always felt we shared the Earth with fellow creatures, and we had no right to dominate them; they deserved our respect and kindness. So, I was constantly upset by the stories I read, from Richard Adams' Watership Down to details of slaughterhouse practices.

While that set the scene and was a backdrop to my life – it was really hard both justifying and being a vegetarian in the 1960s, so much easier in the 2020s – I can't say there was one defining moment that led to 'the' further awakening. More it has been a slow build. In a senior editorial and leadership role in the British Broadcasting Corporation (BBC), the interrelationship between elements such as politics, race, social justice, economic justice and leadership were evident, and I was increasingly aware of what I did in one place affecting the interconnected whole, just like with a balloon animal as you press in one part of the balloon another area expands and changes shape. That has become even clearer as a coach - what you do/coach in one place has systemic ripples.

My learning journey has helped. I am lucky enough to be married to a transpersonal psychotherapist and to have done some of my coach training with Sir John Whitmore's Performance Consultants bringing me into further contact with transpersonal work and the idea of interconnection. Then I did part of my supervision training in Bath, UK, with Peter Hawkins and there was a pivotal moment of understanding that we used more resources than the planet could sustain. Peter called this the challenge of greater demand, higher quality and lower cost, described by him as an 'Unholy Trinity' of:

1. Increasing Demand – population from 7bn in 2012 to 10bn in 2050.
2. Increasing Expectations – the rest of the world wanted what they saw others having, via the spread of the web globally, the so-called developed nations having – so, for example, we now had more mobile phones than people.
3. Decreasing Resources – in 2013 we needed 1.5 globes each year to sustain our current usage of resources, in 2022 it was 1.75 Earths and by 2050, on the same trajectory, we might need 6 (Global Footprint Network, 2022).

I read Margaret Heffernan's Wilful Blindness. Another chink opened. Even then it wasn't a sudden 'awakening.' It was a niggle. I had become increasingly focused on multi-stakeholder contracting, in delivering value beyond the individual, first by considering all the more obvious stakeholders to the work (individual client, boss, team, board, suppliers, shareholders, family, community, etc.). Then finally, something clicked, and this got extended to past and future generations/clients, and to Earth and nature, letting the ecology do the coaching, and out of this emerged what we believe is the first coaching book to specifically consider nature as a stakeholder (Hawkins and Turner, 2020). I was inspired to become even more actively involved by the Open Letter to Coaches, the Coaching & Coaching Psychology Professional Bodies and Coach Educators written by Zoe Cohen, Alison Whybrow and Linda Aspey: (www.linkedin.com/pulse/open-letter-coaches-coaching-psychology-profes sional-bodies-zoe-cohen-1e) (July 2019). This contributed to the birth of the Climate Coaching Alliance.

If I think of Peter's Eco-Phase Cycle, I see it as my journey for life! I may have been slow to awaken, and I appreciate that as a species we are running out of time. The key has been retaining my curiosity, reading, thinking, reflecting, worrying, being anxious – all the points of the cycle on any day, and never fixed! But more than anything as I reach my final decade or two, I have increasingly felt the voice of future generations speaking to me, and asking me 'What did you do when the Earth was in danger?' As I reflect, for me the placard sign now needs to make clear we are responsible ourselves to 'Save Humans' because the Earth may change but she will survive, I understand now we are precipitating our own demise alongside that of other species. Legacy, my responsibility to future generations, is my passion, taking and acting upon my responsibility. So, I do what I can, as I can, in my way, limited as it may be, not think it is for others to act.

CONVENER: Thank you, David and Eve. I really like the time dimension in David and Eve's stories, that develop the understanding of awakening from the earlier stories. Awakening isn't something that happens over night, but over a life. As a colleague said, sometimes you're not aware of that happening, but it happens and then you realise you see things differently to how things were. Walking through a forest every day on a commute, you might find the trees have changed you. I would now like to invite another of the book's early supporters, Zoe Cohen, to tell her story.

ZOE: I think I believe that all human babies have the capacity to be 'awakened', or perhaps we are all born awake and then fall asleep?

As a young child I loved plants and flowers and grew many on my bedroom windowsill. I had a small glass terrarium in which I put some soil, water and one or two small plants, then the lid. I was probably less than ten and at some level I had a knowing that that little eco-system was a microcosm of the Earth.

I'd also always been an animal lover, despite a severely asthmatic father and my own allergies. By 12 I'd decided to never eat animal flesh again after seeing on TV a 'mummy' seal following its newly killed baby, being dragged along the ice by a seal culler. I didn't see the difference between them and us, and I still don't. I've strongly disliked anthropocentrism ever since.

By 14 I was a supporter of the Vegetarian Society, Vegan Society, BUAV (now Cruelty Free International) and was leafleting for the Green Party. I joined Greenpeace and learnt about deforestation, and pollution by manmade chemicals such as DDT. Around the same time, I created my own little teen cosmology of wanting to live my life causing as little harm as possible. I have of course failed, as we all have.

At 15, in 1985, I recall a sense of depression learning that it would be the year when the human population passed 5 billion. I hugely admired the Rainbow Warrior activists trying to stop the whaling ships and other protest action.

I think I went to university fairly 'awake' and then through a process of gradual socialisation, the adult, 'grown up' world of work, jobs, mortgages, career, relationships, compromise, and family, I allowed myself to be partially anaesthetised, although never fully asleep. I think the anaesthetic gradually started to wear off around a decade or so ago, as I got more and more involved in local 'sustainability' in my community, over and above living a relatively 'green' life.

Another level of awakening took place when the UK government drastically cut the financial support for renewable energy, and in one go made the community energy company, that I'd chaired as a volunteer, unviable overnight – along with the rest of the national community energy movement and many of their suppliers.

In 2017 I became my Mum's carer in her painful and tragic last months. It was my Mum who had taught me to love nature, to be fascinated by biology, and who was born in a time (1935) when the hedgerows were plentiful with wildflowers, and when CO_2 was 309.4 parts per million (2 degrees Institute, 2022).

Mum died late summer 2018 – a summer of record heatwaves (again), which also made her last weeks even more miserable. Thirteen weeks without rain made my inner mammal scared – and turned our allotment into semi-desert. I was awake again. The 2018 IPPC report hit the headlines saying we had 12 years to save the planet.

A chance Twitter message from a coach friend of mine introduced me to Extinction Rebellion in November 2018 and I ran towards the movement I'd been waiting for. At last, some people were telling the truth, and for once actually acting like the truth was real.

The last few years since have been an incredible journey of grief, friendship, action and joy – with continual learning, and ever continuing layers of awakening to what is.

CONVENER: Thank you, Zoe, for highlighting the levels of awakening we need to move out of our sleepwalking. I wonder what anaesthetic we each need to clear out of our systems to awake further. Now we will hear from another of our book authors, Alison.

ALISON: In 2013 I co-authored a book about embedding values in organisations (Williams and Whybrow, 2013). Despite writing a blueprint for cultural transformation, I felt deeply unsettled. A background awareness of the unholy trinity of accelerating global warming, overuse of global resources and a growing human population shared by Peter Hawkins was settling in my being (see p25). It led me to question the wisdom of facilitating organisational effectiveness when organisational values might or might not be aligned to a thriving planet. Despite lots of individual positive personal choices, the news was of continuing degradation of our eco-system.

The story didn't start there, it started decades earlier, when as a teenager, I found myself travelling with a group of strangers in a transit van with the purpose of disrupting a fox hunt and saving a life. I always keenly felt the suffering of others so much so that I struggle to witness it. It soon struck me that hunters were not interested in foxes, for many this whole quest was one of power. Decades later, whilst the foxes in the UK had had a reprieve of sorts, systemic power imbalances were increasing around the globe, the book on values written, I started to pay attention again.

I decided I needed to connect to dialogues at a global level to understand the bigger picture. Starting with Naomi Klein (2015) *This changes everything*, I never looked back. Re-connecting with the Pachamama Alliance and their online programmes was a game changer and covered a lot of familiar territory as a coach and psychologist. I became horribly aware of the rigged system and felt the futility of the positive actions I had taken to date. I possibly over immersed and did so largely in isolation, resulting in a rather anxious, despairing and frustrated few years. My family suffered.

I was desperate to do something. I got involved in running the Pachamama programmes, involved in some local community-based organisations, and brought my immediate community together in order to share resources, connect, sensing that both local and global action was required. There was a huge gap still, how could I bring this work into my professional world in a systemic, constructive and impactful way?

In October 2017 Stephen Palmer and I were co-presenting a decade of research on coaching psychology practice at The International Congress of Coaching Psychology. Closing that presentation, I asked: 'What would we be saying in ten years' time about our practice if coaching and coaching psychology were to make a positive difference in shifting the human Earth relationship'? I was thanked for the question by colleagues – but still had no idea what might emerge.

A year later, following a workshop with Leaders Quest, and a conversation with a colleague, I reflected on what kind of leaders I wanted to work with. The idea of regenerative coaching started to emerge. This idea, reframing the purpose of coaching towards a thriving planet rather than just towards a thriving individual or organisation, was a reframe that enabled everything else to come into line. Meeting Zoe Cohen and Linda Aspey opened up further possibilities. Rumi's quote seems most apt 'As you start to walk on the way, the way appears.' And a constant companion along this journey has been Peter Hawkins.

Looking back, I'm struck that I could only hear and do so much, until I was ready to move again. I hope we can each make each other's journey shorter and clearer so it doesn't take each of us as long.

CONVENER: Thanks, Alison, for your story of slowly joining the dots and letting go of 'knowing' and finding alignment. Dominique Barbes can you now tell your story please.

DOMINIQUE: My story is called 'One step at a time'. Since childhood, I have felt close to nature. I sensed the wild beauty of the ocean, underbrush softness, a glance at the tall maple tree around the corner fills me with gratitude. As I grew up, I became aware of the human destructive impact, and how biodiversity was diminishing in the span of my own life. Feeling helpless about it most of the time, I alternated periods of environmental commitment, then giving up, too busy with my head deep in the sand.

I want to share with you how my ambiguous behavior is changing, partly due to my best friend, nicknamed 'self-coach'.

In recent years, there were numerous occasions for this cycle of anger, indignation, sadness, despair, to start over as I heard the news or documented warnings from scientists. But now my 'self-coach' comes to rescue me and brings me back into action.

It all began on one of those environmental tsunami days, some very bad news had broken out. Signing petitions, marching, would definitely not make it for me this time. Self-coach whispered: 'to be stronger, find your allies, do get connected! If you feel this way, others are likely to feel the same near you. Find them!'

I felt an urge to gather, to join consistently other people of my kind, right here in my neighborhood. And it happened. I mobilized all my senses and very quickly located a small group of concerned neighbors, with a project of bringing environmental awareness among candidates for the upcoming elections. This was my first little step, and it was theirs too.

I embarked right away into door-to-door conversations about climate warming impacts of a new pipeline project. Collecting signatures on post cards, we later were able to bring a few thousand of them during our planned encounters with each candidate. Totally ignorant about this field, I began to learn from my buddy citizen on these walks, he happened to be a biologist! I enjoyed also getting more familiar with my neighborhood I knew little about.

Acting for the climate turned out to be both fun and a continuous learning experience. As my self-coach says: 'Connecting, acting and learning together are essential to develop commitment, and influence'.

As a matter of fact, three years later, I am still involved and we share a very special connection. Various initiatives keep emerging from individuals supported by the collective. Amazing actions grow from those seeds, with a very minimal structure: safe bicycle paths, tree plantations, papers on critical environmental issues, networking with other citizen groups over the city and across the country, zero waste actions, inviting people to lectures and movies about current climate issues, and building a partnership with the local government.

Making the decision to connect myself within my community, about such an overwhelming issue as climate warming, has been one of my best moves as a citizen, and as a coach too. This collective teaches me constantly about group dynamics.

Honestly, my contribution over the last three years has been quite limited, due to full-time work. Yet I remained connected and involved as a support for the group whenever needed. My 'self-coach' keeps telling me 'every part of the WE counts'. Dealing daily with bad news is not as hard, since now I have a handle on change. And I know already my commitment is growing, one step at a time.

CONVENER: Thank you Dominque for sharing your need to pause to appreciate the need and desire for community as we awaken. Now I invite the last of the conveners, Peter Hawkins.

PETER: My awakening to the both critical and urgent need to address the ecological crisis and to heal the relationship between humans and the more than human world, has been a long and patchy journey. Looking back over the last 50 to 70 years, I see a patchwork quilt, where parts of me would wake up, while other aspects of my being would stay 'wilfully blind'. Indeed, it has taken far too long to integrate and sew the patches together.

In childhood, like others here, I was an innocent lover of nature, enjoying gardening (my first job), woodlands, mountains and most frequently cycling and sitting on the river tow-paths close to where I grew up. But it was not until my twenties, in the early seventies, I began to awaken to the dreadful havoc our species, and particularly us privileged members of the industrially developed world, were having on the Earth we share with so many other living species.

Rachel Carson (1962) awakened me to the devastation our industrialized world was inflicting on nature. Gregory Bateson (1972) became my greatest mentor, showing how we had misunderstood Darwin, and the unit of survival (and by implication the unit of flourishing) was never the individual, family, team, organisations, country or species, but was any one of these in co-creation with its ecological niche. He also convinced me that without a fundamental shift in human's ways of knowing the world (our epistemology), from linear, atomistic, Newtonian thinking, to a more systemic awareness, we did not 'have a snowball in hell's chance' (Bateson, 1972: 437) of surviving, and he was saying that in the 1970s!

But this was a neo-cortex cognitive awakening and as I often say: 'The road to hell is paved with insight and good intentions.' This was just one square of the patchwork quilt.

In the 1970s, and ever-since, my eco-spiritual awakening came from a universalist Sufi organisation in the lineage of Hazrat Inayat Khan, a great Indian mystic and musician who brought a form of universal Sufism to the west in the early years of the 20th century. This great teacher wrote:

> Anyone who has some knowledge of mysticism and of the lives of the mystics knows that what always attracts the mystic most is nature. Nature is his bread and wine; nature is his soul's nourishment; nature inspires him, uplifts him, and gives him the solitude for which his soul continually longs.
> (Khan, 1972, Vol X1: 169)

For many years his grandson was my personal spiritual teacher, with whom I learnt how the greatest teachings happen through an intense engagement with life in all its forms, human and 'more than human'. I began to learn how nature is our source of energy, the giver of life and our greatest teacher, and how death and life are one. This learning continues for me to this day with his successor Elias Amidon (2012), a non-dual poet, writer, teacher and ecologist.

Spiritual ecology shows the dangers of both a scientific ecology, that sees the environment as something to be analyzed and a problem to be solved by humans and also the dangers of a human-centric spirituality or religion, that despiritualizes all other life forms with whom we are intrinsically connected in sharing this one Earth.

The next patch of the changing quilt came in the 1990s when at Bath Consultancy Group (which I had started, in the 1980s), I began to be joined by colleagues like David Ballard and Robin Coates, who brought ways of moving beyond 'systems thinking' to exploring systemic practices in leadership, teams and organizations. We worked with Jonathan Porritt, as he moved from Friends of the Earth to start Forum for the Future, to explore how organizations moved beyond sustainability policies in a small silo, to more fundamental culture change. We brought in teachers such as Barry Oshry, Bill Torbert and Peter Reason to train us and moved forward our collaborative inquiry. I also travelled to learn from people like Peter Senge, Otto Scharmer, Jo Jaworski, Betty Flowers, Adam Kahane, what others were doing to create more systemic change in organizations and to create eco-systemic ways of being. But I was still flying around the world, not only to meet systemic innovators, but also to work with organizations and give trainings. This was practical awakening through doing, trial and error experimentation, discovering what helped and made a collective difference. I became convinced that the world did not need more psychotherapists, coaches, counsellors, consultants, but was in desperate need of people who could connect in-depth change at the personal, team, team of teams, organization and eco-systemic levels. This led me to develop Systemic Team Coaching, publish widely on the subject and run trainings in it for people from over 100 countries (Hawkins, 2011, 2014, 2014b, 2017, 2018, 2021, 2022.)

But my actions for too long were limited to my professional life. We were helping organisations like Unilever, look at how it could apply ecological and sustainable principles and practices, to not only its direct work, but taking co-responsibility for its whole value chain. Only slowly did I realize that I too should take more responsibility for my whole value chain. My small company Renewal Associates issued an updated sustainability policy on our website. I enacted this policy by having conversations about their ecological practices with the training centres which I use to run trainings, making sure they met my minimal requirements and bringing influence to bear on further changes they could make. For a long time, I had ensured my pension scheme were investing in ethical and sustainable companies, but more recently I have gone further and insisted that the funds investing the money were not only ESG – A rated, but also could report to me quarterly on the positive impact that my investments were having in the world. I firmly believe that we have more influence in the world through our pension schemes and insurance policies than through our political vote.

Even the intellectual, spiritual, and practical I now realize is not enough, for all change is embodied. My love of gardening has never ceased, but it was only when I found myself with responsibility for 37 acres of countryside and became a grandfather that my ecological awakening became rooted in every part of my body. I learned not to plan a garden and then go out and implement the plan, but rather to listen to the land and every aspect of life that is supported by it and discover what is collectively needed and one's own unique contribution to meeting that need. This is an on-going source of constant learning. I worked with the Woodland Trust, the local community and a thousand local school children, to plant a new woodland, and try to listen to what it needs from us to support its growth and development.

Being a grandfather is a great learning gift. The Welsh have a lovely old saying: 'Pure love arrives with the first grandchild'. Discovering the truth of this, of being able to feel the deep embodied connection to each of my grandchildren, without the psychological entanglement that one has with one's own children, is a constant teacher. Age also brings with it a different time-horizon. I was close to my grandmother, who was a child in the 1880s and now I am close to my grandchildren, who may well be alive in the 2080s, so now my time horizon spans 200 years in direct connection. I can feel the gratitude for what I have received from the previous generations and feel the great responsibility for the legacy I am leaving the generations that follow. I can feel being a grandfather and gardener in my bones, in the waters that flow through me, and the oxygen I breathe. I can sit at the meal table and pause and gaze at a plate of food – all of which has grown on the same land I live on. I can look at it and intimately know its origin and story. I can experience gratitude of participating in the life and death of each part of the meal. At the same time, I can sense how in turn I am, and will become, the food of the living Earth and what comes after me.

For too long I have struggled to bring my spiritual, academic, business, political and embodied selves, together into a constant internal living marriage, to turn the patches of awakening into a quilt. But this weaving we cannot do alone, for I am just one small self in a world teeming with interconnected life. To engage in self-making, is to be trapped in egocentricity. As Chief Seattle reminded us: 'Man did not weave the web of life. He is merely a strand within it.' We are both weaved and a very small part of the weaving. Llewellyn Vaughan-Lee (2013: i), a Sufi and eco-spirituality teacher, writes:

> The world is not a problem to be solved; it is a living being to which we belong. The world is part of our own self and we are part of its suffering wholeness.

So to weave these parts of my being together I need to re-join with everything I make 'other', and to awaken to how we are all one; one global family of humanity, one with all of living nature and part of the mystery of oneness. To re-member with every aspect of my being that the eco-system and I are one.

Now, on this journey together, we need to build a bigger quilt together, patch to patch, person to person, nature to nature.

CONVENER: Thank you, Peter.

Exercise

CONVENER: Now we invite you to tell your story. If possible, please divide up into small groups and give an opportunity for those who have not yet told their story to do so. When you have heard from everyone in your group, we invite you to collectively explore what themes, patterns, and threads you have noticed connecting these stories, and to come back and each group to share their top 3–5 themes. If you are working alone, compare your themes to the stories above.

*

When the groups returned each group shared their themes and built on each other's ideas and the Conveners collected the themes on a flip chart as follows:

- We are part of much larger systems.
- We are not 'I' we are 'we', interconnected with a visceral dissolving of the boundary between I and the larger whole.
- We need to see 'ours', not just 'others', role in this and responsibility.

- As we awaken, we see what is within our gift, there are many ways to contribute and together many drops create rivers.
- Agency and hope, a stepping away from, wrestling away from collusion in layers, creating our own paths.
- We are nature.
- Unravelling and re-threading new habits in everything we do, say and think.
- Seeing what is sacred and creating rituals.

CONVENER: Awakenings can be gradual and creeping, and they can be a moment of realization with alarm bells ringing. They can recur and our realizations deepen. They all involve an emotional shift if we have been disconnected, or a deepening if we've never forgotten. Dialogues with self, others, nature herself.

Awakenings enable us to become aware of our unquestioned assumptions and socially conditioned ways of seeing and acting. We become able to see and sense the system we are in.

For these shifts and patterns to take place, first, an openness is required. This seemed particularly striking in Sally Gillespie's narrative. The openness to knowing from more than one place, the openness to feel, the openness to listen, the openness to respond, the openness to learn, the openness to act even though there are no guarantees.

Awakenings can be an opening of a new space in which to explore new possibilities and new ways of being. Journeys of depth, of growth, joy, of love of being able to stand together to look at what is.

I suggest we all have a refreshment break, after which we will explore the collective journey that coaching has been through and the future story the world requires from coaching. Perhaps before we reconvene, take a walk and sit outside if you haven't already.

Just before we break, a word from one of our dear elders in this work, we welcome an observation from Joanna Macy (2014):

The most remarkable feature of this historical moment on Earth
is not that we are on the way to destroying the world –
we've actually been on the way for quite a while.
It is that we are beginning to wake up, as from a millennia-long sleep,
to a whole new relationship to our world, to ourselves
and each other.

Chapter 4

Awakening in the Coaching Profession

Day One Afternoon

CONVENER: This morning we heard many personal stories and explored some of the commonality and patterns that connect them. We invited all of you on this journey because you are part of the wider field of coaching, either as an internal or external coach, team coach, mentor, supervisor, HR practitioner, a leader or a consultant. We have invited you to help us explore together the journey that the young profession of coaching needs to go on to regenerate and make the best possible contribution to the great challenges of our time.

As we become aware of the rapidly changing world in which we are living and working, the role of coaching now is being reconsidered at different levels throughout our profession. Perhaps by no coincidence, all editors of this book were friends of, or acquainted with, Sir John Whitmore and familiar with his work. He often talked about why coaching was emerging in the mid to late 20th century. He suggested (powerfully as John was predisposed to express himself) that coaching was emerging at this time to assist the transformation of human consciousness as we confronted the unintended consequences of economic growth, population growth and a culture that had forgotten the collective.

As we unfold our understanding of the space that coaching might occupy, playing nursemaid to a dying human-Earth story and midwife to a new story, it is worth noticing the phases and phrases that form a map of the dialogue; the phase/phrase space (Kuhn and Whybrow, 2019). Just as the climate and ecological crisis is unfolding at pace around us, insight into the role of coaching is equally emergent.

On this journey, we seek to provide insights into the evolving relevance and influence of coaching; we know this landscape will continue to unfold long after this book is published. Starting the conversation collectively appears to be crucial and we are grateful for those who generously share their emerging thinking here.

We have invited some voices from across our community to share their thoughts on the journey of coaching, starting with David Lane, who has been engaged with how coaching can contribute to the ecological healing for many years.

DOI: 10.4324/9781003153825-4

DAVID: I have called my offering Coaching and Environmental Health – a look back and forward.

The 1960s saw an increasing interest in environmental matters with new college courses on environmental science appearing and modules being added in areas such as social biology to existing disciplines. The build-up to the 1972 UN Conference of the Environment led to a burgeoning range of studies making suggestions for change. A reading of the recommendations arising from that event certainly makes for sober reflection. If only we had taken on board and acted significantly on many of its recommendations. In particular, we had to recognise that we have to act as a community, and that placing responsibility just on the individual will fail (Lane, 1972).

This debate certainly struck a chord with me, and I undertook a series of studies between 1968 and 1972 to look at the education of three groups on environmental matters. This led to a proposed syllabus for education in environmental health and suggestions for work at the level of schools, communities, and professionals, in particular those professionals with the greatest impact on attitudes to the environment (Lane, 1972). We then would have included teachers, planners, logistics, transport, construction, food scientists, etc., but today we would certainly include coaches. Why coaches - because we are working with decision makers in industries that have a big impact on the environment. For example, if we think of the built environment and all the professions involved in its design and development, and the timeframe over which such decisions impact, we are looking at around 25 percent of greenhouse gases (Lane and Malkin, 1994; Samuels and Prasad, 1994). Similarly, in the transport and logistics industries some 24 percent of CO_2 emissions came from transportation (Khan, 2021). In a series of articles on that sector looking at 2021 (CILT, 2021) as a period of recovery or relapse the role of sustainability features strongly.

We must develop a sense of our common future (Brundtland, 1987), and this idea is growing. People feel proud to work for an organisation that contributes to its community and our common future (Rajan and Lane, 2000) and consumers and investors are prepared to reward organisations that are more environmentally friendly (Willmott, 2021).

Professionals are asking the question 'how will my industry be affected by climate change?' and more directly 'what should we be doing to generate sustainability in all our practices?' These are questions that can legitimately feature in a coaching conversation. Enabling sustainable development is now a conversation that any leader needs to encourage throughout their organisation. It is a concern increasingly raised in conversations and once a leader begins to ask, 'how will my industry be affected' and 'what should we be doing', supplementary reflections become possible.

1. What can I as an individual do?
2. What can I as a citizen do?

3. What might I as a member of this organisation do?
4. What might this organisation do?
5. How might we involve our wider community (supply chains through to customers) in building sustainability?

As a coach working across a range of industries, I have seen a change in the extent to which sustainability can feature in leadership conversations. The idea of organisational purpose has increasingly focused attention on longer term goals. Within coaching the joint statement on climate change signed by 11 coaching, mentoring and coaching psychology bodies (Joint Global Statement, 2020) shows the strength of the commitment within our discipline. Some organisations (say Sony or Arup) have always been able to think about horizons stretching to 50 years and many more are now creating plans for zero carbon by 2035. Radical thinking to bring various parts of the economy together in pursuit of zero carbon goals are emerging. For example, the England Economic Heathlands Draft Transport Strategy (see Hickford and Blainey, 2021). Coaches working with such clients can play their part in enabling that dialogue. For organisations beginning to ask the question 'how will my industry be affected' as coaches we can challenge them to go further and develop a bigger perspective on the question.

I have spent 50 years concerned with environmental health. Initially, I had great enthusiasm and belief that we would act decisively as the case was so obvious. That belief has waxed and waned over the years as action in key areas has been delayed and too much weight given to non-scientific perspectives over the agreed science of the intergovernmental panel (Intergovernmental Panel on Climate Change, 2021) coupled with a failure to look at new ideas on economic development that respect the environment (Dasgupta, 2021). Finally, I can see emerging at the eleventh hour the urgency we need. As coaches we have to play our part in generating a sense of our common future.

CONVENER: Thank you, David, you have given us a wide and deep perspective spanning the last 60 years.

Exercise

Just before we go on, as you read this, how did you respond to David's words? What did you feel, as you read of the early enthusiasm, the certainty that decisive action was needed 50 years ago? What is it like for you now, walking this path? Capture what seems important before we continue together.

*

CONVENER: I am now going to ask one of my Co-Conveners, Alison Whybrow, to share her thoughts on how coaching needs to regenerate.

ALISON: Thank you. The idea of regenerative coaching or coaching as a regenerative practice, emerged from various conversations and inputs. No less, the very simple framework from Bill Reed (2007), and his framework from Degenerative to Regenerative, starkly clarifying where we are, yet also, where we need to be.

A pivot is required and as mentioned in our introduction on page 12 many leaders in the coaching profession have called for coaching itself to go through a major transition and transformation in order that it can play an active role in being one of the midwives for the new era (Whitmore, 2017: 22).

The self-inflicted wound is much deeper and the required response more fundamental than a simple shift to low carbon choices. We cannot regenerate our own individual lives, nor regenerate the well-being of our one interconnected human family, if we are not at the same time regenerating the ecology of this precious planet. A rich diverse planet that we both inhabit and share with many other sentient beings who have been here much longer than us human late arrivals. This includes our shared atmosphere, our shared rain and waters, our shared earth and soil. All of these are currently degenerating fast as a result of myopic self-centredness of us human beings.

As Hetty Einzig writes we need to explore, "the potential for coaching to support leaders in transforming the rules of the game itself." (2017: 2). Especially when we as coaches are stuck playing the same game.

A call to step up

Shifting our lives and our practice in a way that responds effectively to our existential emergency is not easy or simple. How do we locate ourselves as a part of nature rather than apart from nature? We have become so deeply disconnected; we struggle to bring this fundamental truth into being. Many coaches I spoke with felt ill-equipped to work with our crisis in their work: 'it goes against my [coaching] philosophy…. bringing something to someone's attention that they haven't chosen or have chosen to avoid.' I'm afraid of 'moving from the traditional role of coach'. For those with a deeply systemic practice, and First Nation coaches I have recently met, the eco-system is always in the room, the skill then becomes one of perspective taking and inquiry, rather than an issue on the agenda.

Shifting the frame for coaching

Shifting ourselves as coaches, requires us to shift the frame for coaching. Currently, the very focus of coaching is often too small, looking frequently at only individual and organizational success. What about planetary success? Kate Raworth (2017) points out we have exceeded the boundaries of our ecological ceiling, and we are putting strain on our social foundations (water, food, housing, social and family networks). Sustaining what we have now is not sufficient – we would remain a planet in decline. Rather, we need to be in the game of restoring and regenerating (Reed, 2007), and apply living systems thinking (Allen, 2019).

Applying a regenerative framework to coaching, we might define a regenerative purpose for coaching as:

> Enhancing the health of our biosphere and life-giving properties of our planet, underpinned by models of coaching, grounded in established and emerging approaches associated with adult and child learning, psychology, leadership, living systems thinking, ecology, eco-systems, economy and regenerative design.
>
> (Whybrow, 2019)

This wide-angled focus does not deny individual and organisational success. They are now aligned within the context of the whole. Focusing on the life force of our Earth which we fully participate in and impact on every day, invites us to coevolve and partner with our Earth. As Kathleen Allen (2019) notes, our planet has 3.5 billion years of research and development know-how about how to create life. Why would we imagine we can do better?

A regenerative frame calls us to look again at what we do with the knowledge that we cannot offset our impact. We cannot externalize the costs that we impose. Our Earth is beyond capacity to hold our poisonous waste. A regenerative frame invites us to shift our awareness and take full responsibility.

What might this start to look like in practice?

The roots of a regenerative practice might keep us connected to what is real and to jumpstart the process of reconnecting what has been lost. From my conversations with coaches, these roots might include:

- Intentional and real contact with the Earth. Whether through gardening, walking, mindfulness or other practice; paying deliberate attention to 'building relationships with elements of the natural world'.
- An openness to allowing the Earth to work on us and through us: 'experiencing nature acting through me, outside my awareness over time'; and 'an open heart enables the work of the soul'.
- Working on how we show up, with self as the site of change, the work is 'in here' rather than 'out there'. 'It's not what we say or do, it's the way we are'. It's how we focus, not what we focus on.
- Letting go, with coaches asking 'How can I be a helper and provider of a more holistic approach'; 'How do I serve the unfolding of the potential around me and manifest that potential within myself?'
- Stepping into a more connected frame and a wider sense of leadership, enabled a greater sense of purpose, an ease with speaking truth to power. Coaches noted a 'blossoming energy and transparency' and 'feeling more me.'

- Deepening connection as a way of seeing one's place in the world: 'everything is alive, everything is connected, we are all interconnected in one conversation'.

Coming into personal integrity seems a necessary part of developing a healthy coaching root system. One coach said: 'I am not a planet or a tree, the depth of my caring for humanity is the only door through which regeneration can happen.' Regenerative coaching is rooted in self work as the foremost site of change enabling us to engage differently with the context, we find ourselves in.

If we act from a place of judgement, anger or rescuing others, it follows that our ability to engage with the wider human and the non-human world will be disrupted, our sense of connection will be diminished. Interdependence and interbeing are part of the wider sense of self and richer experience of community. This critical self-work enables us to access power-with, different to the power-over framework, the latter being enshrined in neoliberal capitalism.

Later in the book we will explore how we might create regenerative coaching approaches (see below and on pages 242–3, Day Six). With a regenerative coaching framework to house our practice, we can rapidly become a forest of coaches, growing ourselves and our Earth-centred practice, able to hold greater depth, challenge and compassion. Our clients and our planet need us to catch up.

A healthy forest root system is deeply connected to the wider forest community, able to draw up more water and nutrients than it would alone and receive greater protection (Wohlleben, 2015).

Is it perhaps time to take a chapter or two out of nature's playbook?

CONVENER: Thank you so much Alison.

Exercise

CONVENER: Let us pause for a moment to reflect on the many concepts shared here. What might a regenerative practice mean in concrete terms for you and us? It simply boils down to every choice made: is it regenerative or degenerative?

Paul Hawken (2021: p249) shares 12 practical questions that expand on this further:

1. Does the action create more life or reduce it?
2. Does it heal the future or steal the future?
3. Does it enhance human well-being or diminish it?
4. Does it prevent disease or profit from it?
5. Does it create livelihoods or eliminate them?
6. Does it restore land or degrade it?
7. Does it increase global warming or decrease it?

8.   Does it serve human needs or manufacture human wants?
9.   Does it reduce poverty or expand it?
10. Does it promote fundamental human rights or deny them?
11. Does it provide workers with dignity or demean them?
12. In short, is the activity extractive or regenerative?

We will work with these questions again on Day Six, on being Eco-Active.

Before we hear from our third contributor, it is worth acknowledging that coaches are often concerned about having permission to bring in certain subjects. They may even feel that their professional body membership and agreement to follow a code of ethics precludes them from doing so. As we shall see in Day Six, Eco-Active, there have been changes in some codes of ethics to make the wider responsibilities of a coach to society, not just an individual or organization much clearer (for example see *The Global Code of Ethics*, 2021). And in 2020 a *Joint Statement on Climate Change* was written, now signed by eleven professional bodies, with at least 250,000 memberships between them. It is explicit in its wording, including this paragraph:

> Coaching, mentoring, coaching psychology and supervision are concerned with developing the potential of human beings, of raising awareness to enable people to take responsibility for their actions and ownership for their contribution. We have a significant role to play in fostering new ways of being in service to a healthy human society and a healthy planet.

Our third contribution is jointly from Joel DiGirolamo, the ICF's Vice-President of Research and Data Science in the USA, and Tina Joanes, from Germany, on coaching, climate change and the evolution of society:

JOEL AND TINA: Our species, *Homo sapiens*, has survived and thrived partly through our ability to band together in tribes for safety and efficient hunting and gathering (Wrangham and Peterson, 1996). This survival of our species has left embedded sequences in our DNA that cause us to consciously and subconsciously favour and trust ingroup relationships and to remain wary of outgroup relationships (Brewer, 1999). So, it is not surprising that relationship with the coach is one of the strongest factors in positive coaching outcomes (de Haan and Gannon, 2017).

When a coach builds a relationship with a client, they are working to establish rapport and get the client to feel that they are together in the ingroup. and is willing to open up, to speak, and work together on whatever it is that the client desires.

Another Homo sapiens propensity is that toward stasis, that is, general inaction toward change. We tend to change only to move away from something less-than-desirable or toward something more desirable – and only when that feeling of the need to change becomes sufficiently great enough to motive us to action (Knight, 2009).

We propose that the use of coaching skills be moved outward, providing more opportunities for coaching skills to be used in dialog for the betterment of society. Figure 4.1 illustrates a sociological view of this schema, wherein one-to-one coaching is akin to the micro level in sociology theory, groups and organisations at the meso level, and communities and society at large as the macro level (Collins, 1988).

Brock (2012) identified the human potential movement as a contributor to the genesis of the coaching modality, discussing the potentiality which coaching brings to individuals seeking a higher level in their lives. This fits in firmly with descriptions sociologists use at the micro level. Collins (1988) and de la Sablonnière (2017) describe the meso, an intermediate level between micro and macro, as involvement at the group and organisation level. For simplicity's sake, we are including personal relationships at the meso level, however, that could be debated. Finally, communities and societies on a country or global level can be considered to be at the macro level.

Coaching is a loosely structured client-centred facilitation of change. Change is ever-present, in each and every moment. Coaching has been a one-to-one coach-client-focused effort for several decades (Grant and Cavanagh, 2004), whereas more recently team and group coaching have been embraced (Hawkins, 2021; Widdowson et al., 2020). The use of coaching skills by managers and leaders has been promoted since the 1950s (Mace and Mahler, 1958) albeit inconsistently and at a low level (DiGirolamo and Tkach, 2019). The idea of a "coaching culture" in organisations has been researched to a

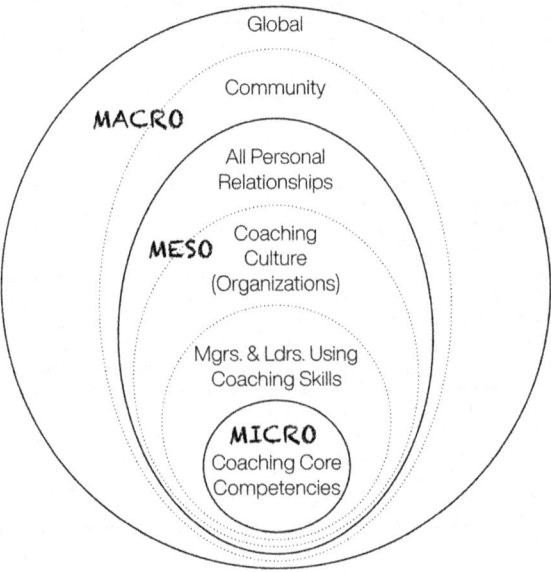

Figure 4.1 Use of coaching skills from a sociological perspective

small extent (e.g. Hawkins, 2012). Moving out into the meso level, community development work has embraced coaching as a way to accelerate and empower community leaders (Emery et al., 2011).

As we consider how the use of coaching skills might help in society even more, we can consider how the ideas of building rapport and evoking awareness could be transformational. When openings arise in which climate change could be a factor, given the existential nature of this crisis, it seems the morally correct thing to do is to raise the issue – within the boundaries of coaching, of course.

So what can we do as coaches?

- Consider how we can be a role model in stewardship of our fragile planet.
- In setting goals and working towards a shift, including climate change as a part of that goal, shift, or the context within which the client is working. Given current scientific data, the coach should raise awareness of the issue and implications of the client's actions or attitudes in this regard.
- Consider that you may have information that the client does not and that this information may widen the client's perspective, foster learning, or provoke an insight.
- Your client may be unmotivated to change or perhaps may not consider the impact of their behaviour on our planet. Through your ongoing work together, this may be an area to ask questions and explore in some level of depth. Such activities may spark some intrinsic motivation in your client that would have otherwise laid dormant.
- Always try to build rapport, an ingroup relationship with your client, colleague, partner, antagonist, and virtually everyone you meet. This is how we can build trust and relationship with all people with whom we enter dialog.
- As you work with all others, use coaching skills in your discussions and be a role model for how the use of those skills can enhance dialog and understanding.
- Whenever possible, promote and educate others in the basics of coaching skills.

Ultimately the adoption of coaching skills in societies across the globe can have a profound influence on the way we interact with each other, engage in dialog – and most importantly – build a better world in which to live.

CONVENER: Thank you Joel and Tina. Josie is now going to take us in a slightly different, but nevertheless related inquiry, to look more critically at the context. Josie will bring a further layer, or two, to this dialogue to help us understand the issues more deeply.

JOSIE: As we sit in circle, I'd like to start by sharing an article recently shared by our fellow convener, Peter, The Omertà of Consultancy by Ed Gillespie (2021). The article pulls at the mask of good and effective work that covers the hidden depths of consultancy. I read this piece quietly. Soaking up the implications of the language of the Italian mafia that he had chosen – a silent conspiracy with menacing undercurrents. Gillespie's article, well worth a read, is directed at sustainability consultancies. I know several that are in the league he is discussing.

The article resonated deeply with me. Employing my systems thinking approach, I have often described how organisational change is so difficult because there is a seeming unconscious conspiracy happening within the organisation. The system keeps everyone very busy doing something. Doing something is better than doing nothing – isn't it? Not when it keeps you too busy to pause. Too busy to think. Too busy to notice the existing patterns of assumptions. Too busy to bring those unconscious assumptions out into the light of day to discuss them. Too busy too chose alternative paths.

But was Gillespie's article suggesting something more sinister? A conscious decision to conspire for ineffectiveness. Claiming to be doing one thing, charging big fees for it, and never achieving the marketing claims. Green washing, as we have come to know it.

I began to reflect on the world of professional coaching. In what ways have we been complicit? I venture to suggest we have all been complicit, I know I have – and am I even more at fault, because I have been conscious of it for some time? So, let me see if I can help us all take stock and count the ways in which we have been complicit with a system that is not serving us, our communities, our society or our planet as we come together here.

My intention is not to whip us all with guilt. Rather to raise our awareness, so that we can make better choices and innovate new ways of doing business together that have a hope of making a real difference. As we step into this exercise, a reminder. Can we be gentle with ourselves? We have been taught to think this way. We have been inculcated – deeply socially conditioned - and trained. We have not questioned deeply enough. We have been in The Matrix. Take the red pill and read on (also see Rita Symons on Day Two, p78). Taking the blue pill, which is a choice, will let you return to dystopian ignorance where the possibilities of awareness appear only in worsening nightmares.

I was told once by a colleague from Scotland, that it's always easier to run into the cold sea hand in hand. So, let's dive in together, swallow the red pill, and I'll start us off with a few examples I have become aware of. And then perhaps you can continue with your own examples to keep us rolling. Please use the sticky notes and pens on the floor. When you think of something to add to the list, I invite you to go to the centre of our circle, pick up a sticky, write your unconscious assumption concisely on the note. Then post it on the wall. We will have quite a diverse collection given that some of us coach teens, mums and dads; some coach small business owners; and some coach in corporates.

My first example is one that I noticed long ago. That the organisations I was working for and with, were exploiting or extracting from their people and the planet. This didn't make the people working in them horrid of bad. They operated within the unconscious assumption that everything could be 'mined' for all that was it possible to extract.

Let me more specific.

- Productivity was about getting as much effort out of staff as possible, for as little as possible.
- Marketing, sales and product delivery was about getting as much profit out of every transaction as possible.
- Procurement was about obtaining all the resources needed for as little as possible, without even a thought given to the impact on biodiversity or natural ecologies.

No thought was given to the follow-on effects of these tactics of doing business.

My coaching indirectly or directly helped my organisational clients do all this 'exploitation' ever more efficiently as I helped to develop high performers. More effectively, as I helped executives understand how to inspire their people to new heights of performance. More quickly, as my coaching helped people endure and devise coping mechanisms for a relentless system of 'more with less'.

Another assumption that I never had the courage to challenge within my blue-chip clients was the assumption of endless growth. Closely aligned to this is the assumption that if you can create a market for something, then it is legitimate to satisfy the demand you have created. (Products that come to mind here are aerated, sugar laden soft drinks and seasonal women's fashion.) This strategy helps endless growth immeasurably! Consumers don't have to identify the need for a product – companies can manufacture the need.

How should we as conscious coaches respond to these realisations?

- Some walk away and work in the not for profit or government sector where economic growth is not an issue. Or is it?
- Some argue that it's better to meet clients where they are and work them through their own realisations as they arise. Do they ever arise?
- Most work within the existing system, replicating the existing patterns of assumptions because we too work within the same economic system that is the root cause. We too need to eat! Do we want to continue doing this?

Morally, can you or we keep doing this?

These are big assumptions or patterns to change. What's needed is a broadening of awareness and responsibility for the impact of our actions but first we need to become aware of what has 'got us'. What unconscious assumptions are we holding that are keeping us as a 'prisoner' of our own thinking because we can't see them, but we could surface into awareness?

It takes courage to look into the mirror. I encourage you to do so. How can we ask our clients to do likewise if we have not done so?

Exercise

I wonder what examples are coming to your minds as you begin to consider these issues? The post it notes are here. Please grab your pen and write.

*

CONVENER: Thank you, Josie, I have really enjoyed this provocation. You've helped me crystallize my own thinking and doing here which I offer in return. It seems obvious now, but, wherever we work in a system, whether it's for profit, not for profit or central/local government, it's the same system. These are three legs of the same stool.

By choosing to work in the not for profit or government sector – because we don't like the values of the 'for profit', then we could easily choose to deny how these aspects interrelate and enable each other, another layer of denial to work through. Once again, radical interdependence shows up. John Perkins (2018), *Confessions of an Economic Hitman* provides some under the hood examination of the design of our toxic system and the deep interconnections between war and disaster, charity and relief, international aid, economic indebtedness and resource extraction. Climate change is feedback for this system.

I am now going to ask Peter to expand on what others have shared to bring us back to how coaching awakens by being aware of its historical roots and how these need rethinking.

PETER: In focusing on the awakening of the coaching profession I want to reflect on how coaching needs to move from being individual and ego-centric to being eco-centric.

For coaching to be really eco-centric, we need to go beyond including the wider environmental issues as one or more items in a personalized coaching agenda. We also need to avoid the trap of thinking that eco-coaching is what happens if, and when, we coach 'out-of-doors', in nature. For Nature is not just 'outside' it is everywhere, and we are part of it. It flows through us in the air we breathe, the food, water, sensations, thoughts that we take in, the chairs we sit on, the heating we absorb and much more. Instead, we need to radically and fundamentally revision the whole coaching endeavour and reshape the foundations of coaching's basic assumptions and plant its roots in different soil. Not only is Nature and the ecology part of us, but we are also inextricably a very small and temporary part of eco-system. And even when we go 'out-of-doors', we carry the closed doors of our mind with us.

Modern executive coaching grew out of humanistic psychology, which in turn emerged from the humanistic and existential beliefs that came to the fore in the 'White Western World' after the end of World War Two (Hawkins, 2017b). At that time these beliefs were liberational and created new hope in human potential, after the dark days of the war and the Holocaust. But they were a child of their times and wider culture, with a focus on individualism, personal growth, humancentrism and limitless development. They were the psychological and interior twin of the white western consumerist and growth addicted cultures, which spread from North America and Western Europe to many other parts of the world. Everything was seen as available and in reach and could be bought and acquired, supposedly including, 'enlightenment'. We were caught in a competitive 'rush to find the fastest escalator up Maslow's Pyramid' (Hawkins, 2017b: 255).

Executive coaching became the richest and most seemingly successful child of humanistic psychology, providing 'very expensive personal development for the already very highly privileged', (Hawkins and Turner, 2020a: 3).

The work was 'client-centric', with the coach a hired supplier, who could be personally chosen from a shelf or panel of prospective possibilities, and the organization paid the cost. There was a contract formed, often starting with asking the 'client' what they wanted from coaching? The goals that emerged were a mix of personal development aspirations and greater performance contribution to the efficiency of the organization and its success.

For coaching to be eco-centric requires a whole new paradigm, based on new beliefs such as:

- Coaching is not something done by the coach, but is a collaborative inquiry and exploration done in partnership between a coachee and a coach.
- Where the coaching agenda is not based on what the coachee wants or needs, or even what their boss, team or organization want and need from them, but what life is requiring.
- Both coachee and coach are in service of what is necessary to co-create a greater beneficial value, with and for, all the coachee's and their organisation's stakeholders (Hawkins, 2021; Hawkins and Turner, 2020a).
- Where both the coach and coachee go to the learning edge and stand together in vulnerable unknowing, deeply listening to what the wider world requires.
- Where service is more valued that personal success.
- Where flourishing and diverse life is more important than growth, whether that is personal growth, quarterly profits or country GDP.

James Hillman taught me many years ago that growth is just one arc of the circle of life. That growth in spring and childhood is to be welcomed, but that growth in middle age is either over-consumption, obesity or cancer.

Coaching needs to see the myths and basic assumptions it is ensnared within. In a world where there are more books on team building than team ending; on on-boarding executives than off-boarding, on success than on learning from failure, and on living abundantly than dying well. Coaching books and trainings have focussed more on ways of creating a successful personality that on processes for 'unselfing' or what traditional Sufis would call 'Fana'; and on finding 'your Self' than on letting go of your attachment to who you think 'Your Self' is.

We cannot fully contribute to the necessary eco-systemic shift in human consciousness, if we do not first shift our own fundamental paradigms and coaching precepts that grow out of it, as well as the language forms in which we practice. In Eco-Engaged on Day Four we will work together to discover new language and forms of engagement, but today we need to awaken to, own and take responsibility for, our inheritance – a profession built on white western capitalism and consumerism, individualism and self-making, and growth addiction.

Professions, like people, can become older and more pronounced versions of their earlier self, or more fully evolve and mature. A number of writers have started to describe the maturational stages of an individual coach's craft (Hawkins and Smith, 2013; Hawkins, 2011b; Clutterbuck and Megginson, 2011; Lawrence and Moore, 2019). Very little has been written on how the profession matures. We can begin to explore here on this workshop how we move coaching from short-term problem centric focus to one which is more relational and development centric which are still higher stages of ego-centric coaching. Then we can explore how we move to coaching to first being 'stakeholder-centric' and systemic' (Goldsmith and Silvester 2018; Lawrence and Moore, 2018; Hawkins and Turner, 2020a) to 'future-life' and 'eco-systemic centric' (Einzig, 2017; Hawkins and Turner, 2020a, chapter 8).

If coaching is going to increase its positive contribution to the great challenges of our time it needs a maturational leap. I do not believe we need to dissolve coaching, which anyway has a life of its own, which we cannot control, but dissolve our construction of it. To recognize that much of it is a home that nurtured and supported us, but in many ways has become a prison that severely limits the difference life is calling on us to make.

EVE: Thank you, Peter, so how do we feed ourselves and our families and live and work as a coach as our climate and ecological awareness grows?

I wonder whether it's as much about where we work and who we work with as also realizing that we are an embedded part of the toxic system we are trying to change.

Being a coach does not absolve us of responsibility for what we see happening around us.

There isn't 'bad' out there and 'good' in here, it runs through everything if we choose to even use those labels.

Exercise

Some questions I reflect on which I now invite us to use in small groups are:

- In what ways, as a coach, am I also enabling the system to perpetuate itself through me?
- In what ways are you enabling this too?
- What are the opportunities I have to breathe life into a different awareness through my coaching practice?
- What might I need to do to steady myself as I look at the matrix and hold that awareness?
- What other levels and layers of the matrix am I not seeing? What else am I also blind too?

*

CONVENER: Welcome back. In the group I was in as we pondered these questions, I was also aware of the extractive nature of our own industry.

As a coaching supervisor I have come across examples where coaches have reached out for supervision, and I have been unable to meet their needs. So overwhelmed was I with the system they are being asked to operate in as one that is toxic and extractive. More self-work required if I am to be of value for these coaches.

New coaches, having invested significant resources (financially and personally) into their coaching qualifications and doing really good work, can end up working in very toxic environments, just like anyone else where their own wellbeing and health is degraded. They feel they have very few choices if they are to feed themselves and pay the rent whilst also following their ambition to coach.

Here I notice my own privilege as a more senior coach, with greater financial security and choice about where I work and who I work with, it's easier to ask these questions and make different choices.

Josie, you've focused on the coaches above, but what are the layers here?

JOSIE: First to be able to allow ourselves to see and accept the matrix for what it is whichever sector we are in – and not to judge ourselves or others for the choices they make about the work they do. We are all supporting this toxic extractive system.

Second, what are the changes we need to make in our coach training programmes to allow recognition of the system, the drivers and to work with it whilst retaining a grounding to allow a better system to emerge. This we will address on Day 5.

Third. What are the changes we need to make to transform our professional bodies that start to really tackle the extraction and privilege in our own profession, both within our profession and through our profession? This we will address on Day 6.

What makes this a particularly imperative examination for coaches and the coaching industry is because we are told that we are agenda-less – we follow our clients' agenda. This idea of agenda-less conversations means without thinking further, we absolve ourselves of what happens next.

Yet the reality is we are not agenda-less. We coach to put food on our table, to build coaching businesses and to have impact. We describe ourselves as catalysts, facilitators, enablers, so we have many agendas. Our agenda is to thrive within the current system, our agenda is to have ease and build our happiness, to be valued for what we do – and that is just the tip of our agendas. We might not be dictating the goals of the coaching conversation, but we are choosing what to focus on and what questions to ask so have many agendas. Those things can get hidden and denied when we describe coaching as in service of the client's agenda as Peter addressed earlier.

In taking the red pill in order to allow ourselves to see the matrix, we must also see ourselves:

- What are your agendas?
- What are mine?
- How might these agendas prevent us from stepping into greater alignment with a flourishing Earth?

CONVENER: Thank you, David, Alison, Josie, Joel and Tina, Peter and Eve, together you have beautifully addressed the Awakening of Coaching and set the context and backdrop for our collective explorations on this journey together. You have each challenged us to think deeply about how coaching can regenerate and make the most beneficial contribution to the great challenges of our time.

I am reflecting on this quote from Margaret Heffernan:

> The next time someone tells you – 'we've cut costs', simply ask them, and where have you moved those costs to?
>
> (Heffernan, 2014)

It has been a rich day of awakenings, but now we need some time to sleep. Tomorrow we will start bright and early and have invited some scientists to travel with us and to help us collaboratively inquire into what is the essential basic science and the key data we all need to know to be Eco-Informed? What is the foundation we can stand on with confidence so as not to be swayed by the latest social media or corporate/governmental propaganda? May you have rich dreams.

Chapter 5

Eco-Informed

Day Two Morning: 'Listen to the Science'

CONVENER: Welcome to Day Two. We are going to start with an original poem from Gillian Walter (2022).

Mother Earth's Dilemma

They grow up so fast,
Just the blink of an eye,
Since they took their first breath,
Since they first saw the sky,
Their mother,
Their father,
This bountiful home.
It was love at first sight,
But my, how they've grown.

Perhaps it's my fault,
Having so much to share,
I spoiled them with love,
Couldn't see beyond their
Passionate,
Curious,
Desire to grow,
Their knowledge, their impact-
How could I say no?

I gave them my all,
Offered life to sustain,
But they wanted more,
Sucked life straight from my veins-
Emptying,
Poisoning,
Air, mountain and shore,
Pawning all that I have,
Desperate for more.

DOI: 10.4324/9781003153825-5

Their Father's frantic,
He's turning up the heat,
Calling in the debt,
Where their actions greet
Destruction,
Extinction,
Death of the seasons.
No thought of family,
No sense of reason.

It pains me to say,
But I know it is true.
Resources are drained,
A reckoning is due-
Pandemic,
Disasters,
Collapsed foundations.
All that will stop them is
Species starvation.

Yes, puberty's tough,
Rebellion's expected.
Tomorrow plundered,
Future lives neglected as
Irrelevant,
Unceasing,
As certain as now.
Great-grandchildren's problems
Aren't theirs anyhow.

My children, the most
Creative inventors,
Can find solutions,
Be the greatest mentors-
Inspiring,
Preserving,
Balance aligning.
Defining, designing,
Insights combining.

Alone, we are lost.
Together there's hope of
A great renaissance,
A new era to love
Unity,
Harmony,
Learning collective.
Essential awareness,
Our new home perspective.

CONVENER: Thank you, Gillian, what a great opening to our second day! We will today explore what we the rich privileged children of the Earth have done to our Earth Mother as well as learn from those who have stayed in closer indigenous connection with their land. But first, we hope you slept well after hearing the great kaleidoscope of individual stories we heard yesterday and the stories of coaching evolution we co-created. Take a moment to reconnect with what stood out the most for you from our opening day.

*

CONVENER: Today we turn to the story of our suffering planet, its Earth, oceans, atmosphere and all the living beings that share this great home with us.

Greta Thunberg tells us so called grown-ups. 'Listen to the Science' and today we will, but we would like to widen this frame further and say let us listen to what the wider Earth our home is saying to us, what is it trying to tell us, let us know, wake us up to?

Our inquiry will be into the question:

> What is it that we each must choose to learn and hold in awareness to be appropriately eco-informed?

To do this we consider, what is the sufficient knowledge we need to have a strong foundation, when the external winds of social media, political posturing and company greenwashing and clients' denial blow around and through us? How do we find a place of balance between those who try and convince us that the ecological crisis is not as bad as we think, or that human ingenuity will solve it, or from the other direction – it is all too late, and we might as well give up doing anything to remedy the situation.

Ed Gillespie (2021) describes how we are all on a tight rope between denial and despair, and tomorrow we will explore how we can face these emotional reactions within ourselves building the capacity within us to hold a space for our clients, friends, family, community to do the same. However, today we will explore how we overcome our own and others' wilful blindness and face what the world is trying to tell us, with our eyes wide open.

To become eco-informed, we all need to be curious seekers so let us start with a few words from Jarid Manos (Manos, 2007: 376):

> Become a seeker, don't be afraid of your journey. A culture of atrocity has become the norm for us. Wherever we look – if we take the time to look – we can see colossal suffering, destruction, waste and killing done if not by our own hands, then in our name, yet we're numb and move on. The violence we do to the Earth mirrors the violence we do to each other and often accept into ourselves. Pathological violence has become Perfectly Acceptable Behaviour.

Today there is a lot for us to face and look at with eyes wide open and as we do this we will need to support each other, encourage ourselves to keep breathing through the enormity of what is happening right now in our world, to stay with it. We intend to not only listen to the science with our conceptual mind, but also listen with our hearts and guts and whole being. And not only to listen to the science but to what the 'more than human' world is telling us and teaching us. We also need to begin to discover how to anchor ourselves, or maintain our sense of balance, amongst the noise of different opinions and what is, what might be and how we need to respond.

We are going to:

- Look at the data and the consensus of the scientists, which we will build together.
- Explore the habits and blocks that get in the way of us really absorbing this and incorporating it into our worldview.
- Engage with other forms of being eco-informed – ways of listening to the Earth and to other sentient beings and life forms and what they can teach us.
- Go on a 'Deep Time Walk' to understand the evolution and history of our shared Earth both through our ears and through our bodies.
- Return briefly to the systemic patterns underlying the environmental science, before examining global inequity as a key aspect of the ecological crisis. After all climate crisis is a symptom of what is going on elsewhere in the global system, not a cause, and is connected to elements such as racial, social and economic inequality (see Thomas, 2020; Mikati *et al*, 2018 and Rita Symons later today).
- End the day bringing our listening to Earth, back into a 'Council of all Beings' a ritual created by Joanna Macy and John Seed which has been used in many countries around the world.

Exercise

CONVENER: But before we hear from those who will share more about what is happening, let's come up with our own list of the ten most important facts or information about the climate and ecological crisis that we think every coach needs to know. You can do this alone, or in a small group and each group will come up with their list and then rank them in priority order. We look forward to hearing and sharing what every individual or group comes up with.

*

CONVENER: Welcome back. Let us hear what you came up with.

We are going to do this as a 'collective build' (Hawkins, 2021: 114–116) so please can each person or group share just the most important issue from their list, so others can build on this, from what they have written down. Together we can co-create a document that is wiser than any of our sub-group lists and more than the sum of all our lists put together. Then we will start a second round and hear issues so far not mentioned.

Together by tapping into the collective wisdom I am confident we have come up with a much better list than any of us could have done alone. Please as we read it out, note down all the feelings and emotions it triggers within you as we will be exploring these tomorrow.

1   The planet's average surface temperature has risen about 1.8 degrees Fahrenheit (1.01 degrees Celsius) since1880 (NASA, 2022a), a change driven largely by increased carbon dioxide emissions into the atmosphere and other human activities. Most of the warming occurred in the past 40 years, with the seven most recent years being the warmest. The years 2016 and 2020 are tied for the warmest year on record (NASA, 2022b). The Paris and Glasgow COP agreed we needed to stay below 1.5 percent but currently we are on track to rise by approximately 3 percent by the turn of the century.

2   To avoid catastrophic climate crisis, we need to halve carbon emissions by 2030 and get to zero emissions by 2050.

3   Fossil fuel use is still increasing by 1.7 percent a year – the growth in renewables is just absorbing some of the increased demand. In 2009 fossils fuels made up 80.3 percent of global energy mix and by 2019 it was 80.2 percent! (Ren21, 2021).

4   Global sea level rose about 8 inches (20 centimetres) in the last century. The rate in the last two decades, however, is nearly double that of the last century and accelerating slightly every year (NASA, 2022b).

5   Since the beginning of the Industrial Revolution, the acidity of surface ocean waters has increased by about 30 percent (NOAA, 2022). This increase is the result of humans emitting more carbon dioxide (CO_2) into the atmosphere, and hence more CO_2 being absorbed into the ocean. The ocean has absorbed between 20 percent and 30 percent of total anthropogenic carbon dioxide emissions in recent decades (7.2 to 10.8 billion metric tons per year) (NASA, 2022b).

6   The science indicates that the climate and ecological crisis is now rather than in the future, and happening faster than the scientists had previously predicted.

7   Loss of species and biodiversity. 'One-third of corals, freshwater molluscs, sharks, and rays, one-fourth of all mammals, one-fifth of all reptiles, and one-sixth of all birds are heading towards extinction' Kolbert (2015). In less than 50 years the population of mammals, birds, amphibians and fish has dropped by a staggering 68 percent (Polman and Winston, 2021: 14).

8   There are already 150,000 human climate related deaths per year and this is expected to rise to 300,000 by 2030 (Kasotia, 2022).

9   Our most significant non-renewable geo-resource is productive land and fertile soil. The world grows 95 percent of its food in the uppermost layer of soil, making topsoil one of the most important components of our food system. But due to conventional farming practices, nearly half of the most productive soil has disappeared in the world in the last 150 years (WWF, 2022), Each year, an estimated 24 billion tonnes of fertile soil are lost due to erosion. That's 3.4 tonnes lost every year for every person on the planet (Global Agriculture, 2022). This threatens crop yields and contributes to nutrient pollution and dead zones as well as erosion. In the US alone, soil on cropland is eroding 10 times faster than it can be replenished (Cosier, 2019). If we continue to degrade the soil at the rate we are now, the world could run out of topsoil in about 60 years, according to Maria-Helena Semedo of the UN's Food and Agriculture Organisation. (Cosier, 2019).

10 Climate change induced floods, storms, forest fires, droughts are increasing each year and are already making people homeless, hungry and furthering ecological migration. In this century "An estimated one to three billion people will become climate refugees" (Polman and Winston, 2021: 14).

CONVENER: Let us now see if we can build on this list by hearing from some of the those who have dedicated much of their working lives to exploring these areas.

Exercise

CONVENER: This is tough and daunting to hear but please again note down what strikes you as critical and then in a column alongside the feeling and emotions this raises for you.

*

CONVENER: The UN Inter-Governmental Panel on Climate Change (IPCC) 2021, which was the world's largest, studying more than 14,000 scientific papers, said that it is unequivocal that human activity is responsible for climate change. United Nations Secretary-General António Guterres (2021) put it this way:

> Today's IPCC Working Group 1 report is a code red for humanity. The alarm bells are deafening, and the evidence is irrefutable: greenhouse-gas emissions from fossil-fuel burning and deforestation are choking our planet and putting billions of people at immediate risk.

So that debate about its cause is over. The climate crisis is not in the future, it has already started and will get worse. Scientists point out that some of the changes that have happened or are now in progress are irreversible such as temperature and sea-level rises. However, they argue that with 'deep cuts' in emissions of greenhouse gasses we could limit the worst excesses of climate change and move towards stabilizing the climate. For this, the period to 2030 is the most critical. The time to act is now.

What is the science telling us? We invite Zoe to share how she might summarise the key messages."

ZOE COHEN: The science tells us that the dominant story we are telling ourselves is a lie.

JOSIE: Wow! That is a powerful and confronting opening statement Zoe…you mean 'we' as in a cultural narrative? And to wake up from what has sometimes been called a trance, we need people to tell the truth as it really is? As we mentioned earlier, that can be very uncomfortable and disconcerting. It's also very different to the way we usually think, or not, about our lives and the impact our lives have on the planet. Maybe this process is like getting ourselves off autopilot; waking up from the trance of the story of progress – and becoming more conscious of our choices, even to the choice of what we believe.

ZOE: Yes, I am discussing what we have come to accept as normal within some of the very big systems here. Recognizing that extractive capitalism feeding individualistic consumerism and never-ending economic growth not only doesn't make us happy but is in the process of annihilating life on Earth and accelerating the irreversible tipping points from what was an awe-inspiring Eden, to an un-survivable hellscape.

The sheer scale, complexity and extent of the harm we've done – much of it in just the last three generations – is hard to conceive and understand, even when we want to… even the experts find it hard (Bradshaw et al, 2021: 1):

> The scale of the threats to the biosphere and all its lifeforms—including humanity—is in fact so great that it is difficult to grasp for even well-informed experts. … what political or economic system, or leadership, is prepared to handle the predicted disasters, or even capable of such action. … this dire situation places an extraordinary responsibility on scientists to speak out candidly and accurately when engaging with government, business, and the public. We especially draw attention to the lack of appreciation of the enormous challenges to creating a sustainable future. The added stresses to human health, wealth, and well-being will perversely diminish our political capacity to mitigate the erosion of eco-system services on which society depends. The science underlying these issues is strong, but awareness is weak. Without fully appreciating and broadcasting the scale of the problems and the enormity of the solutions required, society will fail to achieve even modest sustainability goals.
>
> (Bradshaw, 2021)

I encourage readers/participants to read the full article… in the meantime I invite you to re-read this quoted paragraph and take some deep breaths to help it permeate.

Exercise

- What might be your responsibilities as a fellow human being and as a coach at this time?
- What might be the role of coaches in this reality we find ourselves in?
- What do you notice this provokes in you?

Asking 'what does the science tell us?' cannot be asked in a vacuum, just as there is no such thing as neutrality in coaching.

*

ZOE: Ever since Exxon's top scientists told their bosses as early as 1977 about their (very accurate) forecasts of global heating and its impact (Hall, 2015), many of the fossil fuel companies, and their paid politicians, have been blocking proportionate public communication and action. Right up to the 2021 COP26, where the 500 strong fossil fuel representatives were the biggest single 'delegation'.

There are also other rich and powerful industry lobbies protecting their damaging practices – from the dairy, meat and agribusinesses to the chemical companies.

The biology, physics, chemistry and ecology that govern how the natural world (aka the real world) actually work, tells us loud and clear that we have not only caused over 1.2C degrees global average heating already (compared with pre-industrial levels) and this is unequivocally driving the breakdown of stable weather patterns and more and more extreme fires, floods, typhoons, storms and droughts; but that our actions have actually broken a number of the Earth's Planetary Boundaries (Stockholm Resilience Centre, 2022) (Figure 5.1). Scientists have identified nine quantitative planetary boundaries within which humanity must live if it is to survive and thrive. Crossing these boundaries increases the risk of generating large-scale abrupt or irreversible environmental changes.

JOSIE: Zoe, may I interpret your flow for a moment please? I have worked with scientists in the past too and I notice that they often talk about degrees of changes in temperatures and planetary boundaries – but these terms don't seem to make an impact with people. I was wondering if we could translate the changes you are pointing to into everyday stories and lives? For example, I remember when I was a little girl, I used to go fishing with my Dad. One day we went out in his small boat and it was like going to the shops really! We started by digging up some pippies or cockles on the beach, then we went over to the jetty for crabs, and then, we boated over to the sandy area further out into the gulf for some whiting. This experience is no longer a possibility for people living near my home now. The fish stocks have declined so much.

I'm wondering what our participants here may be thinking of as I share this. In what ways have you noticed changes that are personal examples of the often slow creep of climate and ecological destruction? Perhaps it is in the frequency of fires, floods, typhoons, or snow storms, that cut off and close down entire towns? That cut off access by roads and make water and food supplies something not to be taken for granted. Some ecologists also point to the emergence of Covid-19 as an example too.

Exercise

CONVENER: What have you noticed? Write these down or share with a partner.

*

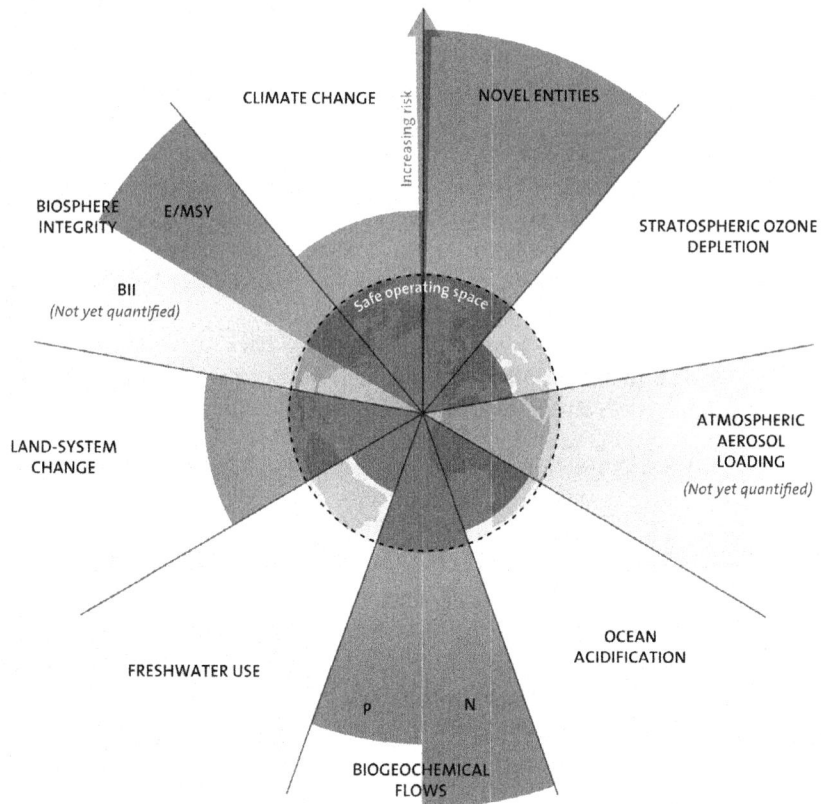

Figure 5.1 Planetary boundaries
Credit: Designed by Azote for Stockholm Resilience Centre, based on analysis in Persson et al., 2022, and Steffen et al., 2015, reprinted with permission

ZOE: Thanks Josie, it's great to recognise this link between these concepts and our lived experiences. It brings us to notice that fundamentally we, our children and our grandchildren are in danger. And being rich and privileged enough to read this book does not mean you and they won't be affected. There is no one and nowhere exempt from the effects of what we have done and are still doing. We are in the 6th mass extinction – around 200 species go extinct EVERY DAY – and 1 million are on a path to extinction in my lifetime (United Nations, 2019). Whole branches of the tree of life being broken off or burnt.

JOSIE: And of course, it's not just about what we destroy to use is it? We also need to be concerned about the waste we assume that Mother Earth will willingly go on absorbing for us too.

ZOE: Exactly! As everyone here today recognised in your opening conversations, man-made persistent organic pollutants (such as PFAS in Teflon and many packaging and clothing products for example), are present in every organism on Earth. They never degrade and they are endocrine disruptors. There are 350,000 manmade chemical pollutants we have pumped out into our planetary home.

Oceans cover 7/10ths of the surface of the Earth, and yet they are so polluted with plastics and a cocktail of chemicals (from agricultural fertilisers and pesticides, to cosmetics, sunscreens and prescription drugs), that there are huge dead zones. This pollution is exacerbated by the massive percentage of excess atmospheric heat (90 percent+) that the oceans have absorbed and the CO_2 all combining to drive acidification.

These factors, plus industrial fishing, linking back to your earlier story Josie, have killed half of all marine life, and risk a total collapse of much of what remains – all within my lifetime (Dryden and Duncan, 2021).

JOSIE: Let's pause and take a breath because we are not done yet.

Exercise

CONVENER: Breath in and out. Notice the emotions you may be feeling as you absorb this information. Are you absorbing it? Do you notice that you are reading quickly to get through it? Or is it sinking in…. deep into your core. If so, you may be feeling anxious, sad, angry, frustrated, hopeless, or numb. All sensations are normal. Just notice them and hold yourself, mind and heart, open as we move on.

*

JOSIE: What else is connected to this seemingly endless line of destruction, Zoe?

ZOE: Well, there is the social justice issue that sits at the heart of the impact of climate and ecological destruction. Responsibility for all this does not sit equally with all of humanity. Cumulative excess emissions data (over 350ppm) show clearly that Global North countries together are responsible for 92 percent of emissions, whilst the Global South are responsible for only 8 percent (Hickel, 2020) despite getting hit with most of the earliest and worst impacts (Figure 5.2).

ZOE: In the present day it is the top 10 percent of the world's richest who causes 50 percent of all consumption emissions (Oxfam, 2015) (Figure 5.3)

ZOE: In 2015, governments agreed to the goal of limiting global heating to well below 2C degrees and ideally to 1.5°C above pre-industrial levels, i.e., the Paris Agreement. In 2022 EVERY government on Earth is failing to even make pledges to hit anywhere near this trajectory, never mind take the actual action needed.

Let's bring this down to a personal level – to perhaps make it more real for us all, when thinking about yourself, your family, your clients.

Responsibility for climate breakdown

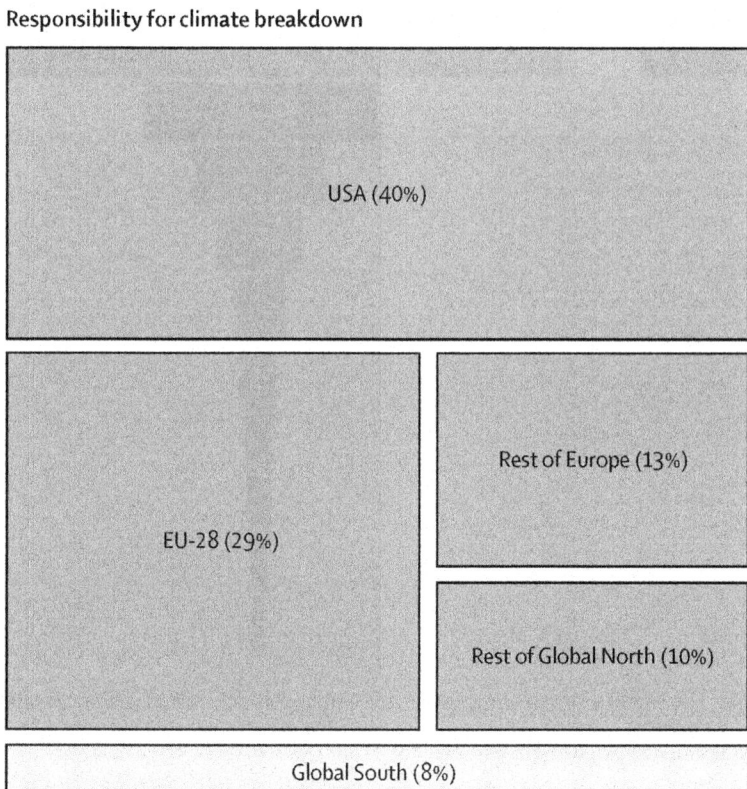

Figure 5.2 Responsibility for excess emissions
For the purposes of this analysis, the term Global North refers to the USA, Canada, Europe, Israel, Australia, New Zealand, and Japan, whereas the term Global South refers to the rest of the world: Latin America, Africa, the Middle East, and Asia.
Credit: Jason Hickel reprinted with permission

To stay within this 'guardrail', every person on the planet would need to reduce their consumption by 50 percent and emit an average of just 2.3 tons of CO_2 per year by 2030 as research from Oxfam (2021) shows. In fact, we actually need to get emissions to zero, and then net negative to get back to safe levels of 350ppm – we are at around 420ppm now. The latest IPCC report (2022) underlined both the scale of the issue and how its impact is unevenly felt: 'how human induced climate change is causing dangerous and widespread disruption in nature and affecting the lives of billions of people around the world, despite efforts to reduce the risks. People and eco-systems least able to cope are being hardest hit.'

Oxfam calculates that by 2030 someone in the richest 1 percent would need to reduce their emissions by around 97 percent compared with today to reach

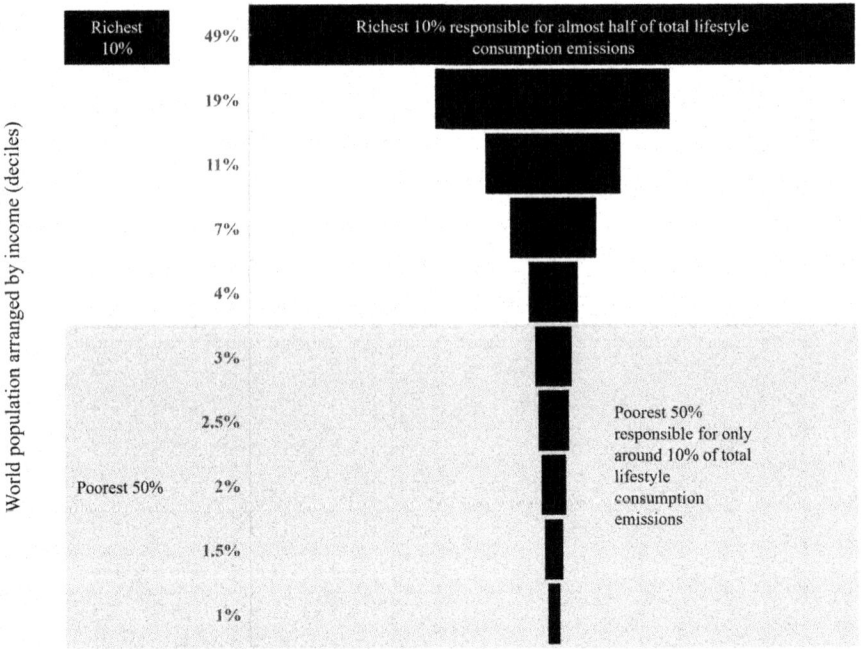

Figure 5.3 The world's richest 10 percent are responsible for half of global consumption emissions

Credit: Oxfam, redrawn with permission

this level (Gore, 2021). (By 2030, you would need an annual income – expressed in $2011 purchasing power parity - of more than $172,000 or more to be in the richest 1 percent).

Science without social, political and economic context isn't fully meaningful or actionable.

Scientific modelling also tell us clearly that the conditions human communities have thrived in for 1000s of years during the Holocene are gone – and that the 'human climate niche' (Chi et al, 2020; Xu et al, 2020) (parts of the world where Mean Annual Temperature is approximately 13 degrees C) is going to shift and diminish so much that 3.5 BILLION people will be in parts of the world which will be uninhabitable by 2070. This includes the Middle East, large parts of Africa, parts of China, Australia, India, Central America, South America, and southeast Asia. This is anticipated as the biggest genocide in human history and will occur without very drastic emissions cuts very soon. If allowed by the world's leaders it will surely result in mass death, suffering, starvation, mass migration, conflict and societal and economic breakdown.

None of this 'news' is really very 'new'... The Limits to Growth (Meadows et al, 1972) report broadly predicted it in 1972. And an updated re-run of their modelling in 2004 (Meadows et al, 2004) and again in 2021 (Herrington, 2021) says we're still on track for overshoot or societal collapse by 2040.

Science also tells us clearly that every single tonne of CO_2 released makes this all worse (IPCC, 2021), and that several of the Earth systems' tipping points are now 'active' (Lenton et al, 2019) – a euphemism meaning they are dangerously near to flipping to their new stable, but catastrophic, state. For example:

- Permafrost across the Arctic is beginning to irreversibly thaw and release carbon dioxide and methane — a greenhouse gas that is around 30 times more potent than CO_2 over a 100-year period
- The Amazon rainforest is at high risk of flipping from carbon sink to carbon source, and then to savanna within the next few years
- 99 percent of tropical corals are projected to be lost when global average temperature rise by 2°C (currently a near certainty based on the low levels of political will and action)
- Parts of the west Antarctic ice sheet are melting and triggering further collapse and more drastic sea level rise
- The Arctic is currently forecast to be ice free in summer around 2034, which is likely to trigger further abrupt extreme weather changes as it is the gradient of temperature difference between the poles and the equator that 'holds' the weather systems in place… they are already 'flickering' due to a more 'wobbly' jet stream (which led for example to the extremes of heat and cold in Canada and USA in 2021).

As we know as coaches, everything is connected – and everything in the field matters and can be having an influence. And so it is with Gaia…
Lenton et al (2019) shows the interconnection of tipping points:

Arctic sea-ice loss is amplifying regional warming, and Arctic warming and Greenland melting are driving an influx of fresh water into the North Atlantic. This could have contributed to a 15 percent slowdown since the mid-twentieth century of the Atlantic Meridional Overturning Circulation (AMOC), a key part of global heat and salt transport by the ocean. Rapid melting of the Greenland ice sheet and further slowdown of the AMOC could destabilize the West African monsoon, triggering drought in Africa's Sahel region. A slowdown in the AMOC could also dry the Amazon, disrupt the East Asian monsoon and cause heat to build up in the Southern Ocean, which could accelerate Antarctic ice loss.

JOSIE: I wonder what 'flipping to their new stable state means'? In complexity theory there is a concept known as strange attractors that a system gravitates towards or around. I sometimes think of it as a sheet of fine wire that someone has punched a dent into with their hand. If we place a ball on the sheet of wire, and wobble it around, the ball, will fall into the hold and wander around it as we move the sheet – but the ball stays in that locale. Flipping to another state – is like punching a second dent into the wire and jolting the sheet strongly enough that the ball moves into that attractor state. There is no way back to the original dent. For people who are used to linear progression this idea of flipping from one state to another is strange to our senses and experiences. It's difficult to believe it could happen. So, what does all this mean for us Zoe? How do you see it?

ZOE: Well, the science, in context, tells us very clearly that we must diminish, very rapidly, our extractive polluting impact on the Earth. We have known we must change since the 1960s and '70s, for example with the publication in 1962 of Rachel Carson's 'Silent Spring' (Carson, 1962), but we have continued to largely ignore our Mother's pleadings…

And now there is more anthropogenic 'stuff' than all of the living world (Pappas, 2020) - we are literally consuming our Mother (Figure 5.4).

Most mammalian biomass (weight) is now us (humans) and the animals we grow for food (livestock) – only 4 percent are now wild animals. Wild animals now mostly exist in picture books, Disney movies or in our childhood memories (Figure 5.5).

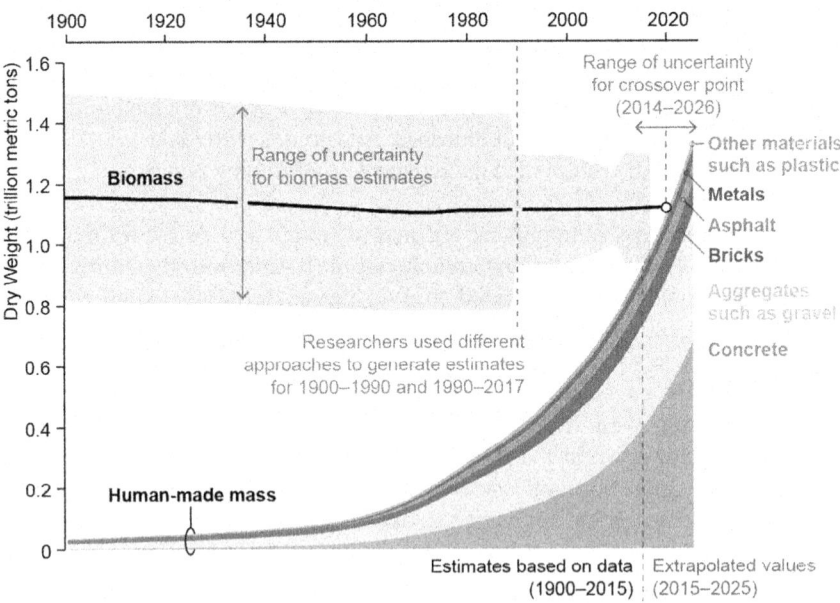

Figure 5.4 Human-Made Mass v Living Biomass, 1900–2025
Credit: Amanda Montañez; Source: *Global Human-Made Mass Exceeds All Living Biomass*, by Emily Elhacham et al, in *Nature*. Published online December 9, 2020, and reprinted with permission

Of all the mammals on Earth, 96% are livestock and humans, only 4% are wild mammals.

60% livestock

36% humans

4% wild mammals

Figure 5.5 Percentage of mammals

The science in context, tells us, because we are alive now, we are the last generation able to change this devouring super tanker's course, before it is truly too late.

But only a 'war-time style mobilisation' will do it (Mills, 2021) – even Britain's then heir to the throne, now King Charles III, says so. 'Quite literally, it is the last chance saloon'. All else is distraction activity.

So, we must produce and consume much less, right across rich nations, to protect our much poorer and predominantly black and brown brothers and sisters – as well as our own children and grandchildren.

There is an urgent need for managed degrowth (Hickel, 2020) of parts of the economy. Gross Domestic Product (GDP) growth drives materials' use and emissions and a recent article reviewing international research shows that 'Green economic growth is an article of 'faith' devoid of scientific evidence' (Ahmed, 2020).

There is no escaping it – we have only one Earth and we must behave as if that truth were true.

As we know we can't deal with an emergency without treating it as an emergency.

But another world really is just about possible…

Scholars of how we reduce growth (degrowth), ecological economists and indigenous peoples know this. For example, Millward-Hopkins, Steinberger, Rao and Oswald (2020) developed a simple, bottom-up model to estimate a practical minimal threshold for the final energy consumption required to provide decent material livings to the entire global population. They state:

> the basic material needs of billions of people across the planet remain unmet…. We find that global final energy consumption in 2050 could be

reduced to the levels of the 1960s, despite a population three times larger. However, such a world requires a massive rollout of advanced technologies across all sectors, as well as radical demand-side changes to reduce consumption – regardless of income – to levels of sufficiency.

So, enough of us need to see it – this kind of different kind of fairer, cleaner, safer future and want it enough to be prepared to step up to make it possible – and fast.

JOSIE: Let's pause again for a moment. Let all of this sink in. Degrowth – that's got to be a HUGE challenge for us personally and in our professional lives. I mean, the western way of life and work is founded on the belief that constant growth is possible and desirable. Can we imagine talking with our coaching partners and suggesting they reduce their budgeted sales? Change their business models completely? A multi-national corporate board shaping direction in ways that makes the world better because their business is in it?
It feels almost impossible!

ZOE: Whilst that's a slim chance, it's not impossible. Just as climate tipping points exist, so do social tipping points Roberts (2020) wrote:

> Will unexpected, rapid changes in coming decades be good or bad, positive or negative? That depends on millions of individual choices made in the interim. Some of those choices, if they happen at just the right moment, could be just the perturbations that spark cascading changes in social, economic, or technological systems. Some of those choices, in other words, will be incredibly significant.

> Which ones? That we cannot know. It could be any of them, any time. Precisely because we cannot know — because any one of our choices might be the proverbial butterfly's wings — we must act. We must take advantage of every affordance, grasp every opportunity. We don't know when history might unlock the door, so we have no choice but to keep pushing on it.

JOSIE: There have been other unexpected, unpredicted social changes of course such as the fall of the Berlin Wall and break-up of the USSR … and further back, the Enlightenment.

ZOE: And really, what else are we going to do? 'There is no lack of money in the world' (Harari, 2022) - it remains about political will and power relations… and that enough of us haven't yet woken up and applied the much-needed multiple sources of peaceful pressure on those in positions of power and influence.
There's no free pass for any of us in 2022, and that includes coaches…

CONVENER: Thank you very much Zoe. This can be very hard for many of us to take in and most human beings do not change by hearing facts. And yet decontextualized, atomistic or reductionistic facts are the diet we've been fed for many years with the hope that we will finally understand. As Joanna Macy so beautifully says:

> Of all the dangers we face, from climate change to nuclear war, none is so great as the deadening of our response.

> (Macy, 2022)

Let's pause for a moment and hear some facts that are placed within the context of a person's life. We will hear what this means from a coach in The Philippines, Jen Horn. Thank you, Jen, for sharing how life is for you and your community.

JEN: While the IPCC predicts the worst impacts of climate change by 2030, we have already been feeling it in the Philippines. Typhoon Rai (locally known as Typhoon Odette) ravaged my country in mid-December 2021. Many of the island destinations affected — including Siargao, Cebu, Bohol, Palawan — had just begun to open up again, after two years of limited tourism due to Covid, and one of the world's longest and strictest lockdowns (Olanday and Rigby, 2020). The fleeting moment of hope and optimism we had for our country's recovery in early December was short-lived. As if that was not enough, in January 2022, the new surge brought about by Omicron has limited mobility and access to health care services, especially for those already living in poverty or have been teetering on the edge of that in the past two years.

Typhoon Rai wreaked havoc in 38 provinces, displacing 475,000 people, injuring 1,100, and killing at least 407. While it was not the most deadly typhoon, it has been the most destructive one since Typhoon Haiyan in 2013. (Sadongdong, 2022). Typhoon Rai was one of about 20 typhoons that hit the Philippines each year. It decimated an estimated P24.5 billion (US$442 million/£365 million, as at August 2022) in farmlands and homes, battered buildings and coral reefs, leaving those who have lost everything uncertain about where to begin picking up the pieces of their lives.

And we're still bracing ourselves for even more climate impacts, being the third most vulnerable country to climate change (Alave, 2011; Flores, 2018) with rising sea temperatures already bleaching our corals and the threat of floods forcing more people to leave their homes in search of higher ground in the coming decades (The Philippine Daily Inquirer, 2019). Thinking of a place's climate resilience has now become a consideration for me as I decide where to live.

And I write this from a place of relative privilege within my country, comfortable in a home in Manila that my family owns, not worried if I'll have food on my table, or if I'll be able to afford health care when I need it. It's a privilege to even have an option to choose if I can live in a place that is more climate resilient. Within my country, I am like what the Global North is to the

Global South. I was born into a life of resources that I did not earn, and I have and use more than my 'fair share'. I am safe, insulated, and can write about this on my Macbook Air while a hot beverage goes cold in an artisanal ceramic mug on my desk. Out of guilt as much as empathy, I donate money, goods and time to different disaster relief and recovery drives, but I know that that is not a systemic or long-term solution.

My research into the motivations of sustainability leaders is what inspired me to pursue my coaching path (Horn and Wehrmeyer, 2020; Horn, 2020). I found that negative feelings like guilt, pain at the sight of social injustice or environmental destruction were strong motivators when my interviewees were able to process that guilt and pain through reflection and meaning-making. This, coupled with self-efficacy, or the knowledge that they had the capacity to do something and possibly make a difference, helped fuel their action.

And while I find myself in the position to support others, I am also grappling with some of the same feelings myself, in processing the difficult emotions brought about by both experienced and anticipated anger, guilt and grief. And the truth is, it feels like there's only so much time to sit around processing our feelings before the next environmental, economic or governmental catastrophe strikes.

Nonetheless, as one of my research interviewees shared, a 'daily reckoning' is important for her to recommit to her purpose – to protect our environment and all that depend on it.

CONVENER: Thank you so much Jen, for that powerful description of what life can be like in the so-called Global South, experiencing the impacts of climate change, and for sharing your own work and practice.

So, as we continue, we might consider, how do we open ourselves to become informed, to take in the data, the facts and allow us to hear, to be impacted and to change? Remember when you first trained as a coach, if you can, one of the first skills you learned was to listen. We need to listen to the science as significant data points that show up the symptoms of the dis-ease running throughout our plant, but then to listen deeper to sense the underlying patterns that are creating the symptoms. Profound listening can be the catalyst for significant change, building a deep connection and relationship.

As we go to lunch, let us pause to reflect.

Reflection

How has your perspective changed working your way through this section to here?

What do you choose now, to accept and to embrace to make the changes the world needs?

In response to all this and more, what it will take:

- to move me
- to move you
- to move us.

Chapter 6

Eco-Informed

Day Two Afternoon: Listening to the Earth

CONVENER: Welcome back and having spent this morning listening to the science and facing up to the terrifying reality or what we humans have caused in and around this planet, this afternoon we are going to listen to the Earth.

So, perhaps it is somewhat comforting to know that profound listening, listening to earth (brought to life in the poem with which we started this morning) and our fellow beings catalyzes our ability to become more fully connected to ourselves, to each other and to Earth. We will address more about deep listening on Day Four.

For a moment – wherever you are, open your window, and pause a moment. Listen to all the sounds or, if you are lucky enough to have a garden, step outside. Notice what you can hear and smell and be aware of what is connecting with you. How might this change at different times of the day – we can experiment!

We have invited Lilith Joanna Flannagan who has practiced deep listening through her life to join us. Lilith Joanna, welcome. We are keen to hear your experiences of building relationship with the more than human world.

LILITH JOANNA: "During a coach training, I facilitated an environmental systemic coaching session. The co-trainer, Meredith Freeman became aware of the presence of Earth and became a channel for its expression. In response, another participant asked, 'How can we help you?'. To which the Earth voice expressed their frustration by saying 'Stop asking questions! You are me! You have it in you!'

The experience I described above, invited me to a deeper level of acknowledgement that being a human I am also the planet, as a part of a whole, and the whole being present in each of its parts. Similarly, as a cell of a human body contains the microcosm of the human body, a human contains the microcosm of planet Earth.

A few years earlier I was doing research for my MSc Holistic Science dissertation. My dissertation was, in essence, about meaningful communication and collaboration in eco-systems. The experiential part of this involved the

DOI: 10.4324/9781003153825-6

River Dart. My dissertation was about intentional listening to non-human parts of the ecology.

I lived by the River Dart in Devon, UK. It is a truly beautiful river that found its place in quite a few artistic expressions over centuries. Through the few days of living, sleeping and being immersed in the wildness of the woods on the bank of the river my human mind quietened down; my guts rested from digestion. I was becoming like a hollow flute, through which the consciousness was coming. I was just being present to what was happening and allowing various perceptions to appear. Sometimes there were pictures as if seen by the riverbed. Sometimes there were physical sensations. At one point a particular awareness emerged. It seemed to take a form of a quite clear message: 'Humans have ambitions, plans, visions. We have been here for a long time and seen a lot. We can support humans in realising the ambitions in the way that it does not harm us.'

That experience had a truly profound impact on me and my work. I realised that if we, humans, consult our neighbours, that are non-human natural systems, about our plans, we may learn how to proceed in a way that is not harmful. If we act in collaboration, we may co-design methods that can be a source of breakthrough innovation. Then perhaps 'thriving' could better describe the result of response to climate change than 'surviving'?

Then I encountered another experience. I observed an act of collaboration between non-human and human colleagues. I participated in a concert at Plymouth University where a slime mould and a human together were co-creating a musical piece. Eduardo R. Miranda and Edward Braund (2015) built analogue computer circuits that include components grown from slime mould, which they call a biocomputer and composed a biocomputer music. For the composition, they programmed the biocomputer, harnessing as they say the slime mould's characteristics to resonate with the piano music and generate musical responses in real-time. These responses were played on the same piano through electromagnets that were setting its strings into vibration, producing a distinctive timbre.

The concert begun by the human artist playing the first few keys. Then there appeared more sounds coming from the body of the piano. It was the slime mould response that created the strings' vibrations. As the process evolved, I seemed to perceive a kind of entrainment emerging between the slime mould and the human pianist. They co-created a piece of art, a moving and beautiful music.

Although it was extraordinary to witness that phenomenon, I was present to a reflection of the slime mould exploitation for the purpose of an anthropocentric experiment. In other words, continuing to treat nature as a resource to exploit, which has resulted in the catastrophic human footprint on our planet and the climate crisis.

I genuinely felt sorry for the slime mould. At some point I realised I was assuming the slime mould's feelings. How about I engage in a dialogue with the slime mould, ask and perceive what it is like from their perspective? So, I did, following the methodology I learned when living by the River Dart. I become present to the slime mould's presence in the room. I greeted them and asked the question. The response I seem to have received was quite clear: 'I made this choice. I am willing to engage with humans and help them learn. I am enjoying this.'

For me that was another example of the willingness of other members of the natural world to engage with humans and collaborate in the cohabitation of our planet. That was also another example of asking and consulting other natural systems rather than assuming, the latter being a focus of human response to the climate crisis.

The consciousness of natural non-human living systems has been a subject of research for many years. In the case of plants, they have evidenced that plants are intelligent, sensitive, exhibit cognitive abilities, communicate between one another and with other life forms, communicate a meaning, oscillate, resonate, entrain, are sentient, conscious, able to generate intention and collaborate. For example, Witzang (2012: 103) finds plants as 'highly sensitive organisms that perceive themselves and can distinguish between 'self' and 'non-self''. While Buhner, citing Volkov (2014: 386) describes: 'the oscillatory nature of plants allows them to establish synchronicity with other plants, and life organisms, including people. Because, at root, plants are a 'vibrating' system that is capable of resonance. Plants can create and can live for years in collaborative relationships.' Buhner (2002: 197) observed 'when some trees were intentionally damaged by scientists, so they could not process nutrients properly, other plants in the range sent them nutrients to keep them alive (radioactively marked for tracking). This kept the trees alive a year longer than the ones not connected to a mycelia network'.

I personally chose to serve a mission of collaborative co-habitation of our planet. I integrated a systemic coaching methodology with holistic science and phenomenology. I developed a methodology of engaging in a meaningful communication with other members and systems of the natural world. I teach it in the ICF accredited coach training programmes as well as offering 'Dialoguing with Earth Certification' for environmental change makers.

CONVENER: Thank you Lilith Joanna, for those profound insights.

As a group, what do we notice about the pace of our own lives when we read these accounts? Listening as a coach requires us to slow down. Attending Earth as a cocreator requires a step change again, we suddenly notice just how much there is to attend to, to connect with, to learn from.

For those of us who may have trained in organizational dynamics we might be reminded that those who come last, need to listen from last place, in order to understand all that came before them. Listening from last place informs us how to act now. What is our place in relation to earth?:

> in Native ways of knowing, human people are often referred to as "the younger brothers of Creation." We say that humans have the least

experience with how to live and thus the most to learn—we must look to our teachers among the other species for guidance. Their wisdom is apparent in the way that they live. They teach us by example. They've been on the earth far longer than we have been, and have had time to figure things out.

Robin Wall Kimmerer (2020: 9)

We are definitely in last place. As Roman Krznaric (2020: 49), quoted John McPhee:

Consider….the distance from the king's nose to the tip of his outstretched hand. One stroke of a nail file on his middle finger erases human history." Let's pause for a moment. I know I needed to allow that to register with me in every part of my being.

Knowing within ourselves

CONVENER: Whilst we can read the sentence above and understand the words, how might we understand what is said in our bodies? How do we engage with the vastness of the life of a planet where our whole history and presence is less than a blink of an eye in our earth's lifespan, and yet every moment we are awake is a fully sentient experience with the added capacity to reflect across time? In contrast, our lives are lived with full intensity at a micro scale. We have developed very short term and self-focused attention spans. The 24-hour news media cycle, business imperatives of reporting short term results, politicians concentrating on being re-elected to another term in say two-five years, have all conspired to keep us distracted from the bigger view. If we gently look, we might see that we and others are absorbed by our daily struggle to survive, and for a growing number, survival is a very real landscape with multiple jobs just to put food in children's mouths. Some dropping dead while at work, from overwork (ABC Foreign Correspondent, 2021). We exist within an economic system that seems to enslave rather than liberate people. Having a short-term focus is all that many feel they can afford.

As we engage with the climate and ecological crisis, we discover ourselves entwined within much longer time frames. Many First Nations people are aware of this reality. There is a principle among these older civilizations that involves the discipline of thinking and acting with at least seven generations in mind. You may have come across this already, if you have, what does it mean to you? If you think about your family, a span of seven generations is about 200 years.

Reflection

CONVENER: Pause for a moment; what are your time horizons? How far ahead do you normally think or plan?

And, how do we lengthen our cultural time horizon?

- Imagine having a conversation with someone in your family who lived 200 years ago? What would you like to say?
- Imagine having a conversation with a relative who lived 200 years into the future? What questions might they ask you? What might the conditions of their life be?

*

JOSIE: And now, imagine a 4.6 billion year time frame.

I struggled with that too, I couldn't make sense of that kind of scale. Fortunately, we have help in the form of a narrated and sound scaped Deep Time Walk (DTW), a rich, experiential exercise that we can engage in together.

If you haven't heard of it before, it is a guided walk that takes people on a journey that reweaves our human family back into the fabric of the living Earth. It steps us back in time to the Earth's beginning 4.6 billion years ago and then, over 4.6km, creates through storytelling, the critical events that have shaped our Earth. It was based on the work of philosopher Arne Naess, and adapted by Dr. Stephan Harding, Deep Ecology Research Fellow at Schumacher College (England) with precious collaborations and inspirations.

The purpose of the DTW is to provide a deep experience that touches 'head, heart and feet' to provoke reflection upon our individual and collective place in the world and to generate a deep commitment to change.

The DTW comes in two different formats. It exists as a phone application that you can download for free and enjoy as a rich, narrated and soundscaped story that provides an experience of how Earth herself evolves. It is a felt experience of the evolution of Earth. It's like a walking meditation on Earth. I have enjoyed the DTW on my own on several occasions and each time I am struck afresh of how significant is the damage we have inadvertently done to Earth. In such a small amount of time, we have interfered directly with Earth's self-regulating forces. For those who may not be able to walk, the DTW can also be experienced as simply a recorded story.

We may also be involved in the DTW, be leading or participating in a facilitated group walk and collective inquiry. Roselyne Lécuyer, is going to share her account of doing this in Paris.

ROSELYNE: I had learned that DTW was coming to France and … we might do something about it at CCA FR (the French speaking community within CCA)! By coordinating and staging a DTW, we'd be joining in a collective effort of 50 other walks taking place around the world on the same day – generating a wave of global action for the 2021 COP26 talks in Glasgow. What an opportunity! It would permit me to build on the enjoyment of my preferred walk – Paris, Ile des Cygnes on the River Seine, between Beaugrenelle and the Eiffel Tower - with an additional layer of knowledge derived from understanding more about the earth's history, but in a highly urban environment.

Our thanks for the possibility of the CCA DTW go to Robert Woodford in Glasgow, Scotland, and Olivier Maurel in France, who respectively ran an excellent online international training program, and conducted a master walk along the River Seine. Together they coordinated a collective marathon effort to translate the materials.

For me, the intersection of coaching with DTW became evident. While the DTW is rich in content in bringing the science and facts of the evolution of Earth together with storytelling, coaching provides an additional dimension with powerful questioning – in connecting people, helping to connect to self, and connecting to Earth while learning its history.

So, with my new CCA colleagues Muriel Huet and Jerome Sussfeld, we got to work organizing the DTW. One of our first steps was to discuss and share our intentions – both globally (to support CCA's efforts in supporting COP26) and personally. We enjoyed adding our coaching skills, and our questioning to the excellent content of the DTW.

On the beautiful day of the walk, 20 people, all new to the idea of the DTW, joined us and we introduced ourselves and set the frame. The storytelling along the walk was augmented with self-introspection, deep questioning in pairs, small group chatting and whole group interaction.

At the conclusion it was evident that all had enjoyed the deep walk, the fascinating conversations, the generous weather, and nature. The last station of the walk ended in deep reflection and discussion about how we felt. It is always a concern that by the end of the walk one might be feeling hopeless in our own individual efforts to combat climate change. However, one of the first of the powerful comments made was from a participant who said: 'Je suis très paisible.' – I am at peace. I better understand the journey of the earth is inevitable and so I can contribute whatever I can, knowing that this is done with the best intention and with grace.

Thanks to CCA and the originators of the Deep Time Walk in providing us with the tools and resources to show our connection to the living and solidarity with climate change action."

JOSIE: Thank you so much for sharing your preparation, observations and reflections from the group you facilitated. I would also like to mention that the DTW website provides guidance about facilitating such a group and holding the space for the conversations along the way. The two documents are Earth Stations: A Resource Guide for Planning a Deep Time Walk (2021) and Holding Space: A Guide to Organising and Facilitating a Deep Time Walk (2021). Both are available and are provided free under Creative Commons License (see References).

I'd like to handout a small section, penned by Stephen Harding (2021), and taken from the second of these two documents – Holding Space. Let's take a few minutes to read it together. It provides an excellent summary of background to the DTW:

We have found over the years that the Deep Time Walk is an excellent way of triggering deep experiences of connection to the vastly ancient life of Gaia, our animate earth. We walk 4.6 kilometres representing the 4600 million years, which is the age of our planet. On this scale every millimetre we walk represents 1000 years of our earth's life, every footstep is 500000 years, every metre one million years. Just contemplating these numbers is often enough to start the process of triggering a deep experience in many people.

Our deep experiences during the Deep Time Walk can open us up to powerful moments of realisation and awakening during which we spontaneously experience a deep love for this ancient turning planet of ours which has given birth to the miracles of her biodiversity, her climate, her rivers lakes and oceans and to her swirling atmosphere, all working together to regulate her surface conditions over vast spans of geological time within the narrow limits suitable for life. By walking we experience how for billions of years bacteria were the only forms of life; how ancient are the first cells with nuclei which arose out of endosymbiotic fusions of bacteria; how recent are the multicellular beings that eventually colonised the land from the oceans, and how we industrialised humans have severely damaged our planet in the last 200 years – merely the last one fifth of a millimetre of the walk.

These deep experiences on the Deep Time Walk can trigger deep questioning and deep commitment. Walking the Deep Time Walk with deep ecology in mind can help to make us whole by waking us up to the wonder that is Gaia, this radiant living sphere, which is our planet, and to the deep meaning of our life and action in her service, for the benefit of all beings.

(Harding, 2021: 5–6)

CONVENER: I am wondering if all this talk about the DTW has whetted your appetite for a walk?

Exercise

Convener: We have allowed time for one now…who would like to join us for our 20 Earth stations facilitated walk? We have facilitators who are walking and also not walking. Please join the group that suits you best.

*

CONVENER: Welcome back together to share experiences and reflections from the DTW. We will chart your reflections against the timescale shown in Figure 6.1.

JOSIE: I've undertaken this walk several times now and each time I am transported into deep time. I become more aware of how I am Earth, and we are all a consciousness of Earth – Gaia. Each time I experience the Deep Time Walk, I am struck by the power of the natural force that life is. And I absolutely love

Deep Time Walk

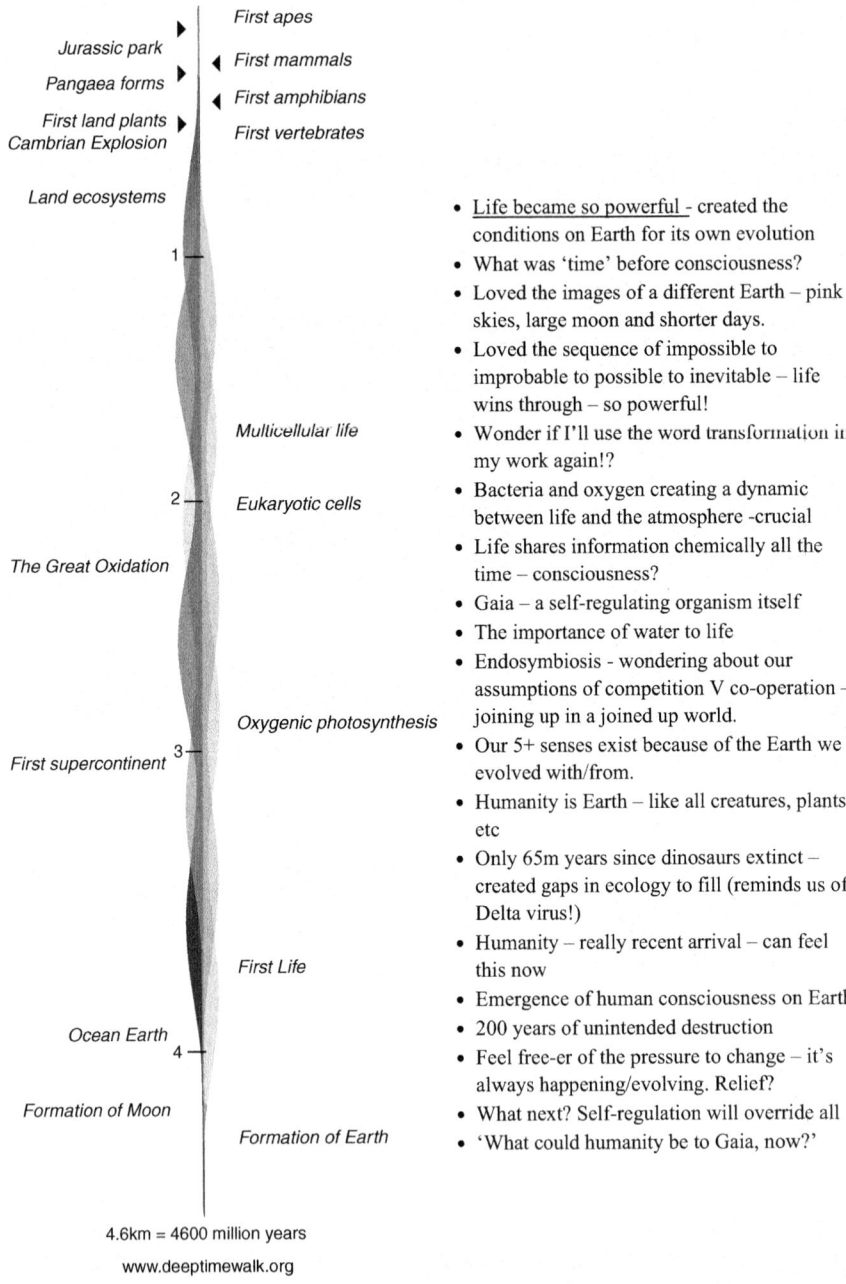

Jurassic park
Pangaea forms
First land plants
Cambrian Explosion

Land ecosystems

First apes
First mammals
First amphibians
First vertebrates

1

Multicellular life

2   Eukaryotic cells

The Great Oxidation

Oxygenic photosynthesis

First supercontinent   3

First Life

Ocean Earth   4

Formation of Moon

Formation of Earth

- Life became so powerful - created the conditions on Earth for its own evolution
- What was 'time' before consciousness?
- Loved the images of a different Earth – pink skies, large moon and shorter days.
- Loved the sequence of impossible to improbable to possible to inevitable – life wins through – so powerful!
- Wonder if I'll use the word transformation in my work again!?
- Bacteria and oxygen creating a dynamic between life and the atmosphere -crucial
- Life shares information chemically all the time – consciousness?
- Gaia – a self-regulating organism itself
- The importance of water to life
- Endosymbiosis - wondering about our assumptions of competition V co-operation – joining up in a joined up world.
- Our 5+ senses exist because of the Earth we evolved with/from.
- Humanity is Earth – like all creatures, plants etc
- Only 65m years since dinosaurs extinct – created gaps in ecology to fill (reminds us of Delta virus!)
- Humanity – really recent arrival – can feel this now
- Emergence of human consciousness on Earth
- 200 years of unintended destruction
- Feel free-er of the pressure to change – it's always happening/evolving. Relief?
- What next? Self-regulation will override all
- 'What could humanity be to Gaia, now?'

4.6km = 4600 million years

www.deeptimewalk.org

Figure 6.1 Deep Time Walk timeframe
Credit: Robert Woodford and Deep Time Walk

one of the closing phrases in the recorded narration where the scientist restates that to life, 'impossibility is a weak wall'. I find that so encouraging because we are an expression of the dynamic of life. I also find it liberating, because no matter what happens, life will find a way.

CONVENER: Thank you, Josie, and Robert Woodford from DTW, who obtained permission for us to offer you Earth's timeline as an image too (Figure 6.1).

Reflection

CONVENER: As you reflect on your own Deep Time Walk experience, please note your reflections:

What thoughts, feelings or ideas emerged during my walk?

How should I lead my life now that I have had direct felt experiences of the immense age and significance of our planetary ecology during the walk?

What sort of positive action and collective advocacy should I commit to?

Let's go and refresh ourselves to sustain us for the journey ahead – perhaps consciously noticing the sources of life that we consume to maintain our own.

With the sound of the drumbeat of the music of the film *Circle of Life* in their heads and laughter, the group share drinks and walk outside.

*

Convener: Welcome back once again, this morning we shared some of the background information and awareness required to become eco-informed. Slowing down enough to deeply listen, starting to strengthen our relationship with earth. We are beginning to find the solid ground on which to plant our feet – from which to develop a flexible stance in a world of different opinions and beliefs. Look around you – notice. Look inside yourself and feel into what is true.

The combination of video and story enables richer sensemaking in some work. Breaking Boundaries (Clay, 2021) is a documentary that builds upon years of scientific research by Johan Rockström, a Swedish professor and now joint director of the Potsdam Institute for Climate Impact Research (PIK) in Germany (referenced by Zoe Cohen earlier). It can help us understand the numerous and inter-related pressures that are being exerted on Earth at this time. Narrated by David Attenborough and Johan Rockström, the film identifies a 'quadruple squeeze' on the planet. These are:

1. The continued growth of human population combined with the desire of the 'developing world' to attain living standards that are enjoyed by the 'developed world'. The 'developed world' represents approximately 20% of global population and has caused most of the planetary demise through over consumption of resources and pollution. The opportunity now exists to eradicate poverty and we can't do it with the same underlying assumptions about endless growth and extracting resources as we have done to date 'as if' we live on an infinite planet.

2. The current, unrelenting, increase in carbon concentration in the Earth's atmosphere. The science indicates that we need to hold that concentration to no more than 450 ppm to limit temperature increases to no more than 2 degrees C. It is currently 419 ppm and continuing to rise unabated (NASA, 2022c, June 2022).

3. Extreme and significant eco-system decline leading to species extinction, ocean acidification, land degradation, loss of soil, and deforestation to name several elements. All of this reduces Earth's resilience and tolerance of the first and second 'squeeze' pressures above.

4. The world system is not linear and predictable. It has the possibility of surprises or tipping points from which we cannot return. Once we pass a tipping point, we can't expect to be able to wind it back and return to where we started. We are already seeing evidence of tipping points at regional levels with the melting of the Arctic, the Amazon turning from a carbon sink into a net carbon emitter and destruction of coral reefs.

Rockström describes science as similar to 'switching on your headlights on a dark road' so you can see where you are going. With the clarity of what we are facing, his optimism comes through because, like many others in this book, he believes that humanity is waking up to what we are causing. And if we are causing it, we can choose to heal it.

Just as we can't talk about climate change without also talking about biodiversity loss and ecological collapse, we can't talk about the ecology and climate change without discussing social inequality. Rita Symons, can you now address the uncomfortable truth about social inequality.

RITA: To awaken and take the 'red pill' is not a place of comfort. This powerful analogy is taken from the 1999 film The Matrix where Morpheus, a character played by Laurence Fishburne invites Neo, the film's hero, to take the red pill to awaken from an artificial reality. In this distorted reality, humans experience an imagined life in The Matrix, whilst in fact being plugged in to giant cells and used as an energy source by machines. The parallels in this analogy to our current situation are many.

To feel the visceral realisation of the deep inter-connectedness of all the challenges facing humanity and the inextricable truth that many of these are created by the choices we have collectively made, is painful but necessary. So, I offer the question: How can we increase our awareness, understanding and, more importantly, how can we begin to address the crisis facing our species and the planet we call home? And what in this, is the role of coaches.

Since the industrial revolution, we have been stuck in a polarised view that the only choices open to us in terms of how we organise society are communism or neo-liberalism. And so, we do what humans do. We form tribes, and we homogenise and 'other' the opposing side. This is eloquently described in Miki Kashtan's book Reweaving Our Human Fabric (2014).

We have normalised a focus on individual needs and desires, rather than the collective good and in doing so have ravaged our precious planet. Neoliberalism and the idea that, in a capitalist society, wealth trickles down, has proven to be a fallacy. We know the gap between rich and poor has become ever greater in the last decade, and we know that one in six children in the world live in extreme poverty (UNICEF, 2022), whilst the richest 1 % own 43% of the world's wealth, with often obscene levels of consumption arrogance, and short termism.

We also know that the poorest communities are bearing the burden of climate change now, not in some distant future, but now. A 2017 UN report (Nazrul Islam and Winkel) described the vicious cycle of the poor being more likely to be impacted by climate events, thereby disproportionately losing yet more. And so, the gap gets bigger.

This exacerbates existing social inequity, created through systemic injustice, often in the way people are governed, by individual governments or by the new power of multinational companies. In more recent times, Russia and China have embraced capitalist ideology, managing the juxtaposition of this with communist ideology, and other states have become more right-wing in their thinking. This has resulted in an increasing gap between those nations who seek to make change versus those who seek to kick any action into the long grass.

And this too plays out in racial injustice and the legacy of slavery, where millions of humans were commoditised, because those with power rationalised this as a justified act in order to create wealth. Is it any wonder that with generations of embodied trauma, under-represented communities continue to have poorer outcomes in key aspects of life such as education, health and wealth?

So, the different types of injustice we see in the world are not distinct but interwoven in a complex web and directly related to choices a small minority who hold power have made on behalf of humanity. We need to fundamentally refocus our priorities and arguably the seventeen UN Sustainable Development Goals are a good place to start, combining a focus on climate action with goals such as eradicating hunger. We each have a part to play in how we drive action in our communities directed at achieving these goals.

What then is the role of coaching? An uncomfortable truth is the fact that in the early decades of the coaching profession, those that had access to coaching tended to be the people who had privilege, and in many parts of the world, this pattern continues. Arguably, therefore, those who systemically are least in need of support, benefit from coaching and mentoring. Has the coaching profession, therefore, increased inequity?

The 2021 Global Code of Ethics (2.8) states 'Members should be guided by their client's interests and at the same time raise awareness and responsibility to safeguard that these interests do not harm those of sponsors, stakeholders, wider society, or the natural environment.'

The dial has shifted and as coaches, mentors, and supervisors this gives a clear steer that our responsibility lies beyond the needs and wants of an individual client. If coaching is about assisting individuals to gain insight and act, there is a huge opportunity for us to take a more systemic view and consider our role in helping humanity to find a sustainable way of living for the future. All change happens one conversation at a time, and coaches are experts in holding the space for generative dialogue. My belief is it is no longer appropriate for coaches to say they need to be led by clients if those clients work in ways which do not consider social value. We have a responsibility to support our clients in the now; but we also have a deep responsibility to future generations.

CONVENER: Thank you Rita. We now turn to Charmaine Roche. Please can you build on this theme from your own research and practice.

CHARMAINE: Since October 2019 my PhD research has been exploring the ethical stance required to coach for social change. My exploration began with inquiries into my own work. Enquiries which I felt were beginning to push against the boundaries of our much-celebrated belief in neutrality and impartiality. I felt as if I was transgressing, violating hallowed ground, when I began to ask whether it was ethical to continue coaching for improved performance in a sector whose business models are experienced as oppressive by many of my clients. As a coach I was experiencing ethical stress and moral injury: the conflict here was not a clash of values between me and my client but between me and some of my professional norms.

I now seek to partner with my clients in critical reflections that make oppressive practices (economic, social and political) visible and historically contextualised. Visibility helps my clients determine how to act in relation to them. A neutral position would have kept the work at the level of individual psychology and personal behaviour change because this is the norm in our industry. This is normativity not neutrality. Neither does working systemically fill the gap if the systemic lens fails to focus on and magnify structural inequalities and the power dynamics and practices that sustain them.

Since the beginning of 2020 I find that I am part of a nascent movement asking critical questions, feeling around, exploring the implications of stepping out from behind the myth of neutrality in favour of transparency and intentionality. We see ourselves not as neutral technical experts but as agents for change in pursuit of what our planet and the human beings on it, especially the most disadvantaged, need for a sustainable future. We call our clients to extend their agenda in ethically mature ways that account for the diverse and possibly conflicting needs of key stakeholders from a social justice, and eco-sustainability perspective. We need to adopt Salmon's (2000) indigenous *Kincentric Ecology* recognising that beyond the stakeholders usually considered, all sentient life is our kith and kin.

We aspire to be transparent about our vision for a more just and equal world and understand that we are challenging the existing unjust distribution of power and ownership over the material conditions of life and the ideologies that justify this.

I would like to offer a framework for coach training, education, development, and supervision that might better prepare coaches for working with an explicit social justice orientation.

This model (figure 6.2) offers a framework with four interconnected dimensions. Each dimension and the overlapping inner connections provide a guide to the knowledge, understanding, ethical stance and embodied presence we need to cultivate as agents for social change.

This is provisional work but is proving useful to me as a tool for practice, thinking, research and dialogue with coaches and other helping professionals.

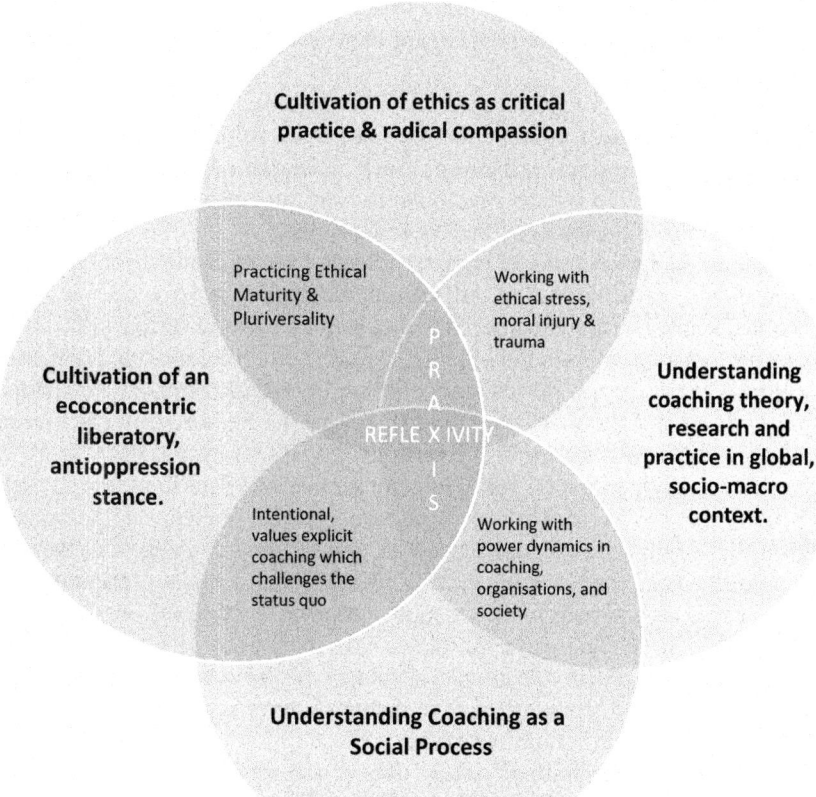

Figure 6.2 Coaching and Supervision for Social Change

CONVENER: Thank you, Charmaine, this is a strong reminder of how things are interconnected inextricably.

Our climate and ecological crisis could be seen as resulting from our inattention, which can become the parent of violence. The bystander who says nothing can also be seen as responsible for creating the violence, as well as the violator, of standing in "wilful blindness" (Heffernan, 2019). We need to attend to our own bystanding, as we can legitimize what we don't speak out on (see David Lane, Day Six, p215). Here Kanishka Sikri invites us to pay attention. As you read on, notice your own responses without being judgemental, allow them to land and accept them all with curiosity and compassion.

KANISHKA: Violence is the space from which I depart in my work. As a writer/theorist/strategic advisor, I am (what feels like constantly) thinking about violability: the practice through which certain bodies, lives, and worlds are marked to the possibility of violence. Can we enter into struggle against climate change from the entry point of violence? In other words, what potentialities for change does reconfiguring the climate crisis from a lens of violation give us?

Violence is often seen as a spectacle, it is the brutality of rape, the grief of hunger, it is murder, torture, terror. Violence is less so seen as a practice, as a gradual, slow, omnipresent haunting. Such a violence is tricky, sly, we cannot see or touch it, but it is there, operating on every surface our hands meet. And what could be the greatest surface violence haunts?

I would like us to think of climate violence through context: the histories, genealogies, particularities of our present. Ecological violence operates in a world of terror—that is patriarchal and heteronormative terror, historically lasting terror of transatlantic chattel slavery, settler colonial and imperial colonial terror, neoliberal and capitalist terror, ableist terror. These terrors, those listed and the many not, exist in a sort of a matrix, they court each other, fortify one another. Our culture of terror permits and calls on us to violate – it gives us the impunity to commit violence. And so we do, to each other and to our world. The logic of climate violence, you see, is not extricated from the continued sexual violence and disappearing of indigenous women across the Americas, it is not much different from colonial extraction that pulls and pulls and pulls from 'the south' – that creates a 'south' in opposition to a 'north' in the first place. We take what's ours, what we want and desire, what we hate and wish to eliminate. Climate violence is a logical progression, it is not a random consequence. The design of terror is the design of a rapeable us, and thereby, a rapeable world.

Terror as a structure and culture does not absolve us. It is through our practices that we may battle violence, that we may live and imagine a world without it. And so our struggle against climate violence is animated by our livingness, our seemingly mundane being: how we work (by work here I do not just mean the capitalist idea of labour, but also the work of our personal lives), how studying and learning gives life to our work, with whom we work and build relationships, what habits and routines ground the work we do, what purposes and visions drive our work, how our work enables us to be more free, and thus, the world to be more free. This may all seem personal—indeed it is. But the personal is the political space from which we enter the revolution. To enter the movement as wholly as we can, we must do the messy work of extracting and then trashing the violences that make their bed in ours. This is not the work we can do; this is the work we must do.

Exercise

KANISHKA: So, I invite you to do this work with me and ask yourself the following:

- What does it look like to question my roles and practices in maintaining the desecration of our world? Reflections here are boundless, but some situating strategies range from introspective journaling, consciousness raising, political organizing, to pedagogical (un)learning.
- How are my strategies of resisting ecological violence bound up and confined by the very structures I am resisting? My hope is that the strategies we use to excavate violence are formed through a recognition of how violence limits and supports certain vectors of struggle. This recognition can lead to non-traditional and alternative modes of resistance to bloom.
- As a coach, how do I facilitate difficult conversations around climate violence from emotional spaces of love and desire and hope, rather than places of hate, apolitical technicalities, and abstract thinking? As bell hooks asks us: 'how do we hold people accountable for wrongdoing and yet at the same time remain in touch with their humanity enough to believe in their capacity to be transformed?' (1998, bell hooks and Maya Angelou).

*

CONVENER: Thank you Kanishka. These can be difficult ideas to hear and digest, especially for those of us who have been raised within the dominator culture. We would like to allow some time for us to all to contemplate the impact of these notions on our lives.

Exercise

CONVENER: We invite you to walk alongside one of your fellow travellers present today. In pairs, consider: How have you moved on your journey to becoming a steward of our planet? In what ways is your path redirected or moved along by the conversation so far today?

*

We've looked together at how things are, we have listened and experienced. We have practised opening our hearts to what the science and the earth are both telling us. I invite you now to listen to three perspectives, each in different ways designed to 'shift' our consciousness.

First we invite Jojo Mehta, from Stop Ecocide International, the charity to which the author royalties from this book are going to be matched by the publisher, to share the work her organization is spearheading. Stop Ecocide International is behind attempts to criminalize the destruction of nature through legislation. Thank you for sharing your work, Jojo.

JOJO: Mass damage and destruction of eco-systems or **Ecocide** has been committed relentlessly and repeatedly over decades in pursuit of economic development and is a root cause of the global climate and ecological crisis. One could think of heavy deforestation to make way for cattle ranching, soil depletion via mass use of pesticides or deep sea bottom-trawling as possible examples of what could be described as ecocide, along with huge pollution disasters such as Deepwater Horizon or Fukushima. And, for the most part, no-one is held responsible.

Stop Ecocide International is developing cross-sector global support at all levels of government and civil society for establishing an international crime of ecocide, creating personal, individual criminal responsibility for key decision makers. The route pursued is that of amending the Rome Statute of the International Criminal Court (ICC), the only global mechanism directly accessing the existing criminal justice systems of 123 member states. Currently there are four international crimes: genocide, crimes against humanity, war crimes and the crime of aggression. Ratifying ecocide as a 5th crime would mean including it in domestic legislation, thus creating a coherent law across borders - important as the worst offenders are often transnational corporations. In this way Earth's most precious eco-systems can be protected and allowed to recover.

Profound changes are needed if we are to move into a safe operating space for humanity. **Ecocide law can support those changes**, levelling the playing field for solutions – regenerative farming, renewable energy, circular economy - while providing a guardrail to ensure best practice. Banks cannot back criminal activity, nor insurers underwrite it. Government ministers can't issue permits for it.

Criminalising ecocide makes serious harm to nature legally and morally unacceptable, preventing finance from flowing to practices that destroy eco-systems. **It stimulates innovation in a healthy direction in all sectors.** Beyond that, it has the power to strongly shift cultural assumptions and our understanding of our place in the natural world and our responsibility towards it, while encouraging governments and businesses to take environmental law and due diligence seriously.

At the time of writing, interest in criminalising ecocide is already a matter of public record at parliamentary and/or government level in the following countries via motions, resolutions, parliamentary questions, petitions, white papers or full proposals of law: Bangladesh, Brazil, Bolivia, Belgium, Canada, Chile, Finland, France, Ireland, Luxembourg, the Maldives, Mexico, Netherlands, Portugal, Scotland, Spain, Sweden, the UK and Vanuatu as well as at the European Parliament, the Nordic Council and the Inter-Parliamentary Union.

This has been actively facilitated by Stop Ecocide International's campaign narrative, alongside legal, diplomatic and grassroots collaborations. Their work sits at the intersection of these and is thus **able to both influence and amplify the global conversation**. The charitable arm of the campaign, the Stop Ecocide Foundation, was the commissioning body for the Independent Expert Panel which in June 2021 announced the Legal Definition of Ecocide:

> Ecocide means unlawful or wanton acts committed with knowledge that there is a substantial likelihood of severe and either widespread or long-term damage to the environment being caused by those acts.

Stop Ecocide International estimates that ratification of this new crime at the ICC could begin within five years, to reach all member states, as well as the potential (via universal jurisdiction) of reaching non-member state perpetrators.

CONVENER: Thanks Jojo. It is easy to think that substantial change is not happening, but these types of legal changes begin to reshape the large economic and legal systems that our existing western ways of life are built upon.

We now invite Sam Suppiah, based in the Philippines, to bring a different perspective on the situation faced. What does the picture look like from those who are often excluded and marginalised from dialogue, discussion, decision making? What does this look like from those at the receiving end of colonialism and colonization? And what is it coaching might offer? Welcome Sam.

SAM: Welcome to the Anthropocene, where one single species has, over a couple of hundred years, built a deep separation via the reductionistic mechanisms of modern civilisation – a cultural evolution begetting potential planetary extinction.

Homo sapiens, representing less than 0.01% of Earth's living biomass, have destroyed 83% of wild mammals, 80% of marine animals, 50% of plants, and 15% of fish (Bar-On et al, 2018). Billions of years of biological evolution disappearing without a mention. Millions more will fade due to accelerated anthropocentric climate change. Playing Jenga and gambling with Earth's complex web of interdependent systems that enable us, and the rest of life as we know it, to live.

Together we have exhausted the planet:

All living beings are relational and part of the web of life. The models and associations we still use to see reality are linear and make no room for what is mystical and unexplainable.

(Parente and Cardoso, 2020)

Despite the United Nations in 2015, releasing a widely acclaimed manifesto entitled 'Transforming our world: the 2030 Agenda for Sustainable Development', humanity has demonstrated great apathy; collectively increasing corporate investment and globalised operations devoted towards accelerating eco-systems' destruction and social oppression, to feed the financial greed of the global white western elite.

A large number of 'Millennials' and 'Gen Zs' worldwide are increasingly committing to veganism and childlessness in response to their entrenched eco-anxiety and socio-economic powerlessness.

We need to wake up and smell the super-typhoons. Collapse is upon us. In these impossible circumstances, how do we head out of the realm of an exhausted planet and into the pluriverse of regeneration, without perpetuating the violence of modernity?

Climate coaches from around the world, you may not be our only hope, but you have your part to play.

Our greatest work in this nascent era of terrible decline is in your own unlearning. It is through deep inquiry of all the stories we tell ourselves - it is through active surrender that we might release the sources of fragmentation within our minds, bodies and souls. It is here we will wrestle with various stages of decolonising our own minds.

At some point, this will naturally unlock each of us and set us on the life-long, cyclical journey of inter- connection - finding new depths. Nature is moving quickly now to ensure we wake up. In our deepening, we can find the parts of us who remember, and remember again, how to embody compassion, empathy, courage and faith. Deeper and deeper, remembering ourselves and each other home. We will then begin to hear nature reaching out from the absolute core of our true inter-beingness.

Lived experience of this immensely challenging process is a non-negotiable necessity in becoming capable of guiding others through their own maze of confusion, discomfort and disarray. Nothing less will do. Beyond these thresholds, the path becomes clearer. All the skills we, climate coaches, have amassed in our work thus far will help.

We need to stay humble. Hold on to your scepticism, your humour, and your passion to act.

CONVENER: Thank you, Sam for that rich and clear perspective. Having listened, watched, absorbed our current state we would now like you to engage in one final exercise today, the 'Council of All Beings'. This allows us to feel into these natural life-giving dynamics as a part of nature.

EVE: As we travel on our journey, how can we bring ourselves to be active participants in what is evolving and to feel that fully? *The Council of All Beings* gives voice to the perspectives of all the Earth's species, and the primary purpose is to: 'enhance human commitment and resources for preserving life upon our planet home'...to enable "a shift from the shrunken sense of self, to which our mainstream culture and its institutions have conditioned us, to a larger, more ancient and resilient sense of our true ecological Self" (Fleming and Macy, 2007: 98–99).

The first exercise of this particular name took place at a riverside wilderness site in New South Wales, Australia in the 1980s, before rippling across the earth. Fleming and Macy recall the co-leader Frank inviting group members to go off for an hour to 'find a place that feels special to you and simply be there, still and waiting. Let ...yourself be chosen by the life-form that wishes to speak through you' (2007: 80).

Reflection

CONVENER: We invite you to empty your mind as far as you are able from all the rich information and conversations of today and go on another walk by yourself. As you wander through nature quietly listen to all that is around you. Let some living being or aspect of nature call you. Discover who or what would like to speak through you, who is choosing you. It might be another living being, animal, plant, or natural feature (swamp, mountain, or desert). We encourage you to go with your first intuition, what comes without hardly thinking. If you are unable to go for a walk, do so in your imagination.

People leave one by one for their walk.

*

EVE: Welcome back. Our council falls into three stages.

1. As participants we speak spontaneously from the perspective of the other life-forms, in the first person, explaining why we have come to the Council, and sharing freely our feelings such as anger, fear, confusion. Those listening can ask questions to allow the speaker to understand and express its Being. Questions we might use include, 'How does it feel to be you? What is happening for you now?' Macy's examples include that of a cow: 'I'm tightly crowded in a dark place, far from grass and standing in my own (excrement). My calves are taken from me, and instead cold machines are clamped to my teats. I call and call for my young. Where did they go? What happened to them?'
2. Next at a signal, five or six of us at a time will sit in the centre of the circle to listen in silence as humans to what is being said. If needed, we might ask of the life-form, 'How has life changed for you in the last x number of years?' 'What would you like to tell us? What are the challenges you face?' Macy (undated) describes this story: 'Humans! I am Mountain speaking.

For millennia your ancestors venerated my holy places. Now you dig and gouge for the ore in my veins. Clearcutting my forests, you take away my capacity to hold water and release it slowly. See the silted rivers? See the floods? In destroying me, you will destroy yourselves.'

3. Lastly each participant, as our life-form, has the chance to offer to the humans (and receive as humans), the powers that are needed to stop the destruction of our world. Here, if helpful we might ask the life-form, 'What gifts do you offer us to help stop the destruction that is going on?' Macy (undated) offers a few examples such as 'I, Condor, give you my keen, far-seeing eye. Use that power to look ahead beyond your daily distractions, to heed what you see and plan.' And 'As Mountain, I offer you humans my solidity and deep peace. Come to me to rest, to dream. Without dreams you lose your vision and hope. Come, too, for my strength and steadfastness, whenever you need them' (Fleming and Macy, 2007: 100–115).

Let me give an example. As I went on my walk, I saw a magpie, and took little notice, but then I kept seeing single magpies and pairs, until there were several ahead of me (a mischief) and I just knew they had 'chosen me' (yes maybe I was a little slow to this realization!). They seemed genuinely inquisitive and allowed me up close. And they were so beautiful with their wonderful purplish-blue iridescence on their feathers and elegant long tails with their greenish hue. But there was still reflection needed, and their more negative reputation considered, so that I could speak as 'I' in these three stages:

1. 'I, Magpie have suffered from a human story that says I steal 'shiny things' and am responsible for the fall in numbers of songbird. It means many humans do not like me. The truth is that I am simply inquisitive, and while I do eat songbird eggs and chicks, and many other things as I like scavenging, in fact I know that songbird numbers in woodland increase most when there are more of us around, so I think that availability of food and suitable nesting sites are the key.
2. I, Magpie have been poisoned in human gardens, and trapped in cages. In some parts of the world, I am treated as a pest and legally shot accused of reductions in numbers of native species. And yet I am happy to eat depending on what is most plentiful at different times of year, from slugs, beetles, flies, caterpillars and spiders to wild fruits, berries and grains, or the household scraps you throw away, alongside small mammals and birds. Yes, when I feel threatened I may attack. Maybe I remind you too much of yourselves humans?
3. I, Magpie, offer my curiosity about all things, and my intelligence to explore objects I pick up without judgement based on colour or material. I also offer my ability to eat whatever is plentiful, not what is in short supply. I offer you my natural intelligence – to recognise yourself as I do

myself in a mirror, as we truly are. I offer too my friendship as I recognise many humans and enjoy their company when I do not feel threatened.'

NEXT IS THE COUNCIL: There are several important elements to holding a Council. Early on, one element is to allow 'The Mourning', 'the pain within us over what is happening to our world. The workshop serves as a safe place where this pain can be acknowledged, plumbed and released' (Fleming and Macy, 2007: 101). This underlines how important our contracting will be and our ability to manage group dynamics, and to hold a safe, contained space. Perhaps tomorrow you can bring the feelings of mourning so we can work through them together. Perhaps it is the ancient forests disappearing so quickly, the loss of coral reefs, of breathable air, or of species that were abundant earlier in our lives. In doing this we are sensing the interconnectedness we share with all Beings.

Then another is termed Remembering, which is in two parts, allowing us to remember our roots. The first takes us 'through the story of the universe from the Big Bang to the beginnings of organic life on earth. The second part is a guided movement meditation on the evolution of organic life from single cell existence through the complexities of form and expression possessed by present-day humans' (Fleming and Macy, 2007: 105–106). This is the remembering we did together as we experienced the deep time walk today.

CONVENER: And so we come to the end of a big day. For some it has been a revelation, and for others an affirmation of what they knew. Either way, we all need to continue to stay informed.

We have identified what we believe is required to know, in our heads, hearts and bodies, to be eco-informed. As important as the science is, our own observation and knowing needs to be added to our more usual western rational ways of knowing. Through centering ourselves in what we know as nature, we can anchor ourselves as the world of opinions rage around us.

As Zoe Cohen has pointed out, science is also only one part of the complex context. There are economic, technological, social justice, worldview, and many other elements that all come together to enable a pattern of outcomes of climate change and ecological destruction to emerge. We are not separate from this complex and wicked challenge. We are a part of it and complicit in it. That realization seems to be the choice point. What do we choose to do now?

Drew Dillinger's earlier words may be ringing in our ears 'what did you do when you knew?' (page 8).

Tomorrow we will begin to answer that question. What different ways of feeling, thinking and being may help us respond more effectively than we have over the past 40 years? What are others already doing as examples of what we might do?

Sleep well tonight.

Chapter 7

Eco-Aware – Processing our Emotional Responses

Day Three Morning

CONVENER: Let us start today together with a poem:

> Thank God our time is now when wrong
> Comes up to face us everywhere,
> Never to leave us till we take
> The longest stride of soul we ever took.
> Affairs are now soul size.
> The enterprise
> Is exploration into God.
> Where are you making for? It takes
> So many thousand years to wake,
> But will you wake for pity's sake!"

> A Sleep of Prisoners
> (Christopher Fry)

CONVENER: Welcome to Day Three. Now we have identified what we need to know to be eco-informed, we will explore the work we have to do to face the deep emotions that each of us feels, in very different ways, when faced with the magnitude of the climate and ecological crisis. Each of us might worry and fear for the lives of our children and grandchildren or have a deep grief for the loss of so many species with whom we share this planet. We may feel guilt and shame that we have been part of the ecocide, and anger and rage at those with more power than us for being wilfully blind and solipsistic. We might also feel overwhelm or hopelessness at the enormity of the challenge.

Today's work will focus on: **How we develop the emotional, cognitive and spiritual awareness needed.**

We will:

1. First explore our human journey and the reality of the struggle we face as we transition towards a deep, unshakeable knowing and being.
2. Then will discover together ways of moving from the Newtonian paradigm through understanding living systems, to systemic awareness and

DOI: 10.4324/9781003153825-7

embodied knowing, to eco-systemic awareness and participation consciousness.

3. And then explore ways of arriving at spiritual knowing – moving from head knowing, to heart knowing, and relational knowing as reflected in Ubuntu and Thich Nhat Hanh's concept of interbeing.

To engage with heart and relational knowing, we will be going back to indigenous wisdom from many traditions both today and on our final day of this journey. So let me share a passage from Robin Wall Kimmerer's excellent book Braiding Sweetgrass (2020) about the prophecy of the eight fires. Accordingly, in our current time, we find ourselves in the seventh fire:

> The people of the seventh fire do not yet walk forward; rather, they are told to turn around and retrace the steps of the ones who brought us here. Their sacred purpose is to walk back along the red road of our ancestors' path and to gather up all the fragments that lay scattered along the trail. Fragments of land, tatters of language, bits of songs, stories, sacred teachings – all that was dropped along the way.
>
> (Wall Kimmerer, 2020: 367–368)

We are all descendants of an ancient indigenous people, that lived in and with the land. What are our indigenous roots and stories, what is the story of the place that we now live, what are our songs, what are our teachings, what do we now need to remember? What is the wisdom we have simply to remember, not from our personal memories, but re-member by rediscovering? Jarid Manos (Manos, 2007: 376) offers this:

> Protecting the Earth and our children's health and future are one and the same. That is our crucible and sacrament. It may be a severe scorching test, but it is something holy. By healing the Earth, we heal ourselves.

Nora Bateson, at the Hawkwood College Regenerative Confluence conference (2021), spoke about how global change doesn't happen through scaling big system change, global change happens through intimate dialogue and exchange. Transformation doesn't happen at a particular time, a particular place, it happens everywhere. System change is in your breakfast bowl, it's in how you speak to your children, it's in every tiny action you take, every thought you have.

The very place where the bigger shift starts is in the tiny repatterning, the focus of attention, the self-love, the self-healing within each of us. We need to focus 'outside-in' to discover what the earth and the world require and become Eco-informed, as we did yesterday, but transform 'inside -out' starting with ourselves and our inner worlds as we will do today to become Eco-aware.

Jarid Manos (2007: 386) expresses it so well and simply: It 'has nothing to do with hatred and killing, but simply people reconnecting with themselves, each other, and our sacred, shattered Earth'.

Embedded above, we have some way finders to help us navigate:

- Be prepared for discomfort and uncertainty, it's part of the territory.
- To create the conditions for future life, we must remember what we have forgotten and piece together ancient stories.
- Healing the earth, heals ourselves.
- The place to start is right here exactly where you are, with everything you feel, think, believe and do.
- Stay as open as you can to connecting with all that comes.

To paraphrase the words of Otto Scharmer (2009) we need to begin this stage of our journey with an Open Heart, Open Mind and Open Will. Open to whatever comes. Be willing to let go of what separates us, closes us off from all that is in us and around us and flows through us. We also need to proceed with Active Hope which for many coaches has been a useful starting place and we will explore this more fully tomorrow (pages 157–160).

As we continue our journey, please spend a minute and breathe deeply, centring yourself. Reflect on what you might need to let go of and leave behind to do today's journey, as well as what you might need to bring with you.

Working through our emotional reactions and responses

CONVENER: I am wondering what emotions arose for you when you listened deeply yesterday to what is happening in our shared world? What is stirred within when you face into the enormity of the crisis and challenges, we face with wide-eyes-open. Looking back at your notes and reflections, from yesterday please reflect on:

- The feelings and emotions that emerged for you as we listened to the science and what's happening in the world.
- What arose for you, when we went for the walk and heard what the 'more than human' world is asking of us.
- Any dreams you had last night.
- Any pictures or music that came to you.
- Any responses that are floating up from your sub-conscious (often the first place where difficult feelings are registered)?

Exercise

CONVENER: If you are able to, please now share these in small groups, using one object from nature that can be passed around the circle as a 'talking stick', the person first holding the object talks. The rest will listen deeply and empathically to the feelings shared by the speaker, noticing our own feeling resonance. This is a form of wisdom council where we listen not to our individual wisdom, but the collective wisdom we can hear through and beyond ourselves.

*

Exercise

CONVENER: In our larger group, you are now invited to each share just one word or phrase, that represents your main emotional response to the climate and ecological crisis.

*

Now look at the collective list that arose from the wider group. Read your list and the longer list below out loud, slowly. See if you can find each of these within you, noticing where they sit in your body.

Text Box 7.1

Fear	Guilt	Loss	Determination
Overwhelm	Shame	Rage	Sadness
Gratitude	Despair	Fury	Awe
Grief	Hopelessness	Anger	Calm
Hope	Numbness	Churned-up	Compassion

CONVENER: Highlight all those you have felt for some time in response to the environmental crisis.

Are there any you haven't highlighted? So many highlight the full list. Are there any you would add?

*

PETER: There are many teachers in this space, and over the years my work has built on Kubler Ross's original Grief cycle and later Otto Scharmer's Theory U and the work of Joanna Macy. We will map some of the emotional stages we go through, not always in the same order, to become more fully Eco-Aware. My wife Judy Ryde (Ryde 2005, 2009, 2019) researched the 'White Awareness Cycle' and she and I built on this in our joint book on *Integrative Psychotherapy* (Hawkins and Ryde, 2019) and particularly in the chapter on 'Ecological and eco-spiritual approaches to Psychotherapy'. This work, I developed further with Eve in applying it to coaches and coaching, in our book on *Systemic Coaching* (Hawkins and Turner, 2020) specifically in our chapter on developing ecologically conscious coaches.

Let me start with some quotes from one of my eco-psychotherapy teachers, the late Jim Hillman:

At some point both psychologists and environmentalists need to decide what they believe our human connection is with the planet our species has so endangered.

(Hillman, 1995: xvi)

To redefine sanity within an environmental context. It contends that seeking to heal the soul without reference to the ecological system of which we are an integral part is a form of self-destructive blindness.

(p: xvi)

At the heart of the coming environmental revolution is a change in values, one that derives from a growing appreciation of our dependence on nature. Without it there is no hope. In simple terms, we cannot restore our own health, our sense of well-being, unless we restore the health of the planet.

(p: xvi)

That air, waters and places play as large a role in the problems psychology faces as do moods, relationships, and memories.

(p: xxii)

In *Integrative Psychotherapy*, (Hawkins and Ryde, 2019: 41-42) we wrote:

The devastation of our shared planetary home has enormous and increasing impact on the mental and emotional lives of each and every one of us, for some consciously and for all unconsciously. This gives rise to direct or indirect feelings of: grief and loss; guilt for one's part in the destruction; fear for oneself, one's community, and for future generations; or anger at others who the individual sees as more responsible for the devastation than themselves. It can emerge through the dreams of the client or their reactions to current events, or even through nameless and unlocatable emotions. Whatever way, the wider ecology and what is happening to it, will arrive in the therapeutic [and coaching] relationship, bidden or unbidden, recognized or unrecognized.

Nature has traditionally been a major place and source of physical and mental healing. Yet with the unprecedented migration of human beings to urban environments and the parallel destruction and humanization of many natural ecologies, this resource is less and less available to larger percentages of the human population. Many have lost their living connection to other mammals and to plants, or ease of access to places of unspoilt beauty away from human noise and interference. Many of us have lost our healthy living connection to our source, and our lives have become plastic- packaged, humanized, noisy and crowded.

PETER: In summary, our planet's health is linked inextricably to our health, and we are destroying the earth's capacity to support human life and our own connection with nature particularly in the west. As well as a mental health crisis across the western world, eco-anxiety is now considered by many of the professional psychology organizations as a recognizable emotional state that many of us suffer from. This is debilitating anxiety about how the climate and ecological crisis will unfold in the future and how it will affect us, our children and our grandchildren.

Glenn Albrecht (2005) usefully recognized and defined another prevalent syndrome which he called *Solastalgia*, which he describes as:

> the homesickness you have when you are still at home and your home environment is changing in ways you find distressing. In many cases this is in reference to global climate change, but more localised events such as volcanic eruptions, drought or destructive techniques can cause solastalgia as well.
>
> (2005: 42)

PETER: We have started to see the scope and scale of the unfolding tragedy, of what we have already lost. But the future outcome is not a done deal. The way forward is one of healing, and the way to that healing is to honour our pain for the world (Macy and Johnstone, 2012).

CONVENER: We invite you to pause for a moment and as you pause, please notice what you are mourning in the ecological loss and what else is stirring and moving, inside of you.

Exercise

CONVENER: Please stand up and go and write some of these on our graffiti wall and read what is written by others or draw a picture that encapsulates your feelings or do movements that express it.

*

'Every day, I feel like I am in a perpetual dance between denial and acceptance. The layers of denial run deep, and every day, a bigger acceptance comes too. I am never still and sometimes I regress ten paces because it's easier than having my face pressed against the window of what is happening in the world.'

'I just want to stand up and shout at everyone who I can reach, shake them awake.'

'I feel frozen in place immobilised.'

'I sense a gut-wrenching hope that it's not as bad as I know it is.'

'I am thankful for my community and for the awakening I see in everyone.'

'I'm scared. What would I take with me if there were flames licking at my house?'

'I fear for the morning I will no longer hear the birds.'

'I am thankful that I'm a coach, I can help and help others learn to help themselves too. I know how to build relationships, to reconnect and earth is bigger than we are.'

CONVENER: We are about to walk through the ecological awareness cycle and later today Mark McMordie will share some inner practices to support this work. He reminds us now that while this cycle offers a helpful framework to navigate emotions there for us all, we can individually and collectively turn toward our ecological crisis in a way that increases our response-ability, rather than becoming overwhelmed by it.

PETER: Now we walk through the stages of the Ecological Awareness Cycle (Figure 7.1).

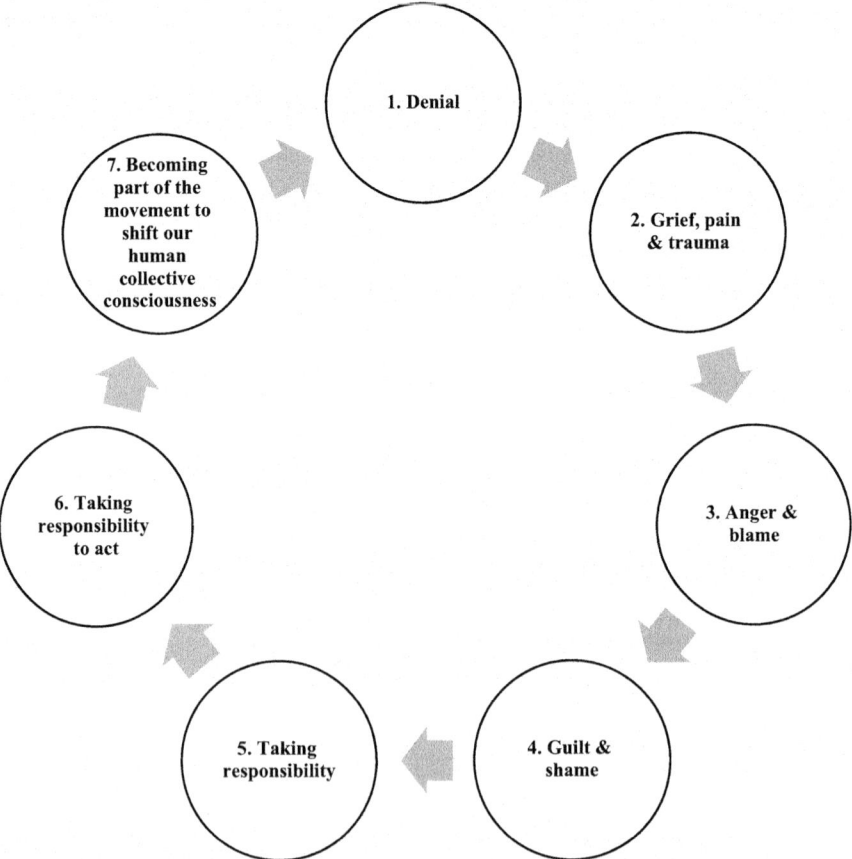

Figure 7.1 Ecological Awareness Cycle

Stage One: Moving through denial

PETER: The first barrier to facing the global ecological crisis is our individual and collective wilful blindness (Heffernan, 2011). This has been described as 'deliberate failure to make a reasonable inquiry of wrongdoing ...despite suspicion or an awareness of the high probability of its existence' (Merriam-Webster, 2022).

Josie McLean (2017: 16) notes that 'the human family are in a state of perpetual denial, unable or unwilling to implement the actions that we know are needed.' Capra and Luisi talk of the illusion of perpetual growth and believe that the 'obsession of politicians and economists with unlimited economic growth must be seen as one of the root causes, if not the root cause, of our global multifaceted crisis...the absurdity of such an enterprise on a finite planet should be obvious to all' (2016: 366). Earth Overshoot Day – the day when we have used more from nature than our planet can renew in a year – moved from 25th September to 1st August in nine years (Earth Overshoot Day, 2018).

> To understand the role of ecology in our work, and therefore in our [coach] training, we need to hold our "varifocal lens" – bifocals and single vision lenses are no longer enough.
>
> (Hawkins and Turner, 2020a: 124)

ALISON: The narratives of denial are many. Joanna Macy and Chris Johnstone (2012: 60–64) share some of the key narrative stories that put us back to sleep, that keep us in denial and running towards ecological collapse.

The seven narratives of denial

1   'I don't believe it's that dangerous', leads to hiding the truth and hiding from the truth. We may find counter stories, no matter how flimsy, that offer an alternative view.

2   'It isn't my role to sort this out' wanting to split from personal responsibility, fragmenting, pointing the finger elsewhere. This is a prolific narrative that allow us to be duped because others are worse and anyway, only those in positions of political influence or policy roles can make a difference.

3   'I don't want to stand out from the crowd' Tapping into our deep need to belong, to be part of a strong social group, we conform to the dominant cultural rules and won't speak until we feel it's safe to do so. We don't allow ourselves to be fully present to the situation.

4   'This information threatens my commercial or political interests' so we may attack the source, often the messengers who are pointing to what we are desperately trying to deny.

5   'It is so upsetting that I prefer not to think about it'.

6   'I feel paralysed. I'm aware of the danger, but don't know what to do'. We are overwhelmed by what we have come to understand and feel unable to respond.

7   'There's no point doing anything, since it won't make any difference' This is 'doomism' and can hold us in a place of full awareness yet excusing our inaction.

Exercise

I invite each of you to notice:

Which narratives resonate most strongly?
Which ones capture your own stories of separation?
How does your story of denial draw from these narratives?

*

PETER: Denial doesn't come just from our own stories, for we are subject to wider influences.

Weintrobe (2012) distinguishes between three types of denial.

1. **Denialism** 'involves campaigns of misinformation about climate change funded by commercial and ideological interests' (2012: 7). It sets out to cast doubt and confusion and undermine belief in the scientific evidence and even in experts per se. Many attribute this to the lack of decisive action both demanded and taken to curb the destruction.

2. **Negation.** Similar to the first stage of Kubler-Ross' stages of grief and mourning, this is a refusal to believe that a person has died and to accept the loss. It is a psychological process of protection from a shock or trauma we are not yet prepared to face.

3. **Disavowal.** Where we paradoxically both acknowledge what is happening and at the same time split off from this knowing. 'This is seeing it but with one eye only' (Weintrobe, 2012: 7). We both know and don't know simultaneously, and Weintrobe sees this as the most prevalent and dangerous form of denial. 'This is because the more reality is systematically avoided through making it insignificant or through distortion, the more anxiety builds up unconsciously and the greater is the need to defend with further disavowal' (Weintrobe., 2012: 7).

As we work through our various forms of denial, we find ourselves having to face up to feelings of guilt and shame in what those of us in the Global North have, largely inadvertently, done to the world. Staying in denial is a defence mechanism against these feelings that are tough to bear, yet moving through denial is essential to find a way to mobilize an appropriate response and develop the skills of resilience to adapt.

Film makers, television producers, authors, journalists, activists and artists are helping by developing well designed 'alarm clocks' to wake us up from our slumbers and denial.

Exercise

PETER: As you listen to this, please write down:

- How do we, as coaches with our clients, explore and unravel the narratives of denial rather than collude?
- How do we find ways to hold the possibility of the truth we are all avoiding, to open a little longer?
- What are the habits you are prone to indulging in to rescue your coachees from the uncomfortable feelings of pain, grief and trauma?
- What might you do to intervene in these habits?

*

Stage Two: Addressing the pain, grief and trauma when we become aware of what we have done to our planetary home

PETER: To wake up to the great destruction the modern western culture has caused on our planet, and to overcome the pervasive denial of the extent of what we have caused, involves going through the cycle of shock, anger, grief, guilt, depression, feelings of hopelessness and despair, before we can truly integrate and become congruent with the wider eco-system of our more than human world.

Eve was struck by the experience of climate scientist Jeff Kiehl, also a Jungian analyst, who understands that the messages from scientists can be traumatizing so when he presents the latest scientific understandings to groups of people, he invites people to speak about their feelings, knowing full well the depth of feeling that this information elicits. He describes one such occasion:

> The moods expressed included sadness, hopelessness, anger, denial, guilt, numbness and fear. We sit together in silence, holding the multitude of moods. Giving voice to the silent spirits inhabiting our hearts brings a certain warmth to the room. In sharing our feelings about these issues, a door opens, connecting us…Our shared feelings evoke within us a profound depth of caring.
>
> (Gillespie, 2020: 30)

Concern about our planet's future is prevalent and is described as 'a fairly recent psychological disorder afflicting an increasing number of individuals who worry about the environmental crisis' (Castelloe, 2019).

ZOE: Peter, if I may add… Rather than seeing eco-anxiety as a dis-order, or a diagnosable illness, we should view eco-anxiety as a right and proper state! It signifies the waves of loss and grieving that are inevitable when we connect with the reality that faces us. As we work through it ourselves and together and keep working with and through it - we empower ourselves to act.

PETER: Yes, pathologizing those who claim eco-anxiety and eco-grief is another form of denial and splitting, allowing a further deepening of the harm and negating of the truth we face.

We need to face the loss of the world as we know it and that 'deep adaptation' (Bendell and Read, 2021) is required. This work, Jem Bendell suggests, requires us to explore:

- Resilience: what do we most value that we want to keep, and how?
- Relinquishment: what do we need to let go of so as not to make matters worse?
- Restoration: what could we bring back to help us with these difficult times?
- Reconciliation: with what and whom shall we make peace as we awaken to our mutual mortality?

In this work, true mourning of what is dying and ending is required.

This stage of pain, grief and trauma is increasingly showing up in the coaching room, sometimes consciously, but more often unconsciously. Some coachees will discuss how upset they have been, or are, at witnessing, directly or indirectly, the tearing down of woods, the loss of many bird species, air pollution, the waste floating in the seas. Others have concerns about what their company is doing, how the supply chain is dependent on non-renewable resources, how their customers are adding to the world's carbon damage, others about their own carbon footprint growing from their enormous travel miles. Some coachees will have free floating grief and anxiety about the health of their children or grandchildren, or focussed concerns such as their children's asthma being caused or aggravated by their school being close to polluted main roads, or even whether it is right or safe to have children at all. In the UK in 2021 a coroner, for the first time, noted that air pollution 'made a material contribution' to the death of 9-year-old Ella Adoo-Kissi-Debrah and called for a change in the law (BBC, 2021b). And the World Health Organisation (WHO, 2022) suggest that air pollution kills an estimated seven million people worldwide every year.

- How might we bear witness to this pain and in what ways can we support our clients experience these feelings without turning away?
- How do we create the space in our coaching work to incorporate this emotional landscape?

We often have a multitude of skills that are more than relevant, and yet, we find our own entanglement with these issues gets in the way of simply coaching.

Exercise

CONVENER: We invite you to come together in pairs, if possible, one listener and one speaker.

The listener listens attentively, without speaking and with full presence. As the speaker, speak the seed sentence starter and continue to say whatever seems to follow naturally, don't censor what emerges. If you get stuck at all, just start at the beginning of the sentence again.

1. When I think about the condition of our world, I would say things are getting…
2. The ecological concerns I have include…
3. Some feelings that come up when I think about these things are….
4. What I do with these feelings is….

Swap roles after each sentence or all four, whatever works best for you. By doing this both of you have heard and witnessed each other.

How might this exercise inform your coaching work?

*

Before we move on, I'm curious to hear some reflections from that exercise. What happened? What moved? What did you discover?

1. Getting better in terms of awareness, but staying the same in terms of concrete, verifiable actions by governments and most large corporations.
2. The extinction of species, the cruelty to fellow sentient beings, the belief that we as humans have the right to do as we like (the anthropocentric way we may operate as humans).
3. Sadness, determination, guilt.
4. Channel them into action, speak to others, remember that any action to improve the situation is worthwhile, believe I can make a difference.

PETER: Now let us look at the third stage

Stage Three. Anger and blame

PETER: Anger can be a healthy response leading to thoughtful activism. Anger can also be an avoidant emotional mechanism, where one's own responsibility is projected on to distant others. At this point, we enter the blame game and 'drama triangle' (Karpman, 2022), where we cast ourselves as victims to external persecutors, and consciously or unconsciously wait to be rescued. Some individuals bypass the stage of grief by going back into 'disavowal' as the grief is too hard to bear, and they lack the support from others necessary to express their grief and be heard. Some go straight from awakening to the reality of what we humans have done to our shared Earth, to anger and fury at those they single out for blame for the growing catastrophe.

We also need to hear the righteous anger towards the privileged white western world, from those who are suffering much more than their fair share of the devastation of the ecological crisis, although as we showed yesterday, had a minimum contribution to creating it. Let me invite Jennifer Uchendu, who supports community-led climate projects in Nigeria that explore eco-emotions with young people, to offer her experience.

JENNIFER: I was eight years old when the Mango tree in our compound was cut down. The act had left me in tears, I found no justifiable reason for this to have occurred. There, as a young child in Lagos, Nigeria, I felt a deep connection with a Mango tree and retained a desire to love and protect nature as often as I could. Little wonder that I grew up to become an environmentalist leading 'sustyvibes', a movement for young Nigerians passionate about the environment, juxtaposed with the recurring development challenges in a country like Nigeria, our passion was pretty niche and so was our style of advocacy. We employ participatory approaches using art, music, poetry and local language to describe the urgency of the climate crisis and our roles as young people.

To effectively guide and coach people to become better stewards of our environment, we need to come into an elevated sense of self-awareness - of our own influence and the impact of our words and actions to the persons who listen to you. So, when I found myself deeply overwhelmed by our advocacy work in Lagos and other parts of Nigeria - especially the Niger Delta region that is currently plagued with an unending soot problem - my work as environmental advocacy was strained by powerlessness and fear, in the face of histories of exploitative systems, generations of vulnerability and the anxiety of what our future holds in Nigeria as we continue to swim in oil and plastic pollution at the expense of health, wellbeing and life. I became acutely aware of my eco-anxiety in 2019 when I no longer felt like leading sustyvibes. I did not think I had what it takes to motivate other young people to be climate activists. Studying about eco-anxiety has, in a way, equipped me with the emotional tools I need to continue my work at sustyvibes, for instance, I am continuously working with "active hope" - which allows me to feel sadness, joy, hope and optimism about our work with environmental sustainability.

I must point out that eco-anxiety is contextual and should not be assumed to be a generic issue for young people. My experience of eco-anxiety has been marked with anger, an emotion that my peers in the western world may not identify with. I get angry at how disproportionately vulnerable Africans are to climate change and how young people will continue to suffer for a long time because the crisis is already here. It is no longer some existential threat. This implies that the climate crisis is in fact a social crisis; it is beyond a degradation of our natural resources, it is also an erosion of people's cultures and ways of living.

In conclusion, may I suggest that the next time you speak to someone about climate change and role, that you consider the intersects of power and privilege and that you deeply seek to connect with their core emotions about the crisis, be it fear, hope, denial, guilt, shame or anger. Another question worth asking is about their vision of the future. This, I belief is critical to whatever support you may be offering. If the future is full of hope and full of solidarity with colleagues in the global south who are already experiencing worse impacts of the climate crisis, then we may all be on a promising path towards living with climate change with courage, love and empathy.

PETER: Thank you, Jennifer. In the privileged North, we need to move beyond blame and anger to taking responsibility, and to be open to listen non-defensively to those who are rightly angry at what costs they bear from our collective actions, on which much of our 'privilege' is built. Once we do this, then we naturally will start to accept our guilt and feel ashamed, and these are the next feelings we will address.

Stage Four. Dealing with guilt and shame

PETER: This stage involves working through our appropriate feelings of shame. Guilt is not necessarily ours to bear, Judy Ryde (2009) notes that guilt is something to be established – am I or am I not partly responsible and should I acknowledge my appropriate guilt? Shame is the feeling we have when we start to take responsibility for how we are part of the ecological destructive-ness that human beings are enacting on our local and planetary eco-systems. Shame alerts us to our destructive actions and encourages us to make amends and repair the damage. It is the work of conscience and looking at ourselves and our actions with honest eyes.

Feelings of shame can become hard to live with if we do not respond to their call for change in our behaviours and choices. Turning away from feel-ings of shame may well lead us back into disavowal. Moving through guilt and shame is part of the process towards a healthy response to the ecological crisis, deepening our awareness of our own part in damaging the ecology and lead to us individually taking more responsibility for what we do and can do.

There is a danger here, that we can flip from blaming others, in Stage 3 to blaming ourselves in Stage 4. The move from 'shame' to 'blame' is rarely if ever helpful, but the move from shame to transparent self-honesty, to the next stage of taking responsibility is a path we all need to follow.

Stage Five. Taking responsibility

PETER: Our responsibility comes from our consumer choices and behaviours, our travel, our use of carbon-based energy, and our collusion in turning a 'blind-eye' to what is happening. Only when we can look with eyes wide open, at our part in the collective destructiveness, and how we are already living new constructive choices, are we able to respond, become responsible. Only then can we develop an appropriate course of action that is neither taking too little or too much but rightsizing our personal share of the collective responsibility and response.

Stages Six and Seven. Taking responsibility to act – finding a sense of greater purpose in your work

PETER: Self-insight and accepting responsibility are both critical but insufficient alone. 'To know and not to act on the knowledge, is to not know.' (Latour, 2017, 140), and on Days Four and Five we will be asking: **'So what can I do that will make a positive difference and how can I contribute to the wider profession?'**.

But let me pause and ask Clover Hogan, climate activist, researcher on eco-anxiety, and the founder of Force of Nature, a non-profit working with young people, to share some of the questions she believes we all need to be asking.

Clover : Here are some questions I ask myself, my colleagues and those I work with:

- If money and time posed no constraints, which problem in the world would you commit yourself to solving?
- Which stories keep you feeling small, anxious, powerless to create change?
- What would you need to do to challenge - and disprove - these stories?
- What comes to you as easily as breathing? How might you show up to solve a problem that ignites a fire within you, with these gifts you've been given?
- Which emotions do you tend to disallow within yourself? How might you begin to create space for them?

PETER: Thanks Clover, a great call to action. And if you want to be further inspired by work with young people, Forde (2022) describes her global movement so that by 2030 millions of children will be equipped with the skills to care for Planet Earth and dare to create a world where everyone can thrive.

Coaching can play a key role in helping us by taking us as coaches beyond our denial, beyond our willingness to turn a blind eye to the impact of our consumption in all its forms and to look squarely in the eye at our own personal responsibility. This requires courage.

We are able to help ourselves and our coachees find a greater sense of purpose in their lives. My favourite strategy question is: 'What can I (and we) uniquely do that the world of tomorrow needs?'. We know that people who have a sense of higher purpose in their work, are more committed, successful and more fulfilled and so are the organizations they lead (Polman and

Winston, 2021: 87; Edelman, 2022; Renshaw, 2018). Deloitte's 2020 survey Millennials and Gen Z showed that their top two concerns were climate change and protecting the environment and this affected what companies they would work for.

As we come to face the ecological breakdown around us, we perhaps need to ask an additional question to bring the future home:

> What is the story that you want your grandchildren to tell their children about who you were and what you did when your great grandchildren sit on their knee?

Our global crisis is such that creating change, one person at a time, will not be sufficient. We are inextricably intertwined and inter-dependent. In this we are like the Bodhisattvas in the Buddhist tradition, who recognise that individual enlightenment is not an end, but a blessing that necessitates us to return and enable others to also open their awareness, liberate their mind-sets and be part of shifting human consciousness.

At each stage we need to ask how we can you use what we have learnt to help others?

As coaches, we are not there to judge, moralize, campaign or convert our clients to our ecological stance or beliefs. We need to ensure that we are neither in denial, disavowal, or into just blaming others for the ecological crisis; that we have moved beyond grief and frozen powerlessness; and that we are constantly looking at the reality of the ecological crisis with our eyes, ears and feeling receptors wide open.

It is essential that we have worked sufficiently through our own ecological awareness cycle – worked through our own despair, hopelessness, anger, grief, guilt, so we can truly listen and be open to the deep ecological feelings and responses of those we are listening to, without becoming reactive, judgemental, or minimizing what is present within us and outside us.

We need to listen with 'ecological ears' to the ecological field within, around and flowing through our clients, to the emotional responses they have in order to enable a constant moving around the ecological awareness cycle and enable a fuller life embracing response at each stage and in each cycle.

The work of coaching includes developing our capacity to acknowledge our own destructiveness, and the pursuit of our self-interest which includes our collusion with ecologically destructive forces.

As coaches we are part of an industry and professions that are embedded within the toxic economy at the heart of fuelling climate change. At times, the coaching industry is unwittingly extractive and degenerative, centralizing power and otherwise replicating the patterns of the system it supports. Those of us who have benefitted most from the coaching profession perhaps have a greater responsibility to ensure that the profession evolves into a healthy profession, one that is focused on creating a healthy planet for all species, is democratized, accessible to all and sustainable in the full meaning of the word.

CONVENER: We would now like to invite Mark McMordie, who has practised and written extensively in compassion and mindfulness work, to share three 'inner practices' that may help us with this challenging work. Thank you, Mark…

Inner practices

> 'The success of an intervention depends on the interior condition of the intervener.'
> (Bill O'Brien quoted in Scharmer, 2009: 7)

MARK: As leaders and coaches, we are all interventionists. Research suggests that our capacity to facilitate transformational change is influenced by our own meaning making and stage of adult development (Rooke and Torbert, 2005). I have written elsewhere about the meta-cognitive and meta-emotional capacities associated with later stages of development and actively developed through regular mindfulness and compassion practices (Chaskalson and McMordie, 2017). What is offered here is a brief signpost to three practices and potential paths for further development. As Peter suggests earlier in this day, it is essential that we have worked sufficiently through our own ecological awareness cycle, so that we can truly listen and be open to the deep ecological responses of our clients. Denial, grief, anger, blame, guilt and shame are all powerful emotions – how can we bring more mindfulness and compassion to these emotions in order to work more skilfully with them? This isn't simply a case of soothing painful emotions away – to coin a phrase from Harvard's Bob Kegan (1994), meta-awareness is a subject-object shift from being 'had by' strong emotions to 'having' them. When we have been able to work with our own strong emotions in the ecological awareness cycle we will be better able to help others to do the same in order to access greater agency and purpose.

Exploring Difficulty

MARK: The most robust and well researched forms of mindfulness and compassion training have one thing in common – they all increase our capacity to be with difficult emotions and move toward what is difficult or unwanted with greater curiosity and kindness. This becomes the foundation for insight and transformation. Perhaps one of the most well researched formats is Mindfulness Based Stress Reduction (MBSR). As eco-anxiety increases in our society one might wonder how more people may reach for MBSR for the practical tools that it offers.

One such tool that sits at the heart of the programme is the practice Exploring Difficulty. If it speaks to you right now you might try it. Do bear in mind that the practice sits in the middle of an eight-week program that prepares the ground for it. For an accessible overview of the full program see Mindfulness (Williams and Penman, 2011).

Loving Kindness Meditation

MARK: When teaching about the inner practices of psychological safety I often refer to the ground-breaking work of Stephen Porges (2017). In pointing to the neuroception of safety, Porges outlines the role of the vagus nerve. This face-heart connection enables us to detect and project features of safety through facial expressions and vocalisations that are simply a reflection of our deeper autonomic state. How we look, listen and vocalise conveys information about whether we are safe to approach. Safe states are a prerequisite for accessing the higher brain structures that enable us to be creative and generative. So if we wish to create safe spaces for powerful inquiry (our own and others) into topics like climate change that can be anxiety provoking, and if we wish to enable generative thinking and adaptive change, the quality of our vagal tone matters. Fortunately, research suggests that you can build vagal tone through regular mental exercise (Fredrickson et al., 2008). After just a few months of practicing Loving Kindness Meditation (LKM) for an average of sixty minutes a week, participants at Fredrickson's lab experienced significant increases in vagal tone and every positive emotion they measured.

If you wish, try this LKM practice and see what impact it has on your state and sense connection with yourself and others who may be experiencing challenging aspects of the eco awareness cycle. For a fuller exploration of LKM and compassion practice see A Fearless Heart (Jinpa, 2015) and the accompanying eight-week Compassion Cultivation Training (CCT) programme.

Anyone exploring the inner dynamics of compassion practice will sooner or later bump into the question of the relationship between compassion and self-compassion. It's not a straightforward one as many of us seem to find it easier to extend compassion to others than we do to ourselves. If you tried the previous LKM practice you may have noticed a subtle but significant difference between extending kindness to the 'easy target' of a loved one compared with yourself. But if we are to deepen our capacity to extend compassion to others in a sustainable way, practicing self-compassion has an important part to play. Interestingly, research suggests that self-compassion actually increases self-improvement motivation given that it encourages people to confront their mistakes and weaknesses without either self-deprecation or defensive self-enhancement (Breines and Chen, 2012).

As we navigate the eco awareness cycle or any challenging aspect in our increasingly complex and unpredictable world, it's likely that we are going to experience moments of falling down and self-judgement. How can we develop a more helpful and resourcing relationship to this? The Self Compassion Break is a powerful pocket-sized practice that can help transform our inner narrative and outer response. This practice sits within the eight-week program Mindful Self Compassion. (If you wish to learn more see Germer, 2009).

As we look to take greater responsibility for engaging with our ecological crisis and to support others to do the same, I hope these practices and paths offer a helpful resource so that we can do so with greater mindfulness and compassion. Ironically, this is likely to accelerate the process of innovation and transformation required.

May we all be safe.

May we all be happy.

May we all be healthy.

May we all live with ease.

CONVENER: Thank you, Mark. It is great to be pointed in exactly the right direction.

We've focused so far on what has been our own work through the cycle of denial and shame and we've explored some ways of dealing with the feelings that arise. What we don't often hear are the voices, unfettered, from those who are not often included in this exploration. What we don't often realize is the systemic depth of turning away and we rarely hear what we are turning way from. There is a collective forgetting that allows us to continue without allowing ourselves to be aware. Because surely, we wouldn't continue if we were aware.

You have worked hard this morning. I can see the journey you've been through written on your faces and in your posture.

I'd like to share a poem here that speaks to the work of becoming aware, the cul de sacs and ditches we can find ourselves in.

Autobiography in Five Short Chapters

CHAPTER ONE

I walk down the street.
> There is a deep hole in the sidewalk.
> I fall in.
> I am lost…I am helpless
>> It isn't my fault.
It takes forever to find a way out.

CHAPTER TWO

I walk down the same street.
> There is a deep hole in the sidewalk.
> I pretend I don't see it.
> I fall in again
I can't believe I am in this same place.
>> But, it isn't my fault.
It still takes a long time to get out.

CHAPTER THREE
I walk down the same street.
> There is a deep hole in the sidewalk.
> I see it there.
> I still fall in…it's a habit…but,
>> my eyes are open.
>> I know where I am.
It is my fault.
I get out immediately.

CHAPTER FOUR
I walk down the same street.
> There is a deep hole in the sidewalk.
> I walk around it.

CHAPTER FIVE
I walk down another street.

Portia Nelson (©1993)
From THERE'S A HOLE IN MY SIDEWALK: THE ROMANCE OF SELF-DISCOVERY by Portia Nelson.
Reprinted with the permission of Beyond Words/Atria Books, a division of Simon & Schuster, Inc. All rights reserved.

EVE: What are some of the holes we fall into again and again? They may be some of the cultural, unconscious assumptions that we make and never stop to question.

Let's take a minute to consider who we are and how we got to be who we are, and how we react as that will influence our eco-awareness, much of which is, of course, socially constructed. We are all the products of our background and all that influences that, so let's discover and acknowledge how we bring our culture to all we do and think, so raising "some of the particular norms, prejudices and assumptions that we can fall into." (Ryde et al., 2019: 41). These can of course be broad and relate to equity, diversity and inclusion, including gender, sexual identity, sexual orientation, age, heritage and disability. But here we would also highlight the assumptions from our geographical location in the world, Global South v Global North for example.

Exercise

EVE: So please take a moment to do this short exercise based on Ryde, Seto and Goldvarg, 2019: 45. We have chosen 15 sentences to complete today, but the sentence starters can alter depending on the context. The important thing is to note the first response that comes to you, your instant reaction, in completing the sentences.

I assume that....

1. Economic growth is....
2. Money is...
3. Time is...
4. Climate change is....
5. My role in the climate and ecological crisis is...
6. The life future generations will have is...
7. Progress is....
8. Living beings include....
9. Human development is...
10. The ecology is...
11. Habitat and species loss is...
12. Change is...
13. The most important thing in life is...
14. I feel ashamed when...
15. People are...

*

As Ryde et al. say, 'we can quickly see what we hold to be truths, even those that are not conscious or are semi-conscious. It can reveal things we would rather not own but also things we think are self-evidently true' (Ryde et al., 2019: 45–6). We have found it an incredibly invaluable exercise to explore deeply held beliefs and even uncomfortable truths both with individuals and groups, changing the questions according to the context. In discussing with others afterwards we may highlight the different assumptions we each hold, but to allow yourself to be honest in your responses, it is important to know that sharing is optional and confidential.

These assumptions are personal and cultural in their origins. We learned these beliefs from our families and they from theirs. These cultural assumptions are brought into view when we take a systemic view – which we will do after lunch.

CONVENER: Let's take a break, and muse for the next hour on what you are harvesting from this morning's labours. We have a packed lunch so you can wander and find the right space to nourish you until we reconvene.

Chapter 8

Eco-Aware – Shifting our Thinking

Day Three Afternoon

CONVENER: Welcome back. Coming into a new awareness once again, we are ready to step into and deepen a wider perspective that will continue to embed our entwined connectivity. Josie and Peter, if you would please, share with us the shift in our basic paradigms of thinking you both see we need to step over and through.

Learning to think, be and do eco-systemically

PETER: Let us start with a quote from the great African-American leader Martin Luther King Jr (1964) that I use at the beginning of many of my teachings:

> All I'm saying is simply this: that all life is interrelated, and in a real sense we are all caught in an inescapable network of mutuality, tied in a single garment of destiny. Whatever affects one directly, affects all indirectly. For some strange reason, I can never be what I ought to be until you are what you ought to be. And you can never be what you ought to be until I am what I ought to be. This is the interrelated structure of reality.

In this he echoes the deep teachings of Ubuntu in southern Africa, Thich Nhat Hahn's Buddhist teachings of 'Inter-being' and many indigenous wisdom traditions rooted in participatory conscious.

We begin by exploring the systemic framework developed by Josie McLean (adapted from the work of evolutionary biologist, Elisabet Sahtouris, 2005). It provides a constant drum beat of the radical interdependence of self, other, and earth encouraging us to foster the depth of re-connection and interbeing that facilitates our ability to co-learn and co-evolve, listening and paying attention. It reminds us of the dynamic flow of relationship between all parts of the larger whole or 'system' and its inherent messiness and uncertainty. Like all models it is a map, not the territory and therefore limited, but it can be a useful way to understand systems thinking and move beyond this into understanding how every system is nested within other systems and to systemic and eco-systemic awareness and eventually participatory conscious. It is

DOI: 10.4324/9781003153825-8

also a reminder of the lesson we learned yesterday on our Deep Time Walk, that Earth itself is a living system comprising multiple feedback loops that have, to date, regulated the conditions for life.

JOSIE: Understanding this model or map (known technically as a holarchy, (McLean, 2020) can be helpful to coaches as they seek to navigate conversations in a systemic way. It can also help us understand that we don't need to bring climate into our conversations – it is always present. And we can appreciate the nature and dynamic of the word 'sustainable' – so long understood as a destination rather than an emergent outcome.

We all know a great deal about living systems (or complexity) because we are living systems and we operate everyday within larger living systems – our families, communities and the landscape or natural environment itself.

Consider your body for a moment. We can use our bodies to illustrate the point about nested subsystems - also known as holons (Koestler, 1976). Imagine a single cell in your heart. It is a 'whole and complete' cell, and it is also a part of your heart.

Let me pause for a moment to explain that when I say it is whole and complete, I don't mean to imply that it is fixed and static in nature. Every living system can replicate and repair itself and respond to the environment in which it is embedded. By 'whole and complete' I mean that each subsystem is able to do these things on its own without other systems contributing their parts – but every living system is coupled to the larger systems of which it is a part – hence all the holons in Figure 8.1 being drawn as coming together at a single point. They are always connected and in constant movement together.

Picking up the example again, your heart is 'whole and complete' in itself, and it is a part of your circulatory system. The circulatory system can be viewed as 'whole and complete' in itself but is also part of a number of different systems including skeletal, muscular and nervous systems that make up

Holons in Holarchy

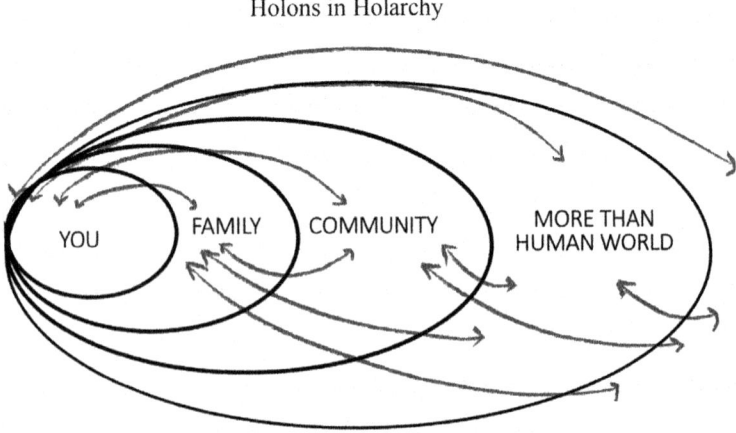

Figure 8.1 Holarchy illustrating the interdependence of us all

your whole and complete physical body. And your physical body, is influenced by your perceptions and emotions, beliefs, and past experiences that in large part are also formed by the perceptions, emotions, beliefs and traumas of your family and the culture of the society of which your ancestors were a part. All of this sits within your nervous system. Now we are beginning to recognize the complexity that exists within each of us.

PETER: Even within you there are more non-human cells than there are human cells. There are over 100 trillion bacterial microscopic life forms in your digestive tract alone (Montgomery and Bikle, 2016). Only recently have scientists realize how our digestion system is so similar to our Earth's topsoil, both dependent on trillions of small microbial beings, for necessary composting and turning matter into life giving nutrients. As a living system we are a community, not a separate entity.

Reflection

CONVENER: Pause for a moment and re-cognize the independence within you. You might like to close your eyes and allow the feelings to emerge and then share what came up for you with someone else.

- Can you sense the generous giving of so many small life forms within you, that are part of the community that is you?
- Can you feel the grace and gratitude in the receiving?"

*

JOSIE: Within the first egg (holon) of the model, the one saying 'you', you could create layer upon layer of feedback loops between the parts within you and seemingly outside of you, to depict this interdependence. Being alive is an inherent quality of a living system. As the parts that make up your system interact, you become alive and you have the capability to repair, replicate, and respond to changes in their environment. We could make it into a movie to appreciate the emergent qualities you have as a living system. Life is an emergent property, as is your ability to learn and adapt from the feedback loops between you and all the systemic levels, that exchange information with you from and to the wider system.

If we kept extending the holarchy from the microbes to your bodily systems to you, and then further on, we see another holon depicting each of your family members, each a system and parts of the larger system that we might term 'your family'. And these families are not islands, for they are formed by, and form, the community in which they live and have their being. And that community, can be viewed as a system, but it is also not an island, but inextricably interconnected to the one human family, where what is done in one community affects the others, as Covid 19 has shown us so well. The human species and community are totally nested within and dependent on the 'more than human' world of this earth we share with zillion other life forms. And so, we can go on. There is really no end, for it is systemic levels all the way up and all the way down.

Maybe as you listen to both of us, you can sense the boundaries we believe in, are those we create to make sense of the world. Where do you begin and end? The edges we think of as hard boundaries are porous. For example, you might consider your skin as the edge of you – but it harbours a very large number of microbiomes that live in a symbiotic relationship with humans. There are no real boundaries or lines in these nested systems except as we choose to draw them.

PETER: Let us pause from this hard thinking and share one of my Nasrudin stories. For those of you not familiar with him, Nasrudin is an archetypal wise fool who teaches wisdom through his foolishness. The original lived in Eastern Turkey in the 12th century.

Leaving his cells behind him

Nasrudin had several friends to stay for a couple of days over the New Year. Over dinner he told them all about a new book he was reading on the 'new sciences', which showed how important bacteria had been to the evolutionary processes of this planet. He continued by saying how we each contained whole colonies of different bacteria.

'What is the difference between bacteria and cells?' asked one of the guests.

Another guest replied, 'Cells are you, whereas bacteria are not.'

Nasrudin was puzzled by this, and even more puzzled when he later read in his book that on average you shed 100,000 skin cells a day, which accounts for most of the household dust. He anxiously telephoned his friends who had returned to London. He enquired of the wife of the second guest:

> 'How is your husband, are you sure he is alright?'
>
> 'Why are you so concerned?' she replied.
>
> 'Because he may not be aware of it, but 200,000 parts of him are still floating around my house.'

(Hawkins, 2005: 48)

PETER: All living systems are interdependent on the systems within them and the many systems that they are nested within. Gaia itself is totally dependent on the Solar system and wider galaxy and universe. We live in constant flowing relationship co-evolving, symbiotically exchanging matter and information, enabling a constant state of fluidity and adaptability - over indeterminable time. The simple holarchy or systemic map above is attempting to point to all this dynamic interaction.

We don't and can't know and be aware of many of the interdependencies. Certainly, in our day to day lives, we are unaware of most of the systemic consequences of actions we take. Josie, can you give an example?

JOSIE: As I got into my car to drive to the shop to buy some groceries. I was not thinking about the carbon emissions that were added to the atmosphere as a result of my trip. Additional carbon molecules that then contribute to further climate change that mean a woman and her children in India must walk further to obtain their daily water supply by bucket. They must walk further because the mountain glaciers in the Himalayas are melting that fraction faster due to my grocery shopping trip. The geographical distance and time involved in feedback loops means that I can remain blissfully unaware of the consequences of my actions. When I sit and reflect deeply however, I am reminded that everything I do has consequences for all the larger systemic levels I am nested within.

And yet, I find the realization that everything I do has consequences a highly empowering notion rather than a source of constant guilt. It's empowering because we are either influencing some beneficial changes within the system or we are choosing to take actions that maintain the status quo. You may never know the 'impact' of your actions because the causation in a living system is indirect and non-linear. But your life, and that of those you work with, is participating in this dance of life continuously. So, the explorations we might consider when we work with coachees, likely go way beyond the stakeholders they first identify. Sharing this simple map can be one way to grasp the enormity of possible stakeholders in the wider systems, while also accepting that we are individually making a difference.

And of course, some players in any system have greater influence than other players in that system. We are not all equal in terms of visibility, authority, and resulting influence. And developing the capability to identify tipping points within the system, places where small actions may result in big outcomes, will enhance your beneficial impact too. The topics of power, position, influence and impact are notions that are not apparent from viewing the map.

PETER: So let us pause to recap and summarize the key steps in shifting our paradigm:

1. Atomistic Newtonian thinking divides the seamless web of life into static separated parts to understand them as objects. This has been very useful for scientific advancement and many worthwhile inventions, but in its wake has, maybe inadvertently, reduced all living systems to resources to be exploited and reduced value to that which can be

measured and acquired. In doing this we are biting off the hand that feeds us (Atomistic Newtonian thinking).

2. Systems are wholes that are more than the sum of the parts and cannot be understood by just studying the parts. It is the dynamic patterns of relating between the parts that give form to the whole to function (Systems thinking).

3. The whole is also formed not just by its parts but by its purpose (Purposive system thinking).

4. But life is more than bounded systems, for every system is nested within and totally interdependent on, many, many larger systemic levels (Nested systems thinking).

5. These larger systems are not just outside us as context, for everything we are part of, is also part of us. That is true of our family, culture, species, and the ecology. This is the principle of reverse-nesting and the beginning of systemic awareness (Systemic awareness).

6. This two-way nesting goes beyond both human and other separate life forms, to understanding the Earth or Gaia as a living system (Eco-systemic awareness).

7. We could go further and apply the same thinking to the solar system, galaxy, universe (Cosmological eco-systemic awareness).

8. Therefore, we are the ecology, and the ecology is us. By polluting, eroding, and despoiling our ecology we are polluting, eroding and despoiling ourselves (Participatory consciousness).

9. We cannot fully see or know any system we are part of, neither can we know any system we are apart from. Therefore, we can only ever know partially. Our knowing is always through participation (Participatory consciousness).

But we cannot shift our paradigm just by thinking about it, as our paradigm is also our epistemology, the frameworks we do our thinking through. We only transcend our own and other's paradigms, by participating in an experience that we cannot comprehend with our current way of thinking, so let me invite you into **an experience, an exercise to transcend paradigm**.

Exercise

PETER: Please close your eyes and think of a coachee or mentee you have worked with. Visualize them in as much detail as you can, what they wear, their gestures, mannerisms, facial expression, movement. Listen to their voice, its tone, rhythm, timbre, pitch. Recall their story and how it has unfolded in your work together.

1. Now listen to how their team and/or family dynamics have emerged through them, its pattern and flow of relating. What might be the music, taste or colour of the family and/or team dynamic?

2. Now sense the organizational culture that has emerged through this person, both the culture they describe, but also the culture that is part of their describing, seeing and framing. How it is creating the stories you have been hearing – both the team story and the individual story.

3. How is this organizational story formed by its purpose, both its founding purpose, and those it serves today? How is it formed by, and forming, the stakeholder world it lives within? The supply chain upstream, the customers and clients and beneficiaries downstream? The financial systems it is part of? The communities and cultures where it operates?

4. Now go further out; What are the flows into this organization from the 'more than human' world – water, air, fossil fuels and or wind, solar energy for heat and light, plants and trees, insects, fish, amphibians, and animals.

5. What is the human consciousness that is present in receiving these generous gifts that are given from life?

6. Now what are the gifts that flow back from this organization into the stakeholder worlds, into the wider cultures and communities and into the ecology?

7. Sense the quality of co-evolution. The flows in and the flows out. What adjectives or metaphors might describe it?

8. Now come back down the levels and see how this quality of participation and co-evolution, plays out in the relationships between the organization and its stakeholders, within the organizational culture, within the dynamics in and between teams, and within the story of your coachee.

*

Exercise

PETER: Please find a partner to share this exercise with, if you are able, perhaps going through it again, one reading it out to the other, and then the other sharing back what they discovered through each step.

*

JOSIE: I would like to say a few words about 'sustainability' before we move on from this exploration of eco-systemic awareness. It's something Peter and I have both been researching for a long time and the use of the word varies greatly. All living systems are interdependent upon each other. Some scientists refer to this as being 'coupled' to their environment. Varela, Maturana and Uribe (1974) and others talk about symbiogenesis (Margulis, 1998). As their context or niche environment changes, information moves through and between the eco-system and the living systems that reside within it, which processes that information with reference to their own DNA. It does so to determine any changes or adaptations that are necessary. These adaptations are experimental in nature. The system making them doesn't know if the experiment will be successful – it just experiments as a response to the changed context.

PETER: As Reg Revans, the founder of Action Learning, said, if we are not learning at the same or faster speed than our environment is changing, we are on the Darwinian road to extinction. This is true of individuals, teams, organizations, communities, civilizations and species. But evolution is not one way, for the living systems are also changing the eco-systems or niches they are part of, it is always a dance between wholes and parts. So here we find a principle for human leadership in complexity too. Learning, meaning making with reference to purpose, and values or the organizational DNA (Wheatley, 1999) and experimentation are central to the task of responding to a changing context.

JOSIE: Taking another step further though, the sustainability of the living system is more of a co-evolving, collective endeavour. One individual's longevity doesn't make a species sustainable. And sustainability is a result of constant experimentation as needed, so it is dynamic. Sustainability is not a destination or a thing. It is a process of constant co-evolution or emergence that reflects the dynamic feedback between the larger wholes and their subsystems. The interests of both must be 'negotiated' (Sahtouris, 2005): 'To serve our self-interest, we must be in service of all the nested systems that are within us and that we are nested within.'

The English language, which is itself an emergent cultural system, that we shape and which shapes, not only our speaking but our thinking, preferences the use of nouns. This makes sustainability sound like a thing. It is not. It is an emergent facet of a living system.

Elsewhere we have both written about these systemic insights and the implications for organizational life, and individual and collective leadership (McLean, 2020; McLean, 2017; Wells and McLean, 2020; Wells and McLean, 2013; Hawkins, 2005, 2015, 2020, 2021, 2022) to shape a sustainable, sustaining (Dunphy et al., 2007) and regenerative organization. You could also watch the short video dialogue between Peter Hawkins and Giles Hutchins (Hawkins and Hutchins, 2020).

What could you do to bring more awareness of eco-systemic interconnectedness into your coaching? What methods, techniques, interventions and questions might be useful?

And of course, responding and choosing our responses is not quite so simple is it? The complexity of each of us humans is huge. It's like we have a whole world within each of us. The voices of our parents and ancestors, of the cultures we were raised within, means different elements of life are more, or less, valued by each of us. In addition, there are the individual triggers we each fall victim to from time to time, resulting from past experiences and even trauma in our lives. Responding sensibly, rationally, is severely challenged by this too. Knowing what to do now, may also require different ways of knowing.

Let us now take a break before we address the third area of Eco-awareness, the spiritual dimension.

Chapter 9

Eco-Aware – Spiritually Connecting

Day Three Evening

CONVENER: We have explored two of the domains of transformation that are essential to our expanding our Eco-awareness, the emotional and conceptual, and now we will turn to the third. This is the need to have a deeper Earth connection and an eco-spiritual awareness, which many might describe as a coming home, a coming back to self; not our little constructed 'self-identity' but our bigger Self, where we are part of everything that is.

A clear view

This message
can never be understood with the mind.
It is not an intellectual grasping.

It is the end
of all grasping,
of all wanting

to get and understand.

The end of trying to reach
something,
somewhere,
some state of being.

And it is the end
of the feeling
that something
is wrong or missing
or that something
is out of place.

It is arriving home.
A place we lost sight of,
looking for something else.

The search and the grasping,
the looking for completion and fulfilment

DOI: 10.4324/9781003153825-9

are inherent to being identified
with this contracted energy
we have mistakenly called me.

The place we never left
is the true sense of I
that shines already
through the stories
that thoughts and emotions
are spinning.

This place
is a placeless place,
home.

This is our felt sense
of who we are.
On top of this open Emptiness,
have accumulated over the years
layers and layers of conditioning
that have condensed
into an energetic construct.

A clear view reveals
this erroneous assumption.

We are only Oneness
looking for ourselves.

Anamika Borst (2021)

CONVENER: What a beautiful poem, with many thanks to Anamika for allowing us to use it. Now we invite Dr Anita Sanchez of the Nawat tribe and a Mexican American consultant, coach, speaker and author to share her perspective, on the union of nature and people, before other voices share their experience of this spiritual connection too.

ANITA: It is in remembering my connection to my own body, to nature, and people that I began to trust that I am part of the One Hoop of Life. It is impossible to be separate.

One of the best ways to reconnect to yourself and your body is through nature. My grandmother from the Nawat tribe, also known as the Aztec, of central Mexico, was the primary teacher of indigenous wisdom in my family. I have a dear memory of being five years old, standing in her backyard with my sisters, Paula and Olivia. Grandmother Medina had woken us up early that August weekend morning; we stood out by her garden. She had the three of us gather near her sunflower plants as the sun started to rise. It felt magical in her garden. For five minutes every hour, until darkness returned that night, she had us go back to the sunflowers and stand in silence, facing the same direction the flowers were facing. Just as the flowers turned, following the sunlight, so did we delight in the sunlight, mimicking the sunflowers.

My grandmother helped me understand my connection to nature with this lesson. With challenges and traumas as a teen and adult, my body, mind and spirit were quick to remember that I wasn't every really alone. Nature is always here – trees, sun, birds, water, Earth, grass. I am a part of nature, and nature is a part of me. We are part of the sacred; everyone is part of the sacred. Yes, that is what my grandmother and other indigenous elders have taught me. And just as nature provides for humans' wellbeing, so must we care for the wellbeing of all our relatives, two legged, four-legged, winged, fist, plants, water, Earth, air.

This simple truth, a worldview of interconnection, invites us to really experience our oneness: we are each members of one Hoop of Life. Just as my indigenous elders taught me, science now shows us that separation is impossible.

As people re-awaken to their connection to all life; they begin to experience the joy and responsibility of their union with people and nature.

CONVENER: Thank you so much Anita, it is such a beautiful story and allows us all to sense the oneness. We are also delighted to hear from Lily Seto and Michelle Degroot to describe how eco-awareness shows up in their work. Michelle is from the Secwepemc Nation and a member of the Tk'emlups First Nation. She currently resides on the ancestral unceded territory of the Tsleil-Waututh Nation; welcome.

MICHELLE: This illustration is the combination of several clients who explored new approaches, or meaning, in their life in a variety of ways. Purpose became a popular theme when each person explored who they were, who they belonged to and what they are a part of. The theme of family as the core of who they are, along with how they serve their family, and in a bigger sense their community, provided a sense of purpose for their life.

Connection to family and community is also a linkage to a person's ancestors, all my relations meaning the land, the four-legged ones, the ones that swim, the ones that fly and the ones that grow from the Earth. The connection with our environment means a connection to the water, land, rocks and everything we come in contact with. Everyone is responsible to all through this interconnectedness.

To illustrate, some clients spoke of their responsibilities as guided by their family teachings, demonstration of cultural protocols within their family. Within a family, different roles are associated with responsibility for names, songs, and ceremonial items. While these specific responsibilities are specific ceremonial protocols, they translate to everyday responsibilities whereby those who hold more responsibility in cultural protocols also hold decision making responsibility for elderly parents, guiding siblings, or as speakers for the family politically. This translates to purpose. When purpose is not defined then it is sought by individuals by other means.

The clients sought purpose by identifying ways to give back to their family, co-workers, or larger community. Their relation to the broader world brought them joy. Coaching offered them the time to reflect on their current relationships and possible future connections. Before considering something as big as a career change, it was important to reflect on their current role with family and community and the impact the change would have. Coaching by grounding each individual in their teachings and protocols was a useful tool for them to set actions to achieve as an individual, and in consideration of the role they had with family and their larger community.

CONVENER: Thank you, Michelle, the connection you make with nature, culture and family is profound and the link back to purpose is one we can all relate to. Lily Seto will now share her experiences from her work with indigenous clients in Canada. She is grateful to acknowledge, with respect, the W̱SÁNEĆ and Coast Salish Peoples whose traditional land she lives, works and plays on.

LILY: To protect identities, this story combines my experience working with two Indigenous client-partners. One lives in her traditional, remote community and one lives in an urban setting. Permissions have been granted by both client-partners to share my experience as their coach especially as it relates to the environment.

Here in British Columbia the Indigenous population is made up of 206 First Nations, as well as Métis and Inuit people. Each community is unique and to honour the space that the client-partner lives and/or works in, it is important to ask about their perspectives of where they are located and where they are from. At times, the client-partner acknowledges that they are on their traditional land, other times they acknowledge that they are guests on other traditional lands and other times, they may be living and working in an urban setting.

These distinctions can be important for the client-partner to make, and the conversation can revolve around how familiar and/or disoriented they are feeling in their surroundings. For example, in one session, the client-partner, living in an urban setting, was missing her community roots and had the urge to walk barefoot in a local park while we were talking so she 'felt more connected to Mother Earth'. And, in this case, I was also happy to walk barefoot in my back yard, which resulted in a very special connectedness. Also, when discussing actions coming out of a coaching session, often there is a revealing of ceremonial and cultural healing that involves the environment. In one case, the client-partner shared that she would approach her local healer and request a cedar-brushing ceremony to help her release some of the tension that she had built up around her situation. She described how she would only be taking one bough of cedar and would honour the tree by giving thanks and acknowledgement.

I notice that it is difficult to separate the connectedness to the environment with other elements of coaching in the Indigenous space. For example, the concept of all my relations, which translates into being related to all things in the environment, as well as immediate family and the honouring of ancestors. I find that as a coach, I need to slow down and walk alongside my client-partner and invite the environment to be a part of the conversation, as natural as the air we breathe.

CONVENER: These short case studies from Michelle and Lily really underpin the connection with Earth and we now invite Tabitha Jayne to step forward to share her experience of 'Earth Connection' and something of the simplicity and transformational impact of awakening to a deeper relationship with Earth.

TABITHA: Earth Connection is a subjective sense of feeling part of the Earth. It helps an individual to gain a deeper sense of who they are within a divine cycle of life. Earth Connection also promotes a sense of a reciprocal relationship between an individual and Earth. While I reached this definition through research, that is now published and evidenced (Jayne, 2020), the motivation behind doing so is far more personal.

As a child, I experienced really bad cold sores that would end up covering half the face. I felt ugly and alone. I read about the Scottish tradition of washing your face in the May Day dew as a way of enhancing beauty. Practiced for hundreds of years, it made sense to follow the path of ancestors and kin.

Getting up early, I went to the back garden and spoke a prayer to the Earth. I don't remember the words I spoke, but I remember the cold dampness of the dew on the hands as I rubbed it over the face. I remember the intensity of hope that this could make a difference. Cold sores covering my face became a thing of the past. This was the start of a lifelong relationship with Earth.

This relationship has led to countless hours spent consciously engaging with the natural world. When I did, I would receive guidance and insight through thoughts in the mind. I don't know if this was Earth speaking or if I was simply hearing the depths of inner nature more clearly. It didn't matter. What mattered was the fact that the insight led to good decisions. These decisions improved the quality of the life I was living.

After the sudden death of a brother when I was 22, I spent even more time alone in nature. It was where I processed grief and where I discovered a purpose in life. For the last 19 years, I've let this connection guide the work I do. This connection inspired a number of Earth connected models and frameworks. It led to the birth of an organisation called Earthself. It inspired the development of a coach training program where I train other coaches to coach with nature and Earth. It birthed the creation of a business model based on how Earth functions. It creates a relationship with something far greater than the human I am and lets this guide the way I live life and do business.

There was a moment after starting Earthself where I doubted this connection and shut it down. Even after all the work I had done to develop it. The quality of life deteriorated and I vowed never to shut it down again. The lived experience I have had, shows me that life is better when it is lived in relationship with nature and Earth. The evidence-base supports this - and highlights that when we love and care for Earth, we take better care of it.

CONVENER: Thank you, Tabitha, as a final contribution here, we hear from Diana Tedoldi, her own discovery of, in her words, biophilic activism. Biophilia for those who have not come across the term before, is the innate human instinct to connect with nature and other living beings. Diana, please do share your piece.

Coaching for Biophilic Activism

DIANA: I've been practising deep nature connection and deep contemplation practices with nature since I was in my early 20s, around 1995.

I've witnessed how immersing myself into feeling one with nature has gradually shaped my identity, the deep sense of who I am and led me to a love of nature that has soon become activism.

My path to environmental activism is not made of knowledge in the traditional sense (learning about nature and the climate crisis, although I'm also really passionate about studying ecology). Instead, it's made of ecstatic and somatic experiences where I feel in tune with the broader Eco-systemic intelligence (learning from nature, directly 'downloading' all that I need to gradually grow my awareness and my calling to become a steward of life).

The best approach to solve a complex problem at work while slowly walking in the wood.

Sudden insights about a challenging relationship during a contemplative swimming practice in the warm waters of the Ocean.

The release of emotional tensions while lying down on the ground and savouring the smell of fresh-cut grass for minutes.

The innovative approach to a new project while hanging upside-down suspended inside my apple tree's branches.

Because of moments like this, I started feeling an embodied interconnectedness, a sensation of a deep dialogue between me and the natural intelligence everywhere around us and within us. Over the years, this embodied dialogue between me and the different aspects of nature has taught me about love, ethics, respect, gratitude, working with what you have, leadership, co-creating in partnership with another being, listening, self-confidence, taking care of self and others, balanced relationships, sustainability, gentleness, silence, surrendering to what you can't control.

And this ultimately led to a strong biophilic sentiment - Biophilia: the love of life, the feeling of affiliation and co-participation with the natural world.

Since we protect what we love, biophilia leads to respecting and safeguarding life in all its aspects, around us and within us, letting a natural sense of ethics emerge.

I call it 'biophilic activism' and coaching in connection with nature (not only 'in nature') has become part of many of my team coaching and executive coaching sessions and the foundation of what I teach to my students at The Nature Coaching Academy.

Facilitating moments of deep nature connection for my clients has so many benefits that it's hard to list them all. Nowadays, a growing body of evidence-based studies also demonstrates what is consolidated in my direct experience over the years (Berto, 2014; Martin et al, 2020; Pritchard et al, 2020; Weinstein, Przybylsk and Ryan, 2009; Whitburn, Linklater and Abrahamse, 2020).

The key aspects I notice are increased problem-solving, innovation and creative capabilities, stress-relief, self-confidence, growth mindset, increased awareness of self and one's potential, reconnection with an inner compass for self-orientation, increased sense of responsibility and care for others (human and non-human), pro-social and pro-environment behaviour, holistic view of problems and long-term thinking.

Contemplative and deep nature connection practices are also the basis of my well-being routines, and it's what I also teach to my clients and students. It's part of what helps me as a leader to hold a space that feels aligned with the force of life and that rejuvenates my energies and motivation to do my part in the world.

We need to develop a systemic and embodied awareness about our impact and how to live on this planet, 'which is ours – but not ours alone' (Janine Benyus). I believe that coaches, trainers, teachers, educators, therapists, facilitators, and each of us can be a force for good in letting a global movement of biophilic activism emerge. We all need to intentionally and consistently weave biophilia and nature-connection in the fabric of our daily lives - because we can only give what we have.

We have everything we need to be happy and impact our social and ecological communities for the better. We just need to trust our belonging to a wider intelligence and naturally let our daily choices align with it.

CONVENER: Thank you very much Diana. We have powerful concepts and patterns emerging that are resonant with others, wherever people really explore what's required. Ancient wisdoms and scientific enquiries meet. Earth connection, the subjective sense of feeling like an interconnected part of the Earth, strengthens the more you pay attention to it (Lumber, Richardson and Sheffield, 2017). When we allow ourselves to be conscious of our relationship to Earth, we move towards deepening it. A choice, not only to 'be' in nature, but specifically, to develop a relationship with Earth and nature, which is ever present within and without. An opening to what might be possible and a turning away from a life of separateness.

Biophilic activism, again along similar lines bringing joy as well as what is sacred and the concept of co-participation in Earth connection. Diana will be sharing more about training coaches with biophilic activism in Day Five.

PETER: We now invite Giles Hutchins to step forward. He has been working facilitating deep Nature Immersions for leaders and Coaches for many years. He uses the term threshold not to describe Earth's limits as Rockstrom did yesterday, but as the transcending of what were previous inner limits.

GILES HUTCHINS: Times they are a'changin. Even the nature of change itself seems to be changing – ceaseless churning change of the catalytic variety concocts this pregnant moment of seismic breakdown and breakthrough. I foresee the next couple of years as simultaneously challenging, transformative and full of potential.

Those of us engaged with transformational leadership work will know that periods of dysfunction and crisis often precede a step-change in evolutionary advancement. Therefore, these are fertile times for an evolutionary shift in leadership consciousness. Let this epochal hour of breakdown-breakthrough not go to waste, but rather spawn a necessary [r]evolution in consciousness. The future of life on Earth depends on it. The future-fitness of our clients' organisations depends on it. So too does the mental health and wellbeing of our client's families and personal lives.

I believe we need to cross a personal threshold and learn 'To die before you die'. The phrase 'to cross the threshold' means to undergo a metamorphic process of 'dying and being reborn'- to endure a shift in 'inner' self-orientation and 'outer' worldview. It's a deep psychological renewal that transforms how we relate to our inner-selves and our outer-world, enabling us to become more in harmony with inner and outer nature. The ancient Greeks used the term 'metanoia' to describe such a shift, 'meta' like in metamorphosis is to 'shape-shift' or 'move beyond' and 'noia' relates to 'true understanding': to transform the understanding of our sense of self and how we relate with the world.

My coaching experience has shown me that for a leader - or for that matter, any adult - to undergo a step-change in psychological evolution, one needs to hold-space for both the 'inner' and 'outer' dimensions of this threshold-crossing: the worldview shift and the inner access to one's deeper truer nature. All advanced adult developmental models, ancient wisdom traditions, and depth psychology approaches speak to this inner reorientation.

Thus, the death and rebirth process happen at both 'inner' and 'outer' dimensions. One might intuit that the inner precedes the outer (and there is much truth in that) yet I have found interweaving both is helpful, especially for the overly rational-analytic leader who may need some 'outer' reframing to help nurture an inner reorientation.

In *Regenerative Leadership* (Hutchins and Storm, 2019), we explore the worldview shift unfolding during this time of breakdown-breakthrough, and its application to leadership and organisational development. Exploring the dominant yet dying **Age of Separation** and an emerging **Age of Regeneration** – where humanity remembers its deep connection with self-other-world. You may have started to sense the word 'regenerative' entering the emerging zeitgeist. It's a word that can be applied to all aspects of life, from regenerative agriculture through to regenerative leadership consciousness, and involves an opening up to life's evolutionary dynamics of relationality, receptivity, responsiveness, rhythm and renewal.

Regenerative Leadership Consciousness, as we define it, correlates with developmental psychologist Clare Graves' (1974) work on Tier Two Consciousness where an embodied cognition of living-systems is activated in our psyche. It also relates to what organisation specialist Frederic Laloux (2014) and integral theorist Ken Wilber (2006) refer to as Teal-Evolutionary. It is an eco-systemic awareness that draws upon the **Logic of Life**, an embodied living-systems way of leading and living that seeks harmony with the way life works. This is not new, by any stretch of the imagination, it's ancient. For instance, over 2,500 years ago Lao Tzu noted, 'Those who flow as life flows know they need no other force' and Confucius noted, 'Those in harmony with nature hit the mark without effort and apprehend the truth without thinking.' Yet this is also cutting-edge contemporary thinking confirmed by scientific findings that show how being in nature, and opening up to nature's wisdom, helps us become more compassionate, creative and connected. The explosion of interest in everything from forest bathing to wild swimming speaks to this rising interest in reconnecting to the rapture of real life.

Today our dominant leadership awareness is largely mechanistic and reductive, tuning-out the rhythmic and relational dynamics of how life works. The result being: linear chains of production that create 'outer' toxicity and silo'ed hierarchies of management that create 'inner' toxicity. The vast majority of people today work in organisational cultures that sap people's creativity, purposefulness, ingenuity, resilience and empathy. Whilst many organisations pride themselves on being efficient and effective the cold reality is that many of our human interactions, decision-making protocols and meeting conventions are woefully inefficient and ineffective. The root problem here is the mechanistic logic. Change the underlying consciousness, and the logic shifts. Shift the logic and the culture can transform in inner and outer ways.

To shift from a worldview of separateness and its mechanistic logic into a worldview of interconnectedness and its Logic of Life, requires a threshold crossing. It's essential for the times we are in. It allows leaders to sense the organisation as a complex adaptive emergent system rather than seeing it as a top-down hierarchic machine. Our understanding of change, systemic interventions and transformation deepens in recognition of how life works. It also helps us perceive the eco-systemic nature of our own presence as leaders, and the presence of the organisation immersed in a sea of stakeholder relations.

This includes the wider social and ecological systems which are all inter-dependent and interwoven in measurable and immeasurable ways. The other side of all this complexity is a beautiful simplicity, found by crossing the threshold.

In terms of the 'inner' dimension, regeneration is a return to our true nature through a journey towards wholeness. This journey is a fresh as it is ancient, and it opens us up to living with more presence and purposefulness; exactly what these times invite of us. Through my coaching working in nature, I blend the shamanic and spiritual with the scientific and sensorial, so all of our nat-ural intelligences (rational, emotional, intuitive, and somatic) are enlivened. Then we unlock human potential, touch our true nature, and connect more deeply with the world around us. As Gandhi noted, 'as one changes their own nature, so does the attitude of the world change towards them'.

PETER: Thank you, Giles. You have echoed the contributions of this morning and taken it to a more holistic level. But now we have listened for long enough and we need to move into embodied action and experientially discover the inter-connection with nature. Giles and I both run nature immersions as part of our practices so together we are going to take you on a mini–nature immersion. So please take these instructions with you as you go for a walk in nature.

Exercise

GILES: Please walk silently for five minutes just noticing, listening and sensing what is around you. Listen to the birdsong, wind and sounds of plants or animals. Notice the clouds, sky and the many life forms you pass. Sense what is under your feet, what you brush against.

PETER: Then pause and ground yourself in one spot. Imagine your roots growing from your feet deep into the Earth beneath you. Take a deep breath up from the Earth and through your feet, legs, your hold body as you stretch your spine and arms up to reach towards the sky above you. Let that breath fill your whole body and then breath out through your mouth, letting your arms gently float down either side of you. Do this three times.

GILES: Now with your eyes wide open, stretch your arms out wide and breathe in everything that is in your 180-degree view, as your arms gather it in, and your hands come to your heart. Then breathe out pushing your hands and arms horizontally out from your body, sending the energy of your out breath to all that surrounds you here. Do this three times.

PETER: Now walk again, for five minutes, noticing any thoughts, that arise and float into your mind, notice them come and pass across your consciousness and then depart. Don't engage with them, just watch them come and go. No judgement, no attachment.

Now notice, all that is around you, and within you as you walk, just as sensations that arrive and leave, without naming or separating what emerges. No judgement, no attachment. Just emergence and disappearing, life entering you and leaving you.

GILES: If you can, please find a partner to share what this experience was like, and how it changed from each stage to the next.

PETER: Now we invite you to go to where there is some loose Earth near you or better still a compost heap and gather a handful of Earth. If you are fortunate enough to have a compost heap nearby, pause and contemplate that this is a sacred space where life and death intimately dance together. A place where that which is dying is transformed by billions of living creatures into new fresh life-giving topsoil.

GILES: Bring your soil or compost back and smell it and feel its rich texture. If you have a microscope or magnifying glass, use it to study the complex richness of life in what you have gathered. In one handful of compost or soil there are in excess of 500 species of fungi and potentially more than 50km of fungal mycelium. There are 10,000 species and a staggering 100 billion individual specimens of bacteria. They are all busy making the soil that will provide the Earth in which nature will grow the food that will feed future generations.

PETER: Try sensing and listening to this great community of life and see if you can feel a response from your heart and being.

*

CONVENER: Thank you, Giles and Peter, for that rich experience. The journey we have travelled today has again been a rich one, through increasing our eco-awareness at the emotional, cognitive and spiritual levels.

Reflection

CONVENER: To capture your reflections on the day, we invite you to complete the following seed sentences:

1. The feelings I need to listen to more in both myself and my coachees are.......
2. The most important change I need to make in my thinking is......
3. The ways I will more deeply connect with nature in my life and work are.......

Now, where possible, we invite you to share this with a partner, so you can feel these statements in your body and have them witnessed.

*

We end the day with another beautiful contribution from American poet and activist Drew Dellinger:

hymn to the sacred body of the universe

let's meet
at the confluence
where you flow into me
and one breath
swirls between our lungs

let's meet
at the confluence
where you flow into me
and one breath
swirls between our lungs

for one instant
to dwell in the presence of the galaxies
for one instant
to live in the truth of the heart
the poet says this entire traveling cosmos is
"the secret One slowly growing a body"

two eagles are mating—
clasping each other's claws
and turning cartwheels in the sky
grasses are blooming
grandfathers dying
consciousness blinking on and off
all of this is happening at once
all of this, vibrating into existence
out of nothingness

every particle
foaming into existence
transcribing the ineffable

arising and passing away
arising and passing away
23 trillion times per second—
when Buddha saw *that*,
he smiled

16 million tons of rain are falling every second
on the planet

an ocean
perpetually falling
and every drop
is your body
every motion, every feather, every thought
is your body
time
is your body,
and the infinite
curled inside like
invisible rainbows folded into light

"every word of every tongue is love
telling a story to her own ears"

let our lives be incense
burning
like a hymn to the sacred
body of the universe
my religion is rain
my religion is stone
my religion reveals itself to me in
sweaty epiphanies

every leaf, every river,
every animal,
your body
every creature trapped in the gears
of corporate nightmares,
every species made extinct
was once
your body

10 million people are dreaming
that they're flying
junipers and violets are blossoming
stars exploding and being born
god
is having
déjà vu
I am one
elaborate
crush
we cry petals
as the void
is singing

you are the dark
that holds the stars
in intimate
distance

that spun the whirling,
whirling,
world
into existence

let's meet
at the confluence
where you flow into me
and one breath
swirls between our lungs

Drew Dellinger © 2003 (reproduced with the kind permission of the author)

© 2022   www.drewdellinger.org   @drewdellinger

CONVENER: This evening we invite you to experiment with creating your own
poem, starting with Drew's opening two lines:
'Let's meet
at the confluence'
and see what comes. We then encourage you to find someone to share your
poem with. We look forward to being with you tomorrow when we will apply
our learnings to our coaching and mentoring practice.

Chapter 10

Eco-Engaged Coaching

Day Four Morning

CONVENER: Welcome to Day Four. We hope you have rested well, yesterday was a roller coaster and the ripples will stay with you. Your body will remember as you repeat the circle at different depths. Today we are on the move into a different focus.

Let's open with *The Wheel*, a short poem from Neil Scotton (2022):

The Wheel

It is empty
You have much
Reaching out. Touch
No attack so no defence
No theory so no argument
Just your truth
The empty receives
Just watch
The separation within without closes
Parts connect
Become one
Whole
Smile. Relief.
Joy!

Everyone has a piece of the puzzle
Part of
Whole but not fully whole
Clear but not fully clear
More pieces than a mind can imagine
How could they possibly come together?
No one could possibly know
Find my piece. Wonder. Wander.
Mine very different to yours?
No problem: just different parts of the one picture
Mine fits yours?

DOI: 10.4324/9781003153825-10

Smile. Relief.
Joy!

Settling into place together
My piece makes more sense
I see myself more clearly
I see you more clearly
I see the emerging picture
You see it too
We see it differently?
It is still the same
The whole picture is more than I can take in
No worry. I trust the artist.

CONVENER: Today, we will address how we take all we have been discovering both in us and around us and apply it to our coaching and mentoring practice. Tomorrow we will spend another day looking at eco-engaged practice, but then our focus will be on eco-engaged supervision and coach training and how we might all begin to develop as more eco-effective coaches and supervisors.

We are all in the process of developing our practice alongside the emergence of our awareness of the climate and ecological crisis. As we take steps other steps appear. Here, we share approaches and practices from those who may be a step or two, or many steps further, down the road. These approaches are in process, and offer insights, and it's worth remembering what we share and what we discover today, is only the beginning – there is so much yet to come.

There are some of you, who have said to us: 'this is all very well to learn the science and explore our feelings about it, but how can we bring this into our coaching?' Those of you who identify with this will be pleased to hear that today is about becoming practical and our key inquiry question for today is:

What can we do in our coaching, mentoring, supervising and training that will make the best beneficial contribution to the great challenges of our time?

Exercise

CONVENER: So let us begin by going into small groups and sharing the most important questions we each have about how to attend to the climate and ecological crisis in all our coaching partnerships. Please listen to each other with open minds, open hearts and open will (Scharmer and Kaufer, 2013) and bring back collective questions that arise. Alternatively, you can do your own reflections and journal your questions.

*

Welcome back. We have put up on the board the top questions that emerged from our coach research (Hawkins, Turner and Roell, in press) and the number of coaches who shared it, and invite you to come and put your initials against the questions and thoughts you share. Then add additional questions and thoughts at the bottom of the list.

- How to focus on environmental issues while staying client centred (29)
- Identifying resources and techniques (18)
- How to convince clients who are sceptical (17)
- How to integrate climate change into coaching (10)
- How to approach this with clients when they have other priorities (7)
- How to gracefully introduce the topic (7)
- How do we start differently with new clients?
- How do we contract eco-centrically?
- How do we build partnership between coach and coachee and a joint collaborative inquiry?
- How do we listen differently?
- How do we move beyond questions?
- How do we move into whole body and somatic engagement?
- How do we do this work virtually?
- How do we monitor our own shadow, projections and where our ego is getting in the way?
- How do we liberate the imagination and the expand the boundaries of what is possible?
- How do we create meaningful change?
- We want to hear positive accounts of climate conversations between coach and client.

CONVENER: What a wonderful set of questions and thoughts.

Let's sort these into a practical flow and I will invite my co-conveners to lead our exploration of different ones of these.

Let us start at the beginning with the question: How do we gracefully start a new relationship with the ecology as a natural part of our work?

Peter, please do share your approach to this.

PETER: The best time to consciously introduce the centrality of the ecology to every aspect of our life is right at the beginning of a new coaching relationship. When I start a new relationship, I invite a client to '*tell me about you*' and from this I get a good sense of how they locate and identify themselves within the larger systems they are nested within (Turner and Hawkins, 2019: 14).

Then I will ask '*what do you most care most about?*' Or '*what makes your heart sing?*' This helps us to establish a heart connection from the very beginning and often brings their wider human family and future generations into the conversation and many times will include the wider natural world.

Then, '*who and what does your work and life serve?*' This I find gradually opens wider and wider connections. I will assist this inquiry in two ways:

firstly, by asking in a tone that shows I am inviting an exploration, not expecting them to have a ready-made answer; and secondly by prompting further and wider exploration, by asking '*who else?.*'

If the coachee's exploration stops at the human stakeholders, I might gently ask: '*and what about the more-than-human world of the wider ecology, that provides more than 99% of what supports our lives?*'. If the response is just focussed on the present, I might ask, '*and what about future stakeholders and future generations?*'

Only when we have as many of the stakeholders and connected systems as we can, named will I ask: '*So, if all these stakeholders were present in the room right now, what would they say is the work you and I need to do together in our coaching dialogues?*'

If it feels appropriate, I will ask them to step into the shoes of each of their stakeholders, often by sitting in different chairs, or finding the right place to stand and speak from, and to speak as that stakeholder to us, the coachee and the coach.

Exercise

CONVENER: We can now try this together. Please divide into groups of three and practice opening a new coaching relationship with each other. One person will be the observer, one the coach, and one the coachee. If you are reading alone, you might like to ask yourself each of these questions, journal your responses, and when the opportunity comes, do this with others.

- Tell me about you?
- What do you most care about? What makes your heart sing?
- Who and what does your work and life serve? And who else? And what about the more than human world? Who in the future?
- If all these stakeholders were present in the room right now, what would they say is the work you and I need to do together in our coaching dialogues?

Please take some time to make notes between each round.

- As Observer, what did you notice? What was the energy like in the dialogue? Paying attention to body language, what did you observe, in others, in yourself?
- As Coachee, what was it like to be asked these questions? What was easy? What was more difficult? How did emotion show up in the dialogue? What did you find yourself censoring? What surprised you?
- As Coach, what was it like to ask these questions? What was easy? What was more difficult? How did the energy flow in the conversation? What did you find yourself wanting to censor? What surprised you?

*

CONVENER: Thank you, Peter. Let us now turn to the second question:

How can climate and ecology be integrated into our practice?

CONVENER: Coming back together and listening to what's emerging it's clear that not knowing how to introduce the ecology within our coaching work in a way that is respectful whilst not being evangelical, or in some way unethical, is a significant concern and block.

There's also a desire for validation, some evidence that coaching in this way would make a difference, a sense of not knowing how to make a meaningful change. One of the other questions was what stories of success are emerging that provide us with the confidence we need to change our practice.

On that note, let's look at the level of importance, confidence and readiness each of us feels to shift our practice to coach from a more eco-centric approach.

Exercise

CONVENER: Spend some time, in pairs if you are able, exploring these questions and whatever thoughts they connect to that are important for you. You might want to walk together, or simply sit, you might want to reflect before speaking, or speak and reflect. Do whatever works for you as you explore this territory within you and between you.

- How are you feeling now about coaching from this different, systemic, place?
    - How important is making this shift to you?
    - How confident are you about making the shift?
    - How ready are you?
    - What becomes more difficult? What is stuck where it previously flowed?
- What becomes easier? Where is the energy flowing where it wasn't before?
- What's less clear? What's clearer?
- What do you need now?

*

We have invited a host of practitioners who have been experimenting and will share their approaches for use in this eco-coaching space.

CONVENER: As Peter shared earlier, the opening conversation is so important. After we have established a real deep connection with the coachee and their world we often then move into wider contracting. So we will now turn to the question: **how might we contract differently in eco-centric coaching?** Eve can you address this question with us?

EVE: Thank you. I have always been aware that contracting is at the ethical heart of a strong coaching relationship. Creating a clear, shared understanding of assignments is challenging at the best of times (Hawkins and Turner, 2020). The impact of contracting is to create the container, creating the trusted space for the work that needs to be done, how we will work together with both the beginning and the ending process clearly signalled.

The contract is 'cocreated between the coach, the coachee (whether an individual or a group or team), and the joint purpose or work they are undertaking together' (Turner and Hawkins, 2019: 10). A question we hold when we engage in multistakeholder contracting is 'Who is the client?'. Peter's and my research at the time (mid 2014–2016) was limited to organizational stakeholder perspectives (e.g. boss, team, shareholders, customers, suppliers), now, we see this as only a partial set of stakeholders.

We now view the multistakeholder lens as much wider, including, but not limited to:

1. Earth/nature/ all the more-than-human world
2. Ancestors
3. Community
4. Family
5. Future generations
6. Business sector
7. Client (individual)
8. Coach
9. Organization - internal: team, boss and peers etc.
10. Organization – external: customers, suppliers, financiers etc.
11. Supervisor
12. Professional bodies

Can you think of others? For example, when Zoe considers the *Nine Domains* of a climate conscious coach in Day Six, she considers political institutions (which could be local, national and global). What about those we bank with, those we have our pensions with, or our insurers? And how might we contract with a wider multistakeholder lens?

The contracting process in coaching needs to establish the three fundamental pillars, the 'Why, What, and How' of the coaching relationship.

The 'why' and the 'what': we have touched on these in the earlier exercise, considering what has brought us here and who our work is in service of. We can expand these questions a little here:

- What do those stakeholders currently value in you and what greater value do they need you to develop and step up to in the future?

- What is the work we need to do together, that you cannot do by yourself or elsewhere? How will that create value for you, your work and our stakeholders?

It is by taking a good amount of time to establish the '**why**' that we are most likely to include all the stakeholders that our coaching practice could impact on. We might draw on something like Neil's Wheel which comes later today.

From there, it is relatively straight forward to deepen and clarify the '**what**' - the broad focus and objectives of our work together. We share with David, Clutterbuck and Megginson (2013) the view of the limitations of pre-determined coaching goals (more in Day Six). We might use a question like:

- How can our work together benefit future stakeholders?

Like other coaches I want my clients to thrive, not just survive. Building on this future view, we can take a question that I have used in the tourism industry among others:

- How will you/your organization (etc.) need to adapt to bring in younger generations as future customers (or as future employees)?

Adding further depth, we might ask:

- In the context of the climate and ecological crisis we are in, what might you regret in five years' time (10 years, 15 years and so on, as appropriate) not having worked on in our coaching together now? What will you be pleased we have worked on today?
- What might be the changes you most need to make?

I might change the initial phrase of the question above to: "Given that your future customers (employees) are more concerned about the impact of climate change on their lives, what might you regret not having worked on in our coaching now?"

It is only at this point that we now turn to the usual practicalities of a coaching relationship, the '**How**' of our working relationship, taking in:

1. Boundary management including that between coaching and other interventions such as therapy; and consider confidentiality and its limits
2. Practicalities such as fees, frequency, cancellation and postponement policies, and contact between sessions
3. Our professional context including the codes of ethics we have as a guide, the standards we uphold, whether we have supervision (also linked to confidentiality)

4. The role of any third-party (line manager, HR) including expectations and reporting

5. Processes from tools and techniques we might use, to preparation and evaluation and work between sessions

Even here though, the 'how' of our coaching relationships may be in flux and we need to be nimble rather than rigid. Bringing in a wider set of stakeholders and supporting leaders to think about impact beyond the boundary of the typical stakeholders may bring a different emotional context and greater sense of challenge to the work. And of course, contracting is constant, not just something we do once! Many thanks to the person who coined the phrase: ABC – Always Be Contracting!

CONVENER: Once we have established a relationship with a new coachee that includes their stakeholders, from their team right up to the ecology and we have contracting with them – we need to find ways of addressing the next question:

How to deepen and broaden the coaching inquiry with our coachees to address the wider issues?

We invite Neil Scotton, to share what he has termed Neil's Wheel, a freely and widely available online tool for the coaching community (Scotton, 2022). It is being embraced enthusiastically worldwide in very different circumstances ranging from private coaching engagements through to participants involved with COP26.

Following a similar structure to the wheel of life, it is designed to open up much broader systemic lenses that go far beyond the individual and reach far into the future. Its simplicity is a key to its success, enabling clients to engage where they feel drawn to and in the way that works best for them. A graceful dance between client and coach, or supervisor and supervisee is enabled, opening a deepening awareness of the elements in their life and work about which they care most.

Welcome, Neil.

NEIL: Neil's Wheel evolved from many places, including my curiosity and frustration around 'Why is change on big issues like the environment, and gender, and colour so slow?' And from observations, such as these issues always being framed as a confrontation – 'The war on...', 'The battle for...' I could understand why there is a confrontational approach. But what if there was a different way? The question 'How come change in coaching conversations happens so fast?' gave many answers that guided the design, and the way in which people are invited to engage with Neil's Wheel.

Buckminster Fuller was inspiring and would often say: "If you want to teach people a new way of thinking, don't bother trying to teach them. Instead, give them a tool, the use of which will lead to new ways of thinking". The wheel is intended as such a tool. John Elkington also influenced the

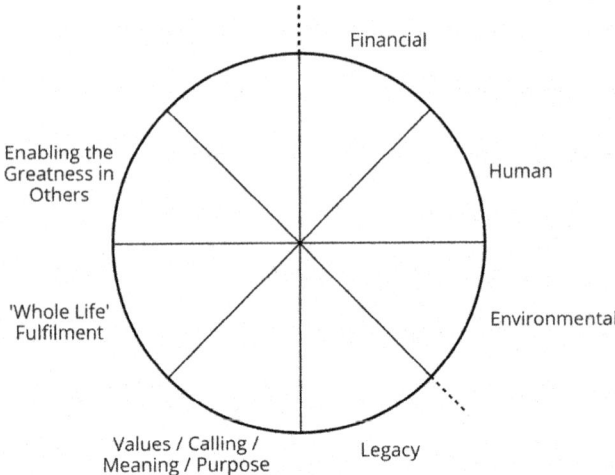

Figure 10.1 Neil's Wheel (2022)

development of The Wheel with his 'People, planet and profit' or triple bottom line approach (cited in Elkington, 2018). This gave a starting point that can be seen in the naming of three segments in the wheel.

The more widely known and used 'Wheel of Life' shows how multiple themes can be tabled at once. It is an inherently self-centred model ('My health, my career, my money…') while Neil's Wheel intentionally invites exploration further beyond self, and deeper into self, to the point where the going out and going in paradoxically connect. When this connection between what's deep within and what is without happens, special things arise for people. People use words like 'revelatory', 'calm', 'peace of mind', 'meaningful', 'amazing', 'energizing' and 'it all makes sense'.

Neil's Wheel is intentionally neutral in terms of politics and persuasion, and its appearance is plain The only known worldview it embodies is that 'All things are connected'. In a world of endless theories and complexity, its power is in its simplicity in unlocking people's innate understanding of how things are connected, without any need for theories or theology – though people are welcome to bring whichever ones they want. They all seem to fit. Often in ways they've never previously conceived or articulated.

As I describe it:

No attack so no defence

No theory so no argument.

So, no explanations, models, research or advocating of a particular subject are needed. Everyone can start immediately from where they are. And the blank section allows them to complete it with whatever would otherwise be missing for them.

Its accessibility is important to me. That's why as many barriers to access (money, time, language, education, identity etc) as possible are being removed, allowing us to act in scale and at pace.

I've been told that this Wheel is at the edge of coaching. I believe it is at the heart of it. For example: staying true to being non-directive; inviting the whole person into the conversation; letting people express themselves in ways that are natural and preferential for themselves; and more. Neil's Wheel enables people to be their own expert and guide, to make their own connections, to make their own meaning, to find their own answers.

There are Five Freedoms for the coach and coachee using the Wheel to remember at all times:

1. Freedom to explore the coachee's own interpretation of segment descriptors
2. Freedom to choose which 'parts of self' to bring to the conversation
3. Freedom to express the coachee's thoughts and feelings in their own way (words, scores, pictures etc)
4. Freedom for the coachee to choose what they want to put in the 'Blank' space, that enables a sense of wholeness and completeness
5. Freedom for the coachee to choose their journey as you explore Neil's Wheel together – where to start, where next, where to finish for the moment, etc.

And there are Four Mantras if you are supporting others to explore their wheel:

1. 'There's no 'One Way'
2. 'Let Neil's Wheel do the work'
3. 'Coach the Person' (the Wheel is simply the catalyst for the conversation)
4. 'Recognise and own your own stuff'

The only times we've heard of people having poor experiences with the Wheel is when they have felt someone else has been trying to take them in a certain direction. Responses to a directive thought-partner are swift and can be raw. If you try using this as a Trojan horse to lead a client to your way of thinking, things will probably backfire. If you trust in the client and are simply alongside them as they explore, following the Freedoms and Mantras, you will probably be surprised.

Here are some examples of people's Neil's Wheels used with permission (figure 10.2).

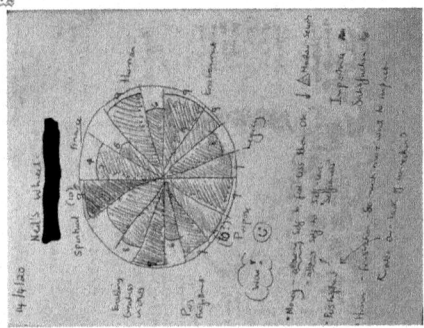

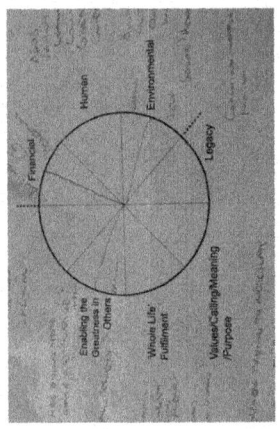

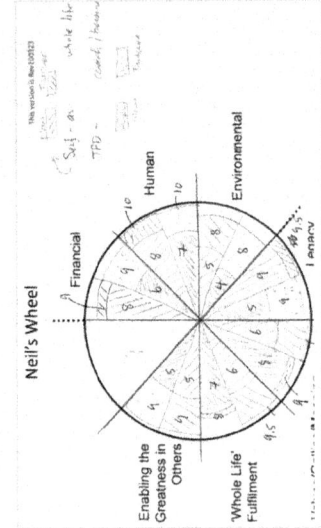

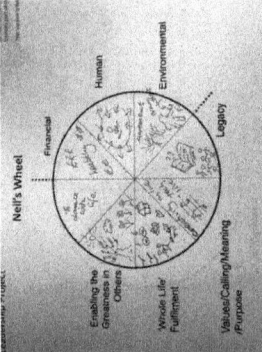

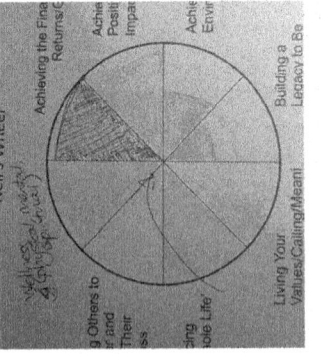

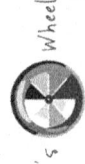

Figure 10.2 Examples of Neil's Wheel

As well as drawing on paper, people have drawn it in the dust of a forest path and in the sand of a beach. People have deconstructed it, placing segments around them and 'stepping in' for an embodied experience. And more, please do play.

I also wish to recognise the beta-testers of The Wheel – coaches from around the world who helped hone its detail and shape the resources and guidance for its use. Their work practicing and playing with the wheel, reflecting and sharing learning and experiences, and rigorously asking 'Is it safe? Is it effective?' was truly invaluable. The 'blank segment' was one innovation that came in through this exposure.

I could explain and share more, but right now this would probably be a distraction to just getting on and using it. Just try it. Go to the website, watch the short explanatory video and get going. Resources are available in multiple languages. If you want to deepen your learning, including learning-by-doing, visit the 'Just for Coaches' page.

CONVENER: Neil, thank you so much for this exploration and making this wonderful tool, freely available. And we would also invite you to look at another tool, also freely available, developed by Linda Aspey. It is called With the Earth in Mind to enable Pathways to Deeper Climate, Environmental and Societal Conversations (Aspey, 2022).

Now we turn to another area of curiosity.

We will address the much-asked question: **how do we address the wider ecology in a way that is still client centric?** I am going to invite Dr Anita Sanchez, whom we heard from yesterday, to share her approach that connects the wider ecology to our direct embodied experience, through our basic human needs of **breath, rest, water, nutrition, and community.** Welcome back Anita.

ANITA: As I worked in one big tech company, in both group and in one-on-one coaching, I began with an essential check-in question, 'How are you doing on the five things all human beings need to thrive?', referring to breath, rest, water, nutrition and community. I described how each is essential to our well-being, our growth as leaders and our capacity to envision and create new organizations, products and services. I also stressed that each is also directly tied to the health of our physical environment; what is good for us is good for the earth, and vice-versa. The key points included:

- We cannot live for more than a few minutes without breath, and conscious breathing is a path toward self-awareness and self-regulation. Consider, as well, the quality of the air we breathe. Billions of people on earth face dangerous air quality that directly imperils their health.
- Every living creature exhibits cycles of activity followed by rest. The greater our stress, the more we require rest to repair and regenerate, to recharge our energy and be open to our creativity. Leaders all too often sacrifice rest, believing they can't afford it. Their efficiency, and their

compassion, suffer. The analogy I used is that this is much like how we are treating the earth around us, extracting to the point of environmental collapse.

- People around the world are exclaiming 'water is life'. We humans are 50–60 percent water. Clean water, readily available, is essential to our personal, community, business, and societal health. As we need to ensure we hydrate, we must also focus on water as the lifeblood of all we do.
- Food that is good for us is inherently good for the earth as well. When we take in nutrition that strays too far from its natural origins, our bodies suffer. Chemically dosed, from exhausted soils, and contaminated, too many foods harm our health, our capacity to lead with vitality and a clear mind.
- Finally, I emphasized that to be human is to be in community. We are held in a web of relationships with other people and the natural world. When indigenous wisdom speaks of 'All our relations', it grounds us in our intimate, inextricable connection to all people, all creatures, the earth, and the cosmos. Our own psychological and spiritual health depends on being in healthy community, and when we recognize and celebrate our place in the largest manifestation of community, we are held, supported, and invigorated to be our very best selves.

One executive particularly stands out. 'Andrea' took in this greater story of her body, mind and spiritual connection to the earth. She began a breathing practice to help her face the systemic opposition she encountered in her work. She began using a sleep app to encourage her to actually prioritize rest, along with her commitment to her work. Andrea also created a regimen of hydration that she found not only kept her energy up during the day but also improved her sleep at night. She either brought a salad to work or had her assistant order lunches full of greens and whole foods. And, Andrea took time to list the many communities she belonged to and what she received and gave to those communities. She began to imagine what else was possible in being a vibrant member of those work groups, church groups, community groups, and the natural world around her.

Throughout all of this, Andrea heightened her awareness and ability to articulate the importance of being a member of the community called the Hoop of Life, where she depended on and appreciated, not only human beings, but all the other beings, water, plants, and earth that supported her in life. Her work became smoother and more productive, her personal life stabilized, and her confidence and sense of agency bloomed in the knowledge of her connection to all life. May it be so for us all.

CONVENER: What I really enjoyed about your story Anita is that there is no hesitancy, there is a confidence – this is how it is. What would it be like for us to adopt this confidence in our own work?

We will now turn to the question: **How do we need to listen differently?** Eve and Peter, can you also address this question please.

EVE: I believe our most fundamental gift is that of listening.

As we awaken ourselves to what is going on, and as we work with our clients (coachees, mentees, supervisees and others) we may feel vulnerable and seek models, tools and approaches to help us. We may hold big questions, yet at the heart of our work as practitioners our most important role remains – creating and holding a safe space, being a 'container' that allows the exploration of our feelings, our thoughts, our concerns, and doing this with love and compassion, free of judgement, expectation, or a pre-defined action plan. This is as true for us as it is for others.

Sometimes we can forget that our most fundamental gift as practitioners is our ability to really listen and to allow the expression of what is held within our clients, sometimes never previously expressed. We can easily underplay the significance of listening. Nancy Kline talks of the power of attentive listening to ignite other people's thinking (2020, 2015, 1999), to generate their best thinking and doing so with 'respect and fascination.' She argues that … giving people 'good attention…makes them more intelligent,' while poor attention 'makes them stumble over their words and seem stupid.' (1999: 37). In the area of ecological and climate crisis, attention is a gift we can all offer to allow us to be our most intelligent selves.

As we grapple with the reality of the 2021 IPCC report (Intergovernmental Panel on Climate Change), describing the damaging impact humanity is having on the climate, that the UN Sectary-General António Guterres called a 'code red for humanity' (McGrath, 2021), we may forget to listen. Our clients, colleagues, family and friends may not believe or ignore the emergency, perhaps because their grief and pain is too profound, as we explored yesterday. In some instances, they may not have heard of it. Yet it's exactly what we need to remember, to allow voice to what is unexpressed. As mentioned in Day Two, Gillespie (2020: 30) talks of the need for climate scientists to connect with the feelings of those they may address and quotes Jeff Kiehl urging us to 'sit together in silence, holding the multitude of moods…our shared feelings evoke within us a profound depth of caring.'

So as coaches, psychologists, supervisors, mentors, we can 'be with' our clients. We can allow silence, we can listen with openness, with our body, with our hearts, with our minds and reach the fourth level of listening described by Peter Hawkins (2022: 357) as 'generative listening' where we are 'connected to something larger than myself' (Scharmer 2016: 12).

In her book *The Promise That Changes Everything – I Won't Interrupt You*, Kline builds on her ten components of the thinking environment and describes the wider system of interruption, giving it four 'main pieces: Conformonomics, Digistraction, Persuasion and Polarization' (2020: 83). To explore one of these, Kline believes conformonomics has a goal: 'to make me and all the other visible 'mes' out there relinquish our discernment, our delight in true difference', which leads to our conformity such as chain store buying patterns. She sees this endless increase and growth as unsustainable, following on from those systems writers who have argued the same (such as Capra and Luisi, 2016) and concludes 'So is this what we mean by a free market? Isn't this kind of conformity-driven market anything but free?' (Kline, 2020: 87).

So, as we sit with our clients, without judgement, listening to their narrative, to their sense-making, as we allow them time to deeply reflect, and based on the contracting work we have done (see – pp. 136-7 and 139–141) we might inquire 'And how do you feel about x?' or 'How might that impact on the legacy you want to leave?' or just 'And what more do you think, or feel, or want to say?' (Kline, 2015: 113; Kline, 2020: 183). If we allow people to truly think for themselves, if we partner with them, we are creating the best conditions for truly liberating them from the constraints that cultural, societal and other pressures put on us all.

PETER: This reminds me Eve, of how Ravi Ravindra (2014: 105) describes spiritual listening:

The requirement is to be present with stillness and a silence of the body, the mind, and the emotions so that one might hear a rose petal fall, the sound of thoughts arising, and the silence between thoughts. The arising of thoughts and emotions is part of the play of nature, and watching this play with complete equanimity, without being disturbed, belongs to the spirit.

In eco-coaching we not only need to listen deeply to the other (empathic listening), but also listen to every person, community and being within their story with 'wide angled empathy' (Hawkins, 2019a), that is, with as much empathy and compassion for them as we do for the person who is present. This takes both discipline and practice and means we are not just listening to the coachee, but also with and through the coachee to their wider world.

The final stage of listening is to listen through and with the person to all the nested systems they are part of, their family, team, organization, community, the ecology. Here we are co- discovering what life is requiring from them and from our coaching work together. (eco-systemic or generative listening).

CONVENER: Can you say more Peter about how coaching becomes a collaborative inquiry?

PETER: We need to realize we as coaches are not the one doing the coaching. Coaching is a partnership where both coach and coachee are inquiring together, both listening deeply to what the coachee's life, stakeholders, and wider systems are requiring them to learn and develop. We move beyond listening to the coachee's conscious agenda, to what is emerging in their lives and deep within them. Both coach and coachee are at the learning edge, co-inquiring and co-discovering what life is requiring. This requires a much wider pallet of coach responses than the traditional coaching questions.

Coaching over the last 40 years has done great work in developing questions that encourage deep exploration in the coachee and many of these we have drawn upon in this book. Many coaches talk about questions as if they are the only form of intervention and that this is what the coach does and the coachee already has the answers, they just need help in finding them. While helpful, questioning is just one form of engagement and dialogue and can at times slow the joint inquiry down, by the coachee having to stop their creative exploration, to understand what and why the coach is asking this question, and what sort of response they are looking for. Also, question and answer interchange can keep the exploration at a cognitive neo-cortex level, rather than a whole brain, whole body, ecological inquiry, where the way forward is discovered beyond our neo-cortex knowing.

If we see coaching as a collaborative inquiry, where both parties are fully engaged, but neither has the answer and the agenda is set by the wider life and context of the coachee, then there are many interventions beyond questions that can be used. In systemic team coaching we spend a lot of time helping team coaches learn how to do *Time-outs* (Hawkins, 2021: 385–387) where they learn a whole range of ways of intervening in the midst of a process or team meeting that moves the inquiry to new levels. Similarly in individual coaching the following types of intervention can be helpful. Here are just eleven I have found helpful and then we can have fun thinking up many others we can use. We have nothing to lose other than our coaching habits and strait jacket!

1. **Offering Reflection and feedback** – that hold up the mirror so the coachee can notice their own process. As with all feedback it is important that such reflection is clear, owned, regular, balanced (positive and developmental), specific. (CORBS see Hawkins and Smith, 2013)
2. **Joining words** – coachees often come up with generalized statements such as: 'I need to communicate more' – 'be more transparent' – 'I need to better engage my team'. Here joining words or phrases are really helpful. So when the coachee says 'we need to communication more'; the coach prompts – 'to......'; 'about.......'
3. **Making action goals specific.** Similar to the above, when an individual agrees an action, but fails to include the 'when' – 'I need to get feedback from our customers' – the coach prompts 'before....'; 'by....'.
4. **Paradigm busting reframes.** When the coach notices a collective limiting mind-set, restricting the individual's options, functioning or impact, they can use simple paradigm busting reframes. An example is when the coachee says: 'You cannot do this in our organisation.' – the coach just adds the word 'yet', turning the situation from a powerless reinforcing belief, into a development need.

Another example is if the individual is stuck in polarized options of an 'either-or' debate, the coach can explore with the coachee, how they find a third option that meets both systemic needs but avoids both the current solutions.

5. **'Outside-in' inquiry.** Coaching can become increasingly inward focussed, and the role of the coach is to regularly open the systemic windows. The coach can bring one or more empty chairs to the table, and invite team members to comment on the issue from the perspective of their client, employees, investors, the wider ecology, etc.

6. **Future-back inquiry.** 'What might you regret in two years' time, not having addressed today?' Can you explore this situation with the voice of your grandchild, or someone in thirty years from now?

7. **Future forward Inquiry.** When a coachee agrees a direction with great optimism of the will, but lacking caution of foresight, the coach can ask them what three or four things might derail their best efforts from being successful. Once identified the coach can ask them to prioritize by rating both the likelihood and possible severity of impact of each derailer, on a scale 1–10, which gives a 'risk score'. Then for high risks to agree how they would handle this situation if it arose.

8. **Turn complaints into requests.** When a coachee complains about somebody in their organization or outside, or some part of the wider system; the coach asks them what request they would like to make to that person. Imagine that person is present and ask them to rehearse asking the person directly.

9. **Connecting separate issues.** It is easy for coachees to bring a list of items, as separate issues and miss the inter-connections. This becomes a form of 'issue siloism'. The coach can invite the coachee to explore what connects these issues.

10. **An invitation to change the form of exploration** – offering a quick method that will move the exploration on – 'let us try a quick brainstorm' – 'try saying that standing up' – 'I think it would help both of us to do a picture sculpt of all the various parties involved in this challenge' (Hawkins, 2021: 378–9).

11. **Celebration.** Positively connoting what has been achieved e.g. 'That sounds like a great achievement!' -opening the space to celebrate what has been achieved.

CONVENER: Peter, you have given talks and written about: **How do we let Nature be part of doing the coaching?** Can you say more about this?

PETER: When we invite nature and the wider ecology perspective of our clients into the conversation, we initiate eco-centric awareness. We can then go to the next step of allowing the wider ecology to have a voice in the coaching room, perhaps an empty chair that the coachee can go and occupy and address the issue under consideration, speaking as the more-than-human world. Then we can go further to step three of inviting nature and the wider ecology to do the coaching (Hawkins, 2020). This can be done simply by walking alongside in coaching, tuning into the pace and rhythm of the other's walking, listening deeply to what they are saying, and then inviting a pause and asking: 'What in the nature surrounding us and holding us right now might have something to offer to what you are exploring?' I might invite them to address their question or issue to the tree, flower, rock, or river they choose and wait to see what response they receive.

Freya Mathews (2021: xix, and 2017) shows how our engagement with the world of nature is through metaphor, imagery, synchronicity and revelation. She has developed the notion of 'ontopoetics' (Mathews, 2017), how we can have deep communicative engagement with nature, but it requires us to listen and engage with a deep humility and openness. The work of the modern approach to Panpsychism (Mathews 2017 and 2019; Reason and Gillespie, 2019, 2021 and Kurio and Reason, 2021), draw on many of the indigenous traditions we have been exploring on this journey.

Reflection

CONVENER: As we break for lunch, we invite you to go and practice some of what we have learnt this morning. You could practice both deep listening and 'invocation' by going and finding a tree, a plant, a river, and asking from your heart for what it has to tell or teach you, remembering that the response may come in different ways of communicating, imagery, sounds, bodily resonance and sensations, or a moment of synchronicity or revelation. Then find some-one to share with, what you discovered.

Chapter 11

Eco-Engaged Coaching

Day Four Afternoon: Stories of transforming our coaching

CONVENER: Welcome back and hope you have experienced refreshment from all the five areas Anita mentioned this morning: breath, rest, water, nourishment and community.

In September 2020 two of our team (David and Josie) invited coaches who self-identified as 'Coaching at the edges' into a dialogue. Exploring what it is like working beyond the previously agreed rules and traditions of coaching as we have known it, they followed their hearts into new coaching practices. Travelling mostly 'under the radar' and on their own, to the beat of a different drum. They are reconnecting to their own values and sensitively bringing these values and perspectives into their work in surprising ways. In the process, they are experiencing a transformation of their own identity as coaches.

First, we welcome Rashmi Shetty, to share her story of her coaching going through a metanoia and her new integration.

RASHMI: "Right now, we are facing a manmade disaster of global scale. Our greatest threat in thousands of years: climate change.' Prophetic words of Sir David Attenborough (McGrath, 2018).

A quote that pre-dated the pandemic, which was also in more ways than one, 'man-made'. A disaster that brought the world to its knees and stopped us from living with complete disregard to Nature. She locked us into our homes and in the name of prevention we caged ourselves while she opened herself up and cleansed a major portion of herself with animals and birds thriving. Mother Nature has given us her warning that if we do not heed to it, she knows how to take care of herself.

For the last few years with climate change affecting the seasons and life-styles in my city of Bengaluru in India, it has made me try to see how I can consciously reduce my carbon footprint with respect to travel, use of plastic, printing only that which is important. I have made sure there is carbon foot-print awareness in my facilitation sessions and consciously made the public transport a part of how I travel. I realised I owed my children a space that they can live in.

DOI: 10.4324/9781003153825-11

My coaching conversations focus on environment. Bringing in this piece in coaching conversations has never been easier with the pandemic and everyone slightly introspective. Reflective conversations especially when they move with the environment open the way they see their entire life in retrospection.

My identity as a coach I notice has altered. Bringing in the environment in the conversation is now a natural process. The first time the conversation started with a client was one where the environment was a casual topic touched only on the surface. It was the subsequent session that changed it completely where my client said that he wanted to share how much spending time with the environment touched him completely, especially after the coaching conversation. He went on to say how it was important for the wider environment to be protected for his child - the next generation – who will inherit the legacy. The vulnerability that came into the conversation opened it up at a deeper level. This I noticed is where I started believing in the power of systems.

Right from our body, to society, to the workplace, to the wider ecology, everything works in systems and this approach adds so much more value to our thinking as they play a big role as our overt and covert influencers. If we are clear as coaches to walk our talk it simplifies a lot of internal dilemmas that we get into. While coaches were once hired only to have difficult conversations that needed to be outsourced, today even individuals are investing in hiring a coach for personal clarity. Every individual values the satisfaction of his stakeholders both personally and professionally and has understood he is no island. The environment has become an important stakeholder in his journey.

I see myself as an Eco-Engaged coach who is not only talking about this with a client but also bringing it in as a lifestyle. In the last few years, I have consciously moved from being an Eco-Curious coach to now being Eco-Engaged. I believe that 'Nature nurtures' is a thought that most clients resonate with. Bringing this awareness of the role of the environment consciously is the need of the hour. 'Climate change is no longer some far-off problem; It is happening here; it is happening now.' These powerful words of Barack Obama (2015) speak for itself. It is now as influencers that we coaches can play a pivotal role.

CONVENER: Thank you, Rashmi, and welcome to Catherine Gorham who is going to share an example of how she continually pays attention to the 3 Cs of contracting, containing, and connecting when coaching people out of doors.

CATHERINE: It is imperative to begin by assessing a client's capacity for psychological containment ahead of an outdoor session as the vastness of the outdoors, coupled with the potentially infinite visual, auditory and kinaesthetic stimulation, can quickly become overwhelming. For that reason, outdoor client work is not recommended for a first session. Once the appropriateness of taking a client beyond a virtual or physical indoor frame has been established, my 3 Cs framework of contracting, containing, and connecting (Gorham, 2022: 83) provides preparatory scaffolding for the client to be held with psychological safety outdoors and enabling 'moments of truth' to arise (The Boston Change Process Study Group, cited in Jordan, 2015: 49). However, an outdoor setting is as dynamic as a client's internal landscape and so it is necessary to continually recalibrate the coaching/supervision process according to how the client is responding. The role of the coach/supervisor is to invite a deep sense of relationship with the outdoor system as a route to the unconscious, thereby facilitating greater awareness in the client and a change in relationship with their inquiry, for example, through metaphor, shifting perspectives, mindfulness. Opportunities for interruption arrive in the moment which may be helpful or unhelpful, for example, an insect landing, a bird singing, and the practitioner has to judge whether that might add value.

Example

My coaching client, Susie, was a newly qualified social worker. We met at her office for her 3rd session having already contracted to try outdoors along the river path next time. I used our walk to the river to indicate the start of our session, that point of entry being an invitation for her to slow her pace and, in silence, start tuning in to her senses. I asked her to notice what she was holding in her body and, as we walked side by side along the river, she connected with her low self-confidence which was an ongoing theme for her. When prompted, she identified a tiny boat bobbing on the water as representing herself, disrupted in its place by passing river traffic. She connected with some raw emotion as she witnessed the boat reacting with no apparent sense of control and even being stuck in a corner of the mooring area, trapped by larger boats around it. At that moment other walkers approached and as contracted in advance for that possibility, I gently offered the option of moving further away from the path to give her more privacy, checking she was OK to continue with the exercise.

This interruption and change of physical position enabled a different perspective of the boat. From there I asked her what she noticed about the larger boats around her – she noticed they were moving too and, in that moment, realised that the larger boats (i.e., her colleagues) were actually providing her with some protection from the buffeting. They were all moving in unison; the movement of the water representing the flow, and whirlpools, within the social care system. She felt comforted to notice that flexing to the system was necessary and part of the work. I then asked her who the swans represented, swimming by with their cygnets. 'They are the families we work with' she said. 'How does that boat that represents you as a social worker serve your families?' I asked. 'I don't want to be big and overwhelm the families, I want to meet them at their level and to find out how I can help them to carry on swimming' she replied. I then asked, 'Are you enough as you are to do that?' 'Yes', she said and smiled. We looked at each other, both feeling emotional at the truth which working outdoors had revealed in a moment of spontaneity and shared intimacy. The client took a photo of the boat in its 'wise' location and vowed to revisit the river when she needed a self-confidence boost.

CONVENER: Thank you Catherine for that wonderful example. Now we will hear from Heather Monro. One thing Heather won't easily share with you is her background as an international orienteering athlete. Coming into coaching, she recognizes the disparity between traditional coaching's more individualistic orientation and the decline of the collective good.

HEATHER: I started coaching about 15 years ago, inspired by the brilliant sports psychologists I'd worked with during my first career as a professional athlete. They had ignited my interest in understanding the psychology of optimal performance and 'being the best we could be.' Fuelled by a desire to do good in the world, I believed I was helping others to lead happier and more successful lives by performing at their best and viewed from this perspective, the first few years of my coaching career were a success.

And yet, as I journeyed further in my development, I was troubled by a growing sense of there being something more. Client organisations were investing heavily in people development and reporting positive indicators of success, whilst at the same time society seemed to be heading further into global crises of social, ecological, and economic breakdown.

And internally I was starting to recognise the darker side of the 'high performance' paradigm, rooted in the age of science and the growth economy's assumptions of ever more and better. My 'never good enough' driver had served me well in terms of achievement, but it came with anxiety and stress, inauthentic relationships, burnout, and depression.

Deep inner work with my own shadow brought into awareness the tendency to 'other' – to try and find a locus of 'blame' for my suffering. Seen through this lens it was all my parents' fault, or maybe the context I was brought up in? And then, ever so slowly, as my field of visioned broadened I realised that my parents had parents and had grown up in a context themselves, and so did my grandparents and great-grandparents before them. With a broadening perspective both in space and time I came to sense that this pain was not rooted in MY shadow or even in THEIR shadow, but a manifestation of OUR collective human evolutionary shadow.

The first changes in my practice emerged with my growing awareness of what science was telling us about the scale and eco-systemic nature of the environmental crisis. As my field of vision widened it became clear that the natural environment could be considered alongside other contributory factors in every 'presenting issue' in coaching. Initially I just held this as a frame of reference in my mind, but very soon a need to be congruent with my values brought it into the coaching conversation. I started to ask much broader systemic questions – beyond organisation and industry to a level of society and planet. And gradually the environment became a 'stakeholder' in every coaching dialogue. We don't ask permission of our clients to inquire into other non-human perspectives (e.g. finance, infrastructure, technology,) so why should planet be any different?

Whilst climate change is the primary focus of this book, it's worth saying that I don't see it in isolation from the many other manifestations of the 'Story of Separation.' The ecological, social and economic crises of our times are all inextricably linked. So, in working with clients I am now explicit about this inter-connected, eco-systemic context right from the beginning. I explain why and how I work with this perspective in all initial exploratory calls. In first sessions I use it to frame the work we will be doing together. I find Neil's Wheel a wonderful framework for these conversations.

The next change to emerge in my work, still unfolding, and perhaps the most profound, is my shift in perspective on the ROLE of the coach. To explain this, I find wisdom that is popularly attributed to Einstein 'you cannot solve problems at the same level of thinking that created them' is insightful. Our global challenges are not simple problems that can be solved with linear, reductionist thinking. We live in a complex, ever evolving web of inter-connectedness and yet our inability to think and act from an eco-systemic awareness finds us standing on the verge of collapse. We need a collective shift in consciousness to tackle these problems. And who better than coaches, whose work is by very definition about developing awareness and new perspective, to expedite this shift?

I, therefore, no longer see my primary role to be helping coachees to achieve their goals (which in most cases, involves solving problems at the same level of thinking); I see it as helping to develop higher levels of consciousness. With this view, there's no longer a struggle between my or the coachee's 'agendas,' rather we are both serving a collective human agenda to elevate OUR consciousness. My own vertical development is inextricably linked to the coachee's. Through the work we are doing we both develop as does humanity and 'Thinking partnership' becomes a much better descriptor than 'Coaching.'

What does this look like in the coaching conversation? Well, with this new perspective I've come to believe it's impossible to stand outside of the client's world and ask the neutral questions that are encouraged in elementary coach-training. Whether we like it or not we ARE part of the client's world and the wider system. This realisation liberates us from the directive vs non-directive coaching debate. As I see it now, the ONLY option is an eco-systemic 'us' awareness where the coaching conversation is a holistic mutual inquiry and coach and coachee are co-collaborators. We are therefore invited to step right into the coaching agenda and sense it from within.

I've found my most valuable 'tool' in this context is simply feeling into and sharing what is arising for me in the dialogue. These are not insights that come from a projection of my thoughts based on accumulated knowledge or patterns of the past (which would be directive). They come, instead, from intuition and the intelligence of the heart. They emerge from more than just listening TO the client; rather, a listening in a deepening and widening field - the space within, between and beyond. Time and again, these 'knowings' that I don't 'make' happen, but just arise from a place of internal stillness, have brought profound shifts in coaching conversations. Still very much emergent and an ongoing journey of discovery, the sharing of insights is profoundly changing the nature of my practice.

And finally, I've also started to experiment with a whole different business model for coaching. If I am espousing 'inter-being' as an ideal underpinning my work, it would seem deeply hypocritical to turn a blind eye to the privilege of our industry and our clients. A privilege reaped from the same market driven growth economy that drives the acceleration of the climate and other planetary crises. Gift Based Coaching is a disruptive experiment in an alternative economy - offering coaching as a gift and inviting clients to pay or gift what they feel in return. Through the lens of the 'Story of Separation' where your gain is my loss, this might be interpreted as altruism (my sacrifice), whereas in 'interbeing' (see p267) your gain is also my gain; I gain through sharing my gifts.

Watch this space as the Gift Based Coaching experiment unfolds!

CONVENER: Thank you, Heather. Your point about the necessary shift in consciousness echoes our exploration of this yesterday (for a summary see pp 115–116) and David Korten's work on The Great Turning (Korten, 2007). He identified the need for humanity to become aware of the ways in which we are collectively harming our planet. Joanna Macy, (2009) one of the founders of the deep ecology movement, took the work deeper. She describes the necessary Great Turning required in our current times as a 'the third major revolution of human times, after the agricultural and industrial revolutions'. Macy's way of being is full of alive-ness and joy, I would recommend her video clips. And indeed, that is her invitation - to join and take action as a way of becoming fully alive, creative humans in this time of crisis and opportunity.

Joanna Macy extended the concept of The Great Turning into a process called 'Active Hope', (Macy and Johnstone, 2012), that carries us through the changes to address our habitual ways of living and working.

We have invited Alice Howard-Vyse to share her work using an Active Hope approach.

ALICE: Being human-centric, I have spent a large part of my career asking people, from all walks of life, how they think and feel about the things that affect them: witnessing, holding and responding to intensely personal accounts of significant moments in a stranger's life, in order to co-create something better. The success of my work is predicated on my ability to stay motivated in believing that positive change is possible.

My optimism was deeply challenged by the side effects of the catastrophic bushfires in eastern and southern Australia in the summer of 2019/2020. In Sydney, we awoke each day to the relentless smoke that challenged the assumption that clean air would always be the norm. Australia wide, social media and the news showed seemingly endless photos of devastation of landscape, livelihoods, homes and lives - human and non-human. This state lasted not for a few days or weeks - but for months. I found it difficult to hold on to hope.

I turned to the work of Active Hope by Macy and Johnstone. I rolled up my sleeves up and convened Active Hope circles (Howard-Vyse, 2020), to support the community amid the collective trauma of the bushfires, and on a personal level, to renew my own sense of direction and purpose.

This has reinforced my connection to personal values and helps me rethink what the role of a coach is now. My experience points to more collective and community-based work, moving away from individually focussed work with traditionally defined 'hero leaders'.

Here, I share my understanding of Active Hope, why it is helpful, what it is and how to get started.

Hope is made through action. It is a practised skill, not an innate quality or character trait. The even better news is that hope follows the law of increasing returns: the more you practise putting your hope into action, the more hope you cultivate in yourself, and the more you enable it in others.

> Active Hope involves identifying the outcomes we hope for and then playing an active role in bringing them about.
>
> (Macy and Johnstone, 2012: 37)

Active Hope is grounded in two concepts: that hope is an attitude that is reinforced by our actions, and resilience is a force of life that happens through us.

Imagining life as a force that happens through us helps to reframe feelings of 'stuckness' or resistance: a reminder that even when we may feel stuck in particular emotions or situations, we're still participants in a much bigger process of life that is constantly shifting, with or without us forcing it.

To take action then, however small and seemingly insignificant, is, as Macy and Johnstone say, to act in service of life, to be motivated by a force far greater than ourselves.

To rage, blame and ignore instead traps us in a sense of hopelessness. And in the middle of the climate crisis - both a physical and existential crisis - we urgently need healthy, collective ways to transform hopelessness to action.

> With Active Hope we consciously choose to draw out our best responses, so that we might surprise even ourselves by what we bring forth.
>
> (Macy and Johnstone, 2012: 37)

As coaches we are perfectly positioned at what Joanna Macy refers to as '**the inner frontier of change**', to the personal and spiritual development that enhances our capacity and desire to act for our world" (Macy and Johnstone, 2012: 32) helping our clients to consciously draw out their best responses in the context of significant challenges. Viewed systemically, the work we do directly influences not only our clients' interior relationship with themselves, but also the communities and eco-systems of which they are a part. The work we facilitate in others, then, sits at the heart of any systemic change. And of course, we know that the quality of the work we do as coaches, or change agents, is dependent upon the quality of our interior too. The work begins inside us all.

Active Hope invites each of us to turn up and turn toward the challenges that confront us, rather than closing down and turning away. And to be courageous in holding space for others to do the same, so that rather than perpetuating the negative cycle of fear, numbing and nihilism we may carve a path of hope through our actions. Macy and Johnstone (2012) explain it as a four-stage spiral.

This process may be used for yourself, with your individual coaching clients or with groups within the community. I will now describe the stages and invite you to engage in each stage experientially.

1 Coming from gratitude

Actively 'Hunt the good stuff', big and small, as often as possible. To connect to gratitude is to notice our place in the Web of Life, and the 'more than human' world that supports our existence. It strengthens positive emotions, increases resilience, cooperation and generosity.

- Please reflect on a memorable experience you've had in nature, or how family or community have supported you in tough times. When did you last feel a sense of awe? Please note this down for yourself.

2 Honouring our pain for the world

As we discovered yesterday it is essential that we respect the intelligence of our emotions. As uncomfortable as experiencing the depths of our fear or worries may be, emotions as our 'survival responses', invite us to 'recognise our pain for the world as a healthy expression of our belonging to life'. (Macy and Johnstone, 2012: 38)

So please pause and reflect on what is troubling you about the world, and complete the sentences:

- The thoughts and emotions that come up for me are...
- My worst fears are....
- I avoid them by....
- I could you use them by....

Make space to listen to your responses with a non-judgemental ear and sit with the emotions that come up. Expressing your concerns through practices, such as journaling, drawing or talking, is the first step in transmuting despair or feeling stuck, into generative action. Doing this with others multiplies the effects. Turning our listening **inwards** to allow the part of ourselves that is feeling unheard, suppressed or unacknowledged – can change our relationship to that feeling and, in doing so, to the circumstances it arose out of.

3 Seeing with new eyes

Please think about a specific situation that's troubling you and ask yourself:
- What would I like to happen?
- How do I see it coming about?
- What might my role be in bringing this vision to life?
- What steps can I envisage for walking these hopes out into the world?

Perhaps your response to the previous questions have helped you to see your role as a coach or business owner in a new light.

4 Going forth

Now focusing on your roles as coach, parent, and/or citizen, consider: 'What story of resilience or hope is happening through me right now?'

Strengthening resilience depends on being able to hunt for the good stuff, even in the middle of challenging situations. In noticing the steps that you are already taking to enact and embody change; you make visible and reconnect with your unique role in the collective story of Active Hope as it is being lived through you. The path to hope is made by walking. I hope to see you there.

CONVENER: Thank you, Alice, that is a profound insight into how hope forms and what is enabled through hope. I too love the book and I recommend everyone to read it.

I would now like to address two other questions that have been posed:

- **How do we need to show up to do this work?**
- **How do we monitor our own shadow and projections in this work?**

And we would like to invite one of our participants, John Wood, to address this.

JOHN: People capable of holding complex perspectives are usually able to see their role as a participant observer (i.e., my 'self' as a coach in this case) in their space of inquiry (i.e. climate change in specific contexts). They are aware of how the act of observing, and the perspective from which the observing is occurring, determines what is observed and the meaning making attributed to it. In addition, the act of observing and the perspective from which the observing is occurring can change what is being observed. The 'self', the observing, the observed and the context become your holding space. And, as a coach, how are you consciously cultivating your holding? What is the holding space you occupy in respect to climate change? What do you allow and not allow in this space? And by allowing and not allowing what are you creating in the relationships and systems of which you are a part?

There are four potential shadow elements in the type of person who 'gets' the importance of working with the ecology and its complexity.

First, this order of mind may not fully understand the workings of shadow and projection and how it may interfere with the important relational and collaborative space needed to advance a global or local movement. Ask yourself, 'what is it that is suppressed or denied in me that disables me in the face of those holding an opposing view on climate change?' Ask yourself, 'how am I doing, in my own way, exactly what I am accusing others of?'

Second, the hubris of 'I know' inferring that 'you don't' which might be translated into 'you're stupid because you don't know'. Or the thinking that my perspective is 'bigger and better' (which it might be) but held in a way that is at the expense of listening, empathy and new action. Ask yourself, 'what assumptions do I hold, and superiority about my perspective, that get in my way of effective action?'

Third, because of these shadows, a failure to inquire into synergistic and integrative solutions that may also hold opposing perspectives. The pursuit of all or nothing objectives that creates its own resistance, the war of opposing wills. Ask yourself whether you are excluding synergistic solutions that do not accommodate opposing perspectives….is this possible? Ask yourself, whether you spend sufficient time inquiring into others' assumptions about climate change, in order to build a bridge of understanding…and then, when this is not possible what can I do?

And fourth, the big picture systems thinking, the wholistic "I totally get it" blinds us to the possibilities of the power of many small things initiated by small communities. When everything is interconnected, where there is Oneness, there are also many small acts like jewels in a larger crown. Ask yourself, am I overlooking these possibilities?

Do you hold superiority and hubris in you quest for climate change action? Do you judge others who 'don't get it' or 'don't care' or see this as a 'hoax'? And how do you hold yourself in the face of what you see as ignorance and stupidity?

Reflection

CONVENER: Take a moment to reflect on those challenges from John and write down your responses.

*

I'd like to introduce another voice in this more provocative discussion, that I discovered on LinkedIn. It's a piece by Tyson Yunkaporta entitled *Integral Theory Thoughts from Land* which I will précis for us. He begins by challenging our addiction to growth and climbing up the vertical development hierarchy. He stresses the importance of looking down as well as up. He invokes how Mother Earth is inviting us to: 'Slow down. Calm down. Scale down. Step down'.

TYSON: **Slow down.** Your life and 'development' is going too fast for you to manage all the knock-on effects and butterfly effects. Each solution, each hack, each fix is killing us all. Slow down. Listen to me. I'm down here, look at the ground, sit down, listen for a bit.

    Calm down. You are too outraged about too much. Make room for other stories. Listen to people you don't like and see their wisdom. You all have the truth, but it's distributed, so you can't see it alone, only together. You can't do that when you're attacking each other.

    Scale down. Stop growing! Stop this bullshit development! You don't need all this stuff, I got everything you need if you'll just live with me and work with me here. Your economy and financial system and supply chains are ridiculous and way too much hard work. Does everything have to be so complicated with you? And your personal development should be as easy as improving your relationships, especially your relationship with me. I am your mother after all! And please don't go to Mars!

    Step down. Your power structures are unnatural and they're killing us all. If you are holding a leadership position, please teach a team of younger people to take that on and distribute it productively throughout that group. You're not smart enough to hold that power on your own. (Used, with permission, from an article published on LinkedIn 19 November 2020.)

CONVENER: This connects to Eve's earlier discussion about listening deeply.

    We have one more voice to hear from before we move to the next question for the day. I'd like to invite Kevin Snorf who has been thinking deeply about the difficulty of 'transforming consciousness' within the time frames we believe we have to make a big difference. Kevin has been experimenting with 'translating meaning' instead. Maybe there are connections to Tyson's thought too?

KEVIN: Climate change is a given - our response to it is not. We are the ones who need to adapt or get steamrolled. There are things we can change and there are things we can't. We have to both respect what we can manage and also what's out of our control. There are technological and ecological management techniques we can use to mitigate and adapt to climate. But let's bottom line this. That takes a willingness and participation on the human side of things. And we don't have a majority opinion. Which, in this interconnected world, could be the death of us. Because getting humans to change is not easy. The current approach, simply announcing from a soapbox that cataclysm is upon us and we need to clean up our act, has never really influenced the masses. Those messages aren't wrong, they're just ungrounded in human psychology.

Knowing that the human side of climate change (influence, buy-in, policy shift, etc.) is the real obstacle to managing our response to climate change, the typical approach is to focus on transformation. In adult-development language, this means trying to get more people to higher and higher stages in developmental psychology models so we will have enough 'woke' people to turn the tide of the human part of climate change. The argument takes various forms. 'As long as more people think like me (holistically, ecologically, compassionately, etc.) then we will have the numbers to chart our way out of this mess and adapt.' I'm afraid not. At best, most adults take two-five years to grow into their next stage of development with great effort and the majority of the world population is two to three stages below world-centric consciousness. Plus, most adults hit a ceiling of development in their mid-20s and don't grow past whatever stage of development they're at unless they have a traumatic event/mid-life crisis or put in prolonged deep conscious practice. So, we are in deep trouble, right?

Not necessarily. It's just that the focus on transformation of consciousness through levels of adult development is a long game. Very long. We don't have time for that.

Instead, we need to focus on translation. Much like translating from one language to another, global leaders and change advocates need to become experts at translating into different value systems than their own. Translating meaning-making looks like taking a truth you hold and reformulating it in a way that is palatable and appetizing to the truths others hold. It assumes a kind of intimacy with, and curiosity about, others that doesn't make them a cog in the wheel, but a living, adaptable bundle of sentience with its own meaning-making that doesn't match ours.

A while back, I realized that people didn't care about the environment and ecological issues like I did. They cared about the environment for very different reasons. Some people can connect more with an eco-warrior's perspective; activism against unjust tyrants, getting out into nature to do my own thing, or exerting my power in nature. Some folks are more interested in how their family, community, and existent socio-centric belief systems can maintain themselves, so translation for them looks like stewardship, conservation, and enjoying and protecting nature. These folks can be convinced of the economic arguments behind climate change, new markets opening up, the thrill of technological solutions, and being cutting edge as climate change increasingly challenges markets to adapt. 'Going green' for others is second nature. This is a simple example of translation looking at levels of development and more applications can follow from using Ken Wilber's AQAL (Wilber, 2007). Translation can be learned relatively quickly by leaders and meets people where they are at, instead of where we need or want them to be.

Exterior adaptation to climate change won't be easy but is possible even based on current knowledge. Interior human adaptation bumps against what Robert Kegan calls 'Immunity to Change' (Kegan and Lahey, 2009). And transformation of human consciousness through the levels is like being in geologic time. We don't have that kind of time. Good translation is being done every day by ad agencies, politicians, and coaches/therapists all over the world. We can get good at this! It takes some practice but it's attainable and necessary. It must be a concerted effort and a unified front. It has to hit the meaning-making of the people we want on board with the climate change agenda, and it needs to be grounded in the 'real-world' limitations of climate change while still honouring the individual and collective perspectives of those we are seeking to 'change'. So, of course, in order to be more effective at change, we have to change our approach. Translation and meaning making is the currency of the future.

CONVENER: Thanks for this explanation, Kevin. I appreciate that the idea of translating also respects all levels of development.

In this part of the day, we have been exploring different approaches from coaches already experimenting at the edge. Some approaches will sit more comfortably with you than others. It will serve us well to remember that these are not either-or solutions – they represent different experiments. And for a complex challenge such as climate and ecological emergency, we need a diversity of responses and collaborative, holistic ways of connecting these. There are no easy answers.

Our next question is **'How do we liberate the imagination and the sense of what could become possible in our coaching work?'** We've been wrestling for a while about how we move with grace and skill, and to address this Zoe interviewed Rob Hopkins, the Founder of the Transition Town network. Rob's approach offers something we can build into our practice. I'd like to invite you to listen to the recording of that interview together.

ZOE: What is the vision of your book 'From what is to what if' with us…?

ROB: The vision was sparked by the fact I kept reading people like Naomi Klein (2014) saying climate change is 'a failure of the imagination'. And I thought 'why are we having a failure of the imagination?'. And then I found some research published in 2010 that said we were indeed having a contraction of our collective imagination, and that that had been the case since sometime in the mid-1990s, and it's been declining ever since.

You have to imagine something before you can build it, and if our imagination isn't up the task then we're really, really in trouble. So, the book explores this… are we living in a time of 'the Disimagination Machine'? It seems we might be. With contributing factors such as the decline of play, the rise of anxiety, the rise of loneliness and inequality, spending less and less time in nature, and the highly addictive, concentration-destroying, devices we're on.

What do we do about it, what would it look like if we had a society-wide push to expand the imagination? If we made that a societal-wide priority, how would that affect politics and economics, and how we felt about things. I play with the idea that being imaginative is an indication of well-being. The point of the book is to put imagination back in the middle and say it really matters. The book (Hopkins, 2021) is full of stories of projects I found around the world.

ZOE: What's your favourite example of the power of story and storytelling in action?

ROB: I love the story in the book about how the Swedish government now have a role of 'Chief Storyteller' and their job is to 'bring together the day to day realities of living in a post-carbon world'. I think we all need to cultivate that capacity.

Another story in the book is about two researchers at the University in Plymouth. They work with people to get them to tell stories about what it would be like in their life if they had made certain changes, so if they weren't drinking so much, if they were exercising more, if they lost weight – the big things they really need to change. They get them to imagine that in a multi-sensory way, to create 'memories of the future'. And they find that when people are really able to imagine it and tell themselves stories regularly about what it would be like if they had made those changes, that those changes stick with surprisingly little intervention.

I think we need to do the same thing with the climate emergency. How do we cultivate longing for a low carbon future? How do we tell stories about what it would be like? The poet Rilke (1929/2012) said: 'the future must enter into you a long time before it happens'. I love that.

ZOE: What have you seen unleashed when human imagination is truly sparked?

ROB: I've been to visit many, many places in Europe where the imagination has led to some incredible projects being unleashed. One of my favourite ones is from Liege in Belgium, where the Transition group had a 'what if' question – they said 'What if in a generation's time the majority of food eaten in this city came from the land closest to this city? And then they did it – they created a project called 'The Liege-Foodbelt' - which has now created 25 cooperatives. They've raised 5 million Euros of investment from local people. It's extraordinary, but it started with a really good 'what if' question.

I met the Mayor of the City who said, 'we used to say we want to be a smart city, now we want to be a Transition City'. This is now the story of this place. There you could see something that started in the imagination of citizens and grew and grew to the point where the municipality are saying 'how can we help? How can we remove the blockages to this happening?'. For me that's a beautiful example.

ZOE: How can coaches help revive or harness the collective imagination to help create a different future?

ROB: One of the things coaches can do is to be explicit about imagination. Coaches also need to be explicit about the climate emergency. We are in a climate and ecological emergency, and we need to be looking at all the work we do through that frame. And the beautiful thing about imagination is that when you impose limits, we're able to be so much more imaginative within limits rather than without limits.

Firstly, I would be hoping that coaches would not be coaching clients in the belief that there's a business as usual future, because that's not going to happen, and can't happen. Coaches have to be coaching for a low carbon, zero carbon future, and embracing the possibilities of that.

Coaches need to recognise that imagination needs certain things, it needs space, it needs certain practices that we can do, it needs us to intentionally make room for it. It needs good facilitation when we do it with other people. I would hope that coaches would be working with their clients to help them to really see quite differently their capacity and potential to be imaginative beings. To bring practices in which move us out of our heads and into a more creative kind of a space.

ZOE: How do you get people to use their imagination?

ROB: One of my favourite exercises that I do is called 'potato monsters' – it's a really good way of getting people into a more imaginative space. At the beginning of the course, I say to people 'here are four potatoes and some cocktails sticks. You can go outside and use anything you find; you've got 20 minutes and I want you back here with a monster, and I want to know what it eats, what its mating call is and what its name is. Go!'

They go off and giggle for 20 minutes and come back with something ridiculous, but it puts them in a very different headspace than they were in before, and much more imaginative, much more playful, much less in the headspace of 'I have to get the right answer, I can't take any risks'. Because you can't make anything sensible with potatoes, whatever you make is going to be ridiculous, and we need to be able to put ourselves in the space of being more playful.

ZOE: You talk about asking better questions – what are some 'better questions' that coaches could be asking themselves and their clients?

ROB: What I mean by better questions is questions that start with the words 'what if'. I'm a great believer in the art of a really good 'what if' question, which opens up so much possibility. One of the people in the book talks about a good 'what if' question being like writing the first half of a really audacious sentence on a black board and respecting their capacity to write the second half, which I think is great.

I am always really inspired by some of the amazing political movements - particularly black political movements in the US, who use really big 'what if' questions like 'what if we ended mass incarceration?', 'what if there were no police?' - and are able to sustain and nourish those big 'what if' questions over a long period of time. I get so much inspiration from that, and from their writings.

One of the ways you create good 'what if' questions is by creating a space which is a 'yes and' space, which is something I learned from doing improv training – the difference between the 'yes but' and the 'yes and'. So 'yes but' is what we all experience when you suggest an idea and someone will always come up with some idea why that can't happen, it's too expensive, too late, just not possible, we tried that before and it didn't work. When you replace that with saying 'yes and', where you build off each other's answers, you create a very different culture. Any training around 'what if' has to include some 'yes and' work, and really getting people into that kind of headspace. And then creating a framework about 'ok, how do we actually make these things happen?'.

Good 'what if' questions are like creating a new constellation in the sky above us. Again, it's about creating memories of the future, they open up, they give us a taste of how the future could be and a really strong invitation to make it a reality.

ZOE: A final wish for the global coaching community to take on board?

ROB: You have often a privileged degree of access to people of real influence. And in this time, you need to be really, really honest with those people about the challenges of now and about the fact that the climate and ecological emergency means that business as usual is completely finished. This is a time for really radical, brave, bold new thinking and that that really radical, brave bold new thinking needs to be rooted in values of equity, social justice, resilience, more local economies and that represents the most enormous opportunity for innovation and imagination, and it needs them to be part of making that happen.

If we're just coaching them to rearrange the deck chairs on the Titanic, if we are coaching them to be better at accelerating the demise of eco-systems and social systems, and human and climate wellbeing then we're really, really failing them. I would say that ideas that feel really radical in 2022 will not feel radical at all by 2024 and that this is really the time in the position you're in to be inviting and encouraging those people to be really bold and playful in the steps that they take.

CONVENER: Great practical ideas about how to help people be more imaginative and of course, wonderful encouragement to us all to step out of business as usual. Business as usual is not going to serve our coaching partners very well, even if it remains comfortable for them to believe that its possible.

Exercise

CONVENER: Let's play for a moment with some "What If…" questions ourselves. Take a moment to write a few down and make them as creative as you like.

*

Let us share what we have come up with.

To coaching bodies/trainers

What if all coach training had sessions on social equity?

What if competencies for accreditation/credentialing in coaching included bringing in social issues ethically?

To coaches

What if all coaches asked their clients questions about future generations such as 'What would your future employees expect from you in 2030?' Or 'How will future customers relate to your use of finite resources?'

What if people did want to hear about climate change?

To coachees

What if this does matter to your staff/family, customers/shareholders…?

What if there were no barriers, what would be the first thing you would do?

What if you were fully confident you could make a difference, how would you take this forward?

*

CONVENER: Thank you, that is a great list we can all keep adding too and some of which we will pick up on in Day Six.

So today we have explored many ways of coaching 'eco-engaged', right from the very first meeting, in how we contract and listen deeply and eco-systemically, how we bring the voices of all stakeholders including the wider ecology and future into the room so we can coach the connections not the parts, how we can broaden the focus of the coaching, and invite 'Nature to do the coaching'. Tomorrow we will explore how we do all of this, and more, in the context of supervision and coach development and training.

Exercise

CONVENER: To end today we invite you go back to the questions that were important for you at the beginning of the day and see how they have moved on. Then to collect up your key learnings by completing for yourself the following key sentences as personal commitments:

1. How I will start the next new coaching or mentoring relationship differently is....
2. Approaches and techniques I will build into my work with a current coachee or mentee or group are......
3. My biggest encouragement to myself is.......
4. I could be more courageous by.........

If you can please share these with a colleague – we find voicing our commitments to another takes them into our bodily knowing, and makes them much more likely to happen.

Let us end with two lovely quotes from the poet Maya Angelou (2014):

Courage is the most important of all the virtues, because without courage you can't practice the other virtues consistently.

If one has courage, nothing can dim the light that shines from within.

Chapter 12

Eco-Engaged Supervision

Day Five Morning

CONVENER: Welcome to Day Five in which we will explore eco-engaged coach development. This morning we will explore how supervision can not only move from being problem, coachee or even supervisee centric, to being eco-centric, but also a key resource for developing the eco-systemic literacy of both the supervisee and the supervisor. Then this afternoon we will look at coach training and how eco-literacy can become a core part of all trainings, from short courses to master's programmes to CPD offerings.

Much of what we explored already, particularly yesterday when we look at eco-engaged coaching and mentoring is relevant to both coach training and the practice of supervision. For instance, you could reread yesterday's programme and simply 'swap' the terms coaching practice and coach for supervision practices and supervisee.

Today is not just for coach supervisors and coach trainers, for supervision is not something that is done by the supervisor, but a collaborative endeavour between a supervisor and one or more supervisees. Also, the quality of a training is co-created by the pro-active involvement of the trainees. How do we all take responsibility to ensure we receive the development in working eco-systemically that we and our world needs us to acquire?

So let us start by asking, how many of you have regular supervision and attend some form of coach training every year?

Now, a follow-up question, how many of you use the supervision and the training to expand your own capacity to work more eco-systemically?

Whatever the answer to the last two questions , now ask how you could use both the supervision and any CPD coach learning, including attending conferences, to more effectively develop your ability to work in ways that co-create greater beneficial value for the earth and all your direct and indirect stakeholders?

DOI: 10.4324/9781003153825-12

Reflection

CONVENER: I invite you to go into small groups, or do your own reflections, and share your answers to these questions.

*

Welcome back, and please share some of your responses from your small groups.

'I could contract with my supervision group that someone also takes the role of the wider ecology and someone the role of future generations, when we feedback on each other's case situations.'

'We discussed setting up a peer learning group after this programme finishes where we specifically contract to keep challenging and supporting each other in being more courageous in how we open the open up the wider scope of our work.'

'I decided I need to change my supervisor to someone who is further along this journey than I am.'

'Two of us here are on a master's programme in coaching together and we decided we are going to lobby the course organizers to have a specific module on the ecological dimensions of coaching. If they will not teach it, we will! So, we are eagerly looking forward to this afternoon's session.'

Exercise

Please add your comments here.

*

CONVENER: Thank you for getting us started so now we turn to our focus on:

The role of supervision in moving from ego to eco coaching

CONVENER: We invite Eve and Alison to share their exploration of this topic.

ALISON: Let us start our exploration together by looking at how coach supervision has developed over the last thirty years. Listening to your answers to my colleague's questions I am aware that some of you have never had supervision and some only as part of your training, and it was more a form of coach mentoring.

Supervision in coaching has developed very quickly in the last 15 years after a very slow start in the early days of coaching (Hawkins and Schwenk, 2006; Hawkins and Turner, 2017). As in some professions, early on, it was thought of as something you only did when you were in training, and it was about senior practitioners helping trainees apply their training correctly and effectively. It would often be problem centric, focusing on the coachee and the problems they were bringing. Or it might be intervention-centric, focusing on how the coach intervened and how else they might have coached.

Our fellow convener Peter Hawkins had already spent 20 years developing wider systemic approaches in supervision across many helping professions before applying these approaches to coaching and mentoring from about the turn of the millennium.

EVE: We wanted to explore how supervision practice might respond to the climate and ecological crisis to become a clear partner towards a flourishing earth. Supervision lends itself naturally to a very wide angled perspective; being in service of multiple stakeholders. Systemic supervision frameworks encourage taking perspectives that move beyond the focus on clients, supervisees and their immediate context. So rather than aiming to establish what supervision is or its importance as others have done (e.g., Hawkins and Smith, 2013; Bachkirova, Jackson and Clutterbuck, 2021; Hawkins and McMahon, 2020; Hawkins, Turner and Passmore, 2019; Turner and Palmer, 2019), we wanted also to consider what it might be.

We invited eight coaches from the Climate Coaching Alliance (CCA) to participate in a co-inquiry, in a unique supervision group. With the purpose of experimentation, we explored the role and practice of supervision with an intentional ecological perspective. What might happen if we allowed the ecology to 'do' the supervision?

From late 2020 we, together with our eight fellow travellers, met once a month for three hours. We met four times in total. What happened? Read on to find out.

We are so grateful to: Gosia Henderson, UK; Jaime Blakeley-Glover, UK; Janet Mrenica, Canada; Katerina Kanelidou, Greece; Dr Kenza Khomsi, Morocco; Musa Nxumalo, South Africa; Stephan Ulrich, Vietnam; and Vaishnavi Viswanathan, India. Each of our collaborators have written reflections on their experience, some of which we share here with permission.

As part of our reflection, we share how we created the container and contracted together for the work as we began. As co-conveners, we invite you to review the approaches we experimented with, read some of the impact they had and to test these out for yourself.

Creating the container for supervision is akin to weaving a rich tapestry. Including Alison and myself, there were ten threads and enormous diversity, and the tapestry was inevitably going to be colourful and intricate. The point of similarity was that all were members of the CCA. Some had experienced many years of supervision previously and others had never even come across it as a concept.

Our guides were living systems principles including:

- Emergence, the pattern of what we were making was something we could only see when we looked back at what it was. No plan was followed.
- Inclusivity, everyone shaped the container for how we worked together and how the group unfolded and came together.

- Experimentation and 'play' were at the heart. Coming from a place of not knowing and in some way allowing the journey itself to be the teacher, using what was emerging and experienced to guide the next step.
- Collaboration and cocreation. We used a collaboration of approaches from old favourites to provide a sense of foundation and comfort and new approaches to stretch the edges of awareness facilitating new insights.
- Unlearning. To allow all of this flourishing to take place, it was important to be prepared to 'unlearn'. To leave some ways of being and doing behind. In particular, moving from experiencing and knowing through 'head' alone, to one where we experienced through head, heart and being.

The frame, holding this colourful and intricate weaving needed to be clear and firm, to prevent the threads coming apart and unravelling. The frame was created in multiple layers between:

- Alison and Eve.
- Alison, Eve and the group.
- Each of the group as co-inquirers.

We also had a sub-group that developed, working on the 'turtle' story, as we will find out.

A safe and confidential space was created that also held our hopes and outcomes for our work together. This was done in the knowledge that we would share our findings in the book, and everyone would have a chance to see the contribution and make changes and suggestions.

In her reflections Katerina observed, 'I loved that you were experimenting with us, and I wish that was standard practice. There is something when we try things for the first time. I believe we become more open to anything that can emerge, as at least I do not feel the pressure of expectations of how it should be, how to do it right.' Janet's discoveries included 'a world of experimentation, safe containers, new beginnings, friendships, professional relations that spanned the globe, a new outlook on coaching that depends a lot on place'. She saw in supervision 'shared experiences for coaching considerations and like-minded hearts and minds who are guardians of Mother Earth wishing to create safe containers to hold individuals during uncertainty in their lives.'

On diversity and inclusion, Musa noted: 'Our combination of personalities; cultural backgrounds and areas of specialization gave this experience an added advantage in that every moment was different, unpredictable, and thought-provoking. In a way, this enabled us as participants to see what is possible when people operate without the narrow and shallow limits of borders.'

Gosia shared, 'In my view the diversity of the group significantly contributed to the depth and richness of the learning experience.'

And Janet added, supervision groups should always be 'global, which allow for discussion from different lived experience voices. Ensure more participants from the south than the north in each cohort. Due to the coaching profession, we each have different training, different cultural interpretations of supervision, and are facing different realities in our daily lives due to the changes in climate that vary across the globe'.

The contracting questions, shared below, guided the depth and breadth of the container we created. Within these questions, you may see the bigger question of who supervision serves (Hawkins and Turner, 2020a: 150–151). In this specific supervision group, we wanted to really extend the stakeholders to wider society, Earth, and our ecology with our understanding of kincentric ecology, to recognise that all that life are our kith and kin (Salmon, 2000).

- What is the purpose of our work together? How can we co-create this?
- How do we need to work together to deliver value for us and for our stakeholders?
- Who/what are the stakeholders to our work? Who/what else?
- What will success look like for us and for our stakeholders?
- What might earth notice because of our work together?
- How do we stay at our learning edge? How can we be comfortable with our discomfort?!
- How might we combine lightness of touch with depth of learning? How can we share our intuitive and collective wisdom?

In addition, we used three 'simple, transformational questions' (Hawkins and Smith, 2013: 44) which the group found helpful:

- What is the truth that needs to be spoken?
- What is the shift that needs to be enabled?
- What is disconnected that needs to be connected?

In the context of the climate and ecological crisis, this reconnecting what has become disconnected is key. As Jarid Manos notes, the map towards the best of the future is beguilingly simple, it involves 'people reconnecting with themselves, each other, and our sacred, shattered Earth' (Manos, 2009: 386).

Jaime appreciated the power of the third question describing it as: 'a question that works at so many levels and it straight away made me reflect that our collective disconnection from the natural world is a root cause of much of what we face.'

Creating the container continued throughout our work together. We touched on our working contract at each session to ensure it remained fit for purpose. We closed the group after four sessions, leaving time for a gratitude circle to give appropriate time and attention to closing well."

CONVENER: "Here, we invite you to pause for a moment.

Exercise

As you listen to Eve describe these reflections, please complete the following sentences:

1. One practice I could adopt for contracting in my supervision is....
2. One additional area I could add to what Eve and Alison did is....

*

The Experiments

EVE: In supervision we find we are often bringing ways of being or perceiving through a wider set of lenses to look afresh at the work we are engaged in. Over the course of the four supervision sessions, we introduced a number of approaches to playfully experiment with, that were aligned with a strong systemic and ecological focus.

Here, we share four of the models and approaches that we worked with. At the end of each description is an invitation to engage in a reflective exercise, please select which one most appeals and go from there. You may use these exercises individually to reflect on your practice, or you may want to use them in your supervision with others, either way the learning can be so much richer and deeper.

Inquiry using living systems principles:

Kathleen Allen's work (2019) identifies five shifts to embody a living systems approach to leadership. These provide a deep and underpinning focus on how we worked together as a learning group, that would also be easily transferable into our coaching facing practice. The shifts move us from the world of control, looking for a 'who', avoiding resistance, seeking individual influence and change to something more dynamic, collaborative and interconnected, capturing something of the practical essence of 'interbeing'. We did this by changing supervisory questions we used (Table 12.1):

In all of these five shifts, we are asking a fundamental question: 'How can we use nature as a model to help redesign the questions we ask and the way we run organizations?' With billions of years of life on this planet, there is a lot we can learn from wider nature, about the conditions that create life, and the principles for flourishing. Those principles will lead us to better ways of living and working across our lives and societies. The five shifts house elements of these principles.

During our supervision sessions, these questions and reframes were available explicitly to work with.

CONVENER: Let us pause again and work with those reframed questions that Eve and Alison experimented with.

Table 12.1 Changing supervisory questions

From	*To*
What do I need to control?	What could we unleash together?
Who can make this work?	What interactions will facilitate this?
How do I avoid or overcome resistance?	How might we welcome or play with any resistance, with loving curiosity, rather than resist it?
How do I influence what individual action?	How might I influence the field, to unleash its potential? What is the culture and cohesiveness that will enable this organization to thrive?
How do I create change?	How do I release, transform and connect the energy that already exists in the organization?

Exercise

CONVENER: In threes, have the coachee bring a situation where they want to, or are struggling to, bring about change. The coach will experiment with versions of the second column questions above. The coaching will last for 15 minutes and then the observer will reflect with the other two on what they noticed was most effective and together you can explore what else might have been helpful.

*

Neil's Wheel

ALISON: You will remember we introduced this yesterday (page 141). In our supervision inquiry, we decided to explore what it was like to stand on the wheel and move around it, experiencing what emerged somatically, through sensations and awareness of words, stories or emotions that arrived. Each member of our group was invited to prepare a physical representation of the wheel that they could step on.

We began by presencing and centring to get in touch with our intuitive selves. During that moment of breathing, we were guided in an exercise which started with thinking of a question we were holding, about life, our practice, or a client, a training programme, or something else we were engaged in. Then using the physical representations of the eight segments, we were invited to step into whichever segment was most appealing at that moment and pay attention to the physical sensations in our body. Were there any thoughts, phrases, ideas or images that came to mind? How did this segment make sense in relation to the question we were holding?

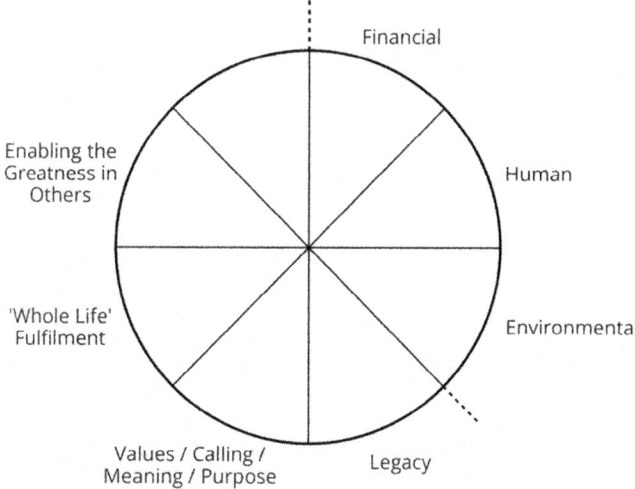

Figure 12.1 Neil's Wheel
Neil Scotton, 2020

ALISON: Rather than walk around all eight sections in one go, we walked through three or four, each time pausing to recentre, and consider which segment was inviting us next. Asking, how did any one segment feel in relation to another? In each segment we also had the option of exploring from a future perspective, in answer to the question we were holding, what needed to happen now?

We took a note pad and pencil with us as we explored the wheel, noting and reflecting in silence, with the odd prompt and the odd giggle. Not everyone had the space to create a 'standing' Neil's Wheel, in which case, using our fingers to do the walking worked equally well.

KATERINA: The experimentation in the group setting with the wheel was very helpful. I felt confident to 'skip' the guidelines and instead of having one question for the sections, I went to the sections and the questions found me there – different in each one. The fact that I felt I had permission to do that, is a result of how comfortable I was feeling in our group' ....in a space 'that allowed me to follow my instincts and in a way experiment within an experiment.

CONVENER: We invite you to explore Neil's Wheel to reflect on either the supervision you give or receive. If you have no experience of supervision, then think about the supervision that would be of most benefit for your work. For each segment, please score what percentage of the supervisory time is spent here and then write what are topics that are covered.

Now you might use the following questions to reflect on what you have written:

- Which segment(s) are you most drawn to?
- What segment(s) are you less comfortable with or avoid?
- What happens when you consider more than one segment?
- Which are easy to engage with? Which are 'empty'?
- What do you notice about the way you are engaging with your wheel as a whole?
- What are the patterns?
- What have you found in the blank segment? What does that segment represent?
- What questions come up for you in relation to your supervision?

Now you might like to do a new wheel on how you think this supervision could be more beneficial to all the stakeholders it serves. What would be the new percentages and what would be the areas covered in each segment?

Nature metaphors

EVE: There are so many ways we can bring metaphors from nature that provide new perspectives. Metaphor is such a creative medium, allowing us to express what's hard to clarify through other language, so we can discover what might be hidden, and inquire into a situation with the playful knowledge that we are speaking indirectly, 'as if'. In the full knowledge that metaphor is make believe, we give ourselves more freedom and can sometimes say what can't be said directly. The work of Srivi Kalyan (2015, 2019, 2020a, 2020b, 2022 video) exploring painting, music, poetry and movement and ecology, is another source of creative ideas to connect us with the natural world.

We used a particularly fun approach where we used nature metaphors for our clients. Each one of us was invited to bring a client to mind, someone we were currently working with, or had done very recently. Someone where it would be useful to explore the relationship between our client and ourselves (Hodge, 2013).

Exercise

EVE: The next step needs to be done quickly. If you would like to have a go with us now, bring one of your clients to mind that fits the description.

Quickly, imagine your client as an animal. Take the first image that comes to mind, that first animal that pops into your head.

Now we invite you to share your image with a partner or in a small group and here are some questions we invite you to play with, pausing every so often to hear from the group about emerging feelings, regarding their own and the menagerie of other animals.

- Tell me something about this animal (or figure, image etc.)? What are its characteristics?
- How does it behave? How does it interact with others in its area/community?
- What are its strengths and weaknesses? How do these affect them?
- What does it do that might prevent it from succeeding? What patterns of behaviour makes its life difficult?
- What other living beings might help it in its quest?
- Who are the stakeholders this animal is not taking into account? Who or what might be the '13th' fairy/fairies? (These are the stakeholders we may forget at our peril, rather like the 13th fairy not being invited to Sleeping Beauty's party who then cast a spell) (see Hawkins and Turner, 2020: 186–7).
- So, what changes could this animal (figure/image) make that would help it survive more effectively and succeed? What does it already know or do that will help it? Who or what else can it involve?
  And then considering the client…
- What insights have you gained that might help you to understand or interpret what is happening with you/your client and in the space between you?
- Knowing what you know now, how might you engage differently with this client?

*

EVE: This exercise allowed our group to be curious and withhold censorship of what was emerging, to step back and create space, to become more aware of masks worn, providing a different level of engagement and curiosity about the client and the relationship with the client. More specifically, Jaime noted that the client in mind seemed comfortable being detached, showing little emotion and seemingly superficial engagement. This exercise provided another way of considering how he might provide a safe space to allow the client to share more of himself.

Vaishnavi noted the feeling of stuckness with her client prior to the exercise, yet after the visualizations and discussion 'I stepped into coaching them differently, with patience. After two to three sessions, we entered a huge milestone that shifted in their way of living and being. There was a renewed sense of living with purpose which we were unable to tap into earlier despite several processes. It has also impacted them in being a better leader at work and share a deeper connection with their family at home.'

We specifically used animals as the metaphor, but there are many possibilities such as: a character from history; an image from the natural world; or a choice from a collection of cards, cut-outs from magazines, a bowl of figures, or a collection of objects. I (Eve) shared that I have a collection of small objects typically used to make children's jewellery. This collection houses a wonderful array of possibilities from butterflies and other insects to flowers, chairs and castles, I add to these objects such as small coins, minerals and crystals. All of these ways of representing our clients can bring new insight, freedom from self-censor and assist in our moving through hidden barriers.

We also used **the Seven-eyed model** in this group.

The Seven-eyed model, familiar to many in the coaching and psychology fields, was developed by Peter Hawkins (Hawkins, 1985; Hawkins and Smith, 2013). The model combines personal and social systems perspectives, focusing on seven lenses to ensure every angle is covered. Whichever 'eye' we look through we will see a different aspect of the whole.

EVE: Questions for each of the seven lenses or modes can be found in Hawkins and Schwenk (2021: 152–158).

Working through each of the modes in supervision, particular modes were highly fruitful providing useful insights. Mode three, the conscious and unconscious relationship between the coach and client, the 'dance' we are engaged in, is often drawn on as a way of reflecting on the relationship, using metaphor.

Jaime brought his work with a client to the group:

'...the nature of the contracting with that client already provided an ethical foundation to bring the ecology into our conversation and to challenge our work to keep at the 'development edge' of humanity. We had all the ingredients to 'dance with the earth in mind' but I discovered, the only thing holding me back was permission from myself to take the next step with confidence and a freedom from fear of being judged.'

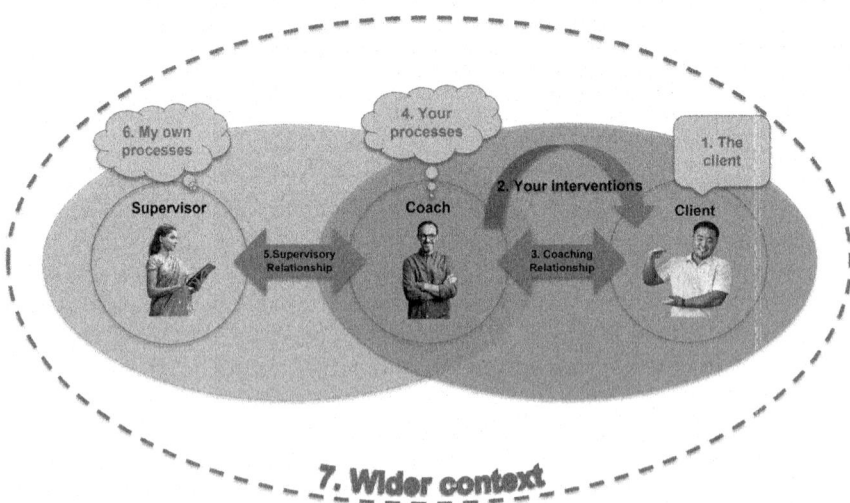

Based on the seven-eyed model by Hawkins and Smith (2013)

Figure 12.2 The Seven-eyed model

We explored many dimensions of Eye 7 and the expanded systemic levels (see Hawkins and Turner, 2019: 155–156). As we expanded outwards from the level of the organization it revealed hidden patterns sitting outside everyday awareness that can shed new light on what we and our clients are experiencing.

- What is the physical space like in the organization in which the client works? What does it say about the way the client values themselves, what is important to them and how they are valued?
- How do people treat each other in this organization?
- Who are the people in the organization that care about the client's success?
- Who and what are the communities that rely on this client?
- What do their families need? What other pressures exist?
- What are the concerns and challenges facing the sector they are in?
- What other countries/sectors are part of their context?
- What are the wider connections which the client might need to discover – political, social, environmental – that are shaping the background that is creating challenges for the client and their organization?
- What are the global dynamics that might impact on the challenges your client is facing?

Whilst these systemic levels are not new, we altered how we approached the exercise to tell the story from that part of the wider system we were in. This approach 'giving voice to' different levels and stakeholders led to the case study below, The Tortuga Voice, where the story is told from the turtle's perspective. This approach of providing a living voice to all elements is described in the afternoon of Day Two for example and the Council of All Beings (Seed et al., 2007: 35 and Hawkins, 2022: 328–348).

This approach takes us beyond the normal limits of our anthropocentric perspective. Seed et al. talk of the need to move 'Beyond Anthropocentrism… the idea that humans are the crown of creation, the source of all value, the measure of all things…deeply embedded in our culture and consciousness.' (2007: 35). The voices we spoke from were:

1. Coachee
2. Earth/nature
3. Community
4. Family
5. Future generations
6. Sector
7. Organizational stakeholders - such as internal stakeholders: e.g., team, boss and peers, external stakeholders: such as customers, suppliers, funders/investors

So, as we pause here, you might like to add your own.

Exercise

EVE: What other stakeholders to our work do you think we might need to give a voice to?

*

Subsequently as we have been developing this with groups, we draw from many other potential voices as you may remember from yesterday's Day Four on coaching (see p139), including our ancestors and future generations. But there are always more - for example participants have mentioned training schools/bodies and educators, the banks we have a mortgage with, pension funds, our accountants/business managers, researchers, scientists, governments and other policy makers (locally, nationally and internationally), unions, NGOs and sponsors.

GOSIA: I was surprised how different perspectives can be weaved into the coaching space, and how much more depth, breadth and complexity can be uncovered within a short space of time, through stepping into this systemic view, where Nature is at the core. For me it demonstrated the importance of connecting with the story that is beyond the story which we can hear from our client. It is this stronger connection between head, heart and gut, created through bringing into the awareness the inner body sensations, that shines the light on the unconscious knowing, and gives a deeper understanding of the client's perceiving of their world.

VAISHNAVI: My commitment moving forward is to continuously be conscious of how I can look at the impact at a systemic level and not just the individual. How I can look at the stakeholders beyond family and team. How I can bring the ecological perspective into coaching conversations. Of how I can balance the focus on individual growth along with growth for a larger system like community or planet and the role they play.

Janet shared her meaning of coaching with the ecology in mind:

JANET: First, respect for how we speak of nature. I have become very aware that the English language only allows inanimate terms for nature and all is relations – 'it'. Coaching with this awareness provides for the opportunity to shift and re-kindle the one being with nature and to support our clients in discovering what possible meaning it has and how to change our language to reflect the lifeforce of nature, who we are as a presence and our mutually enforcing relationship.

EVE: The story of *The Tortuga Voice* narrates the lived experiment of supervision with the earth in mind - letting the ecology do the coaching as experienced by, and then retold by, Janet, Kenza and Stephan. The supervision exercise applied the different voices or perspectives of ecological awareness that are mentioned above and challenged the group to open up to all these perspectives of awareness. The story is retold with permission.

The Tortuga Voice

KENZA: One day (2020) before the Covid-19 partial lockdown in Morocco I led a child-oriented training that aimed at making children understand our connection, as humans, to nature, how we can learn from it and why we need to care about it. The trainees' group had ten teens, aged between 10 and 15. It was a beautiful Sunday, under a clear blue sky and yellow coloured sunlight. In the early morning, we left the crowded urban city and drove towards the countryside where we planned to spend the day. I was the coach-trainer and prepared an education session for the morning and an outdoor excursion for the afternoon. Ali Salama, aged 11, was one of my trainees. He was such a happy, active and smart child. Ali was full of life and actively engaged in our programme. All the children showed interest in the concept of environmental intelligence and how connected humans are to nature, yet Ali was proudly putting forward his love for animals, his dream to be a vet, and how he was breeding birds, and had a chicken, a dog and a cat at home. It was clear to me that these creatures play a meaningful role in Ali's life.

The morning education session prepared the children for the afternoon excursion. We headed into the woods. It was a typical spring day, and the children were enjoying their time practicing some of the morning learnings, observing trees and soil animals. Some were classifying trees, others were attracted to ants and their movement, others were wondering why the river was dry and of course I was explaining the water cycle and the impact of climate change on water resources scarcity in our country.

Suddenly, I heard Ali shouting as if he had found a treasure. 'Teacher, look what I found, a turtle!', he said. 'I want to take it home,' he added. Ali was so passionate in his asking. His love and care for animals were shining in his eyes. 'Do you think that the turtle would be happy to leave its home Ali?' I asked. 'Why not?' Ali answered. I was challenged by the strength in my student's demand and simultaneously reminded of my role as a coach and trainer that understands what it means to break an eco-system.

I explained the turtle's life cycle and how it may be dangerous for the turtle to leave its environment. Ali stopped the negotiations and seemed convinced. At the end of the day, we returned to the city and said goodbye. Ali came to me, gave me a hug and whispered: 'you are a great teacher, but it is a pity you don't like turtles.'

JANET: Kenza brought this situation to the supervision group and the supervisor suggested we apply the seven perspectives (see p181). In this exercise, other members in the supervision group represented the seven different perspectives, from which to observe this coaching situation as described above.

Through this exercise, all these levels of awareness are given a voice played by a member of the supervision group. The child Ali in the story became 'the client', Nature was represented by the turtle, and other voices included the family, future generations and education representing the 'sector.' Group members 'became' the different voices. After an opportunity to hear an outline of the case, and ask clarifying questions, each member spoke, asked questions and shared their intuition and feelings from that position.

In the supervision group, the child was convinced that the turtle would have a better life with him at home than in the 'dangerous' open nature. He is not aware of the role and place of the turtle in the eco-system. From his individual perspective, he simply observes that he loves animals and how he keeps them at home and cares for them as others have done before him.

Nature was represented through the voice of the turtle. In response to the child's wishes, the voice of nature said at different times 'I am a turtle; I am a very ancient soul living in the water and on land. I walk very slowly. I come from my ancestors, who have inhabited this land before me. I am a living being'. 'As a living being, I am ki. I am also kin with those beings who are one with the land. I am perplexed that one would want to take me away, one would own me, as I am not one to own. Who is determining how I might be best living with you?'. 'Are you aware that in the English language, animals are known as objects, as the pronoun that references them is inanimate, one such as 'its'. Once upon a time, we were respected, we were referenced by our names and the lands where we came from. The settlers changed that. The settlers came and took our lands and labelled my ancestors, and in turn I, as an 'it' - a non-living thing. Child, is it possible for you to see that I am a living being that is unable to be owned? That I am one with nature? Is it possible to see that I am 'kin'? Could we be 'kin', if we learn to walk together to discover what this land, the nature has to bring?'

Other voices spoke: The level (perspective) of the community asked what language a turtle speaks and how we can see the world through its eyes. The family observed whether the child could imagine being taken away from his family. Future generations wondered where all the turtles had gone. The education sector wondered how learning with and from nature could be amplified.

STEPHAN: The seven levels exercise opened up new ways of looking at the situation for Kenza, in assigning nature its voice through the turtle. She considered that this voice is a call for connection and is powerful for awareness of nature in our coaching conversations. The other members of the supervision group could equally observe that they each held one perspective and could contribute important insights from their unique position. Respecting nature through its voice turned out to be transformational, as it became clear that we mostly speak for nature and to nature but perhaps not often enough as nature.

EVE: Thank you Kenza, Janet and Stephan.

I am now going to invite members of the group to share their reflections and learnings from the supervision process across our sessions, including the Tortuga story. Some of these were provided some months afterwards when their learning had been assimilated.

JAIME: As coaches we need to provide nature with a clear voice if it is going to be truly heard. My learning is that whilst I have asked my clients what nature means to them, this was probably listened to but not heard and certainly not felt. Had I been working with a different topic, there is no doubt I would have provided more challenge to my client. There is something about the fact that climate and ecology has been so polarized that makes it tricky to navigate in a balanced way. It requires, experience, knowledge, skill and the sort of safe space that this supervision provided to learn and practice.

There are some key learnings that have continued to benefit my practice. The first was the awareness I developed that I was waiting for permission to bring the earth into my coaching, but it was my own narratives that were holding me back. This question is now firmly embedded within my reflective practice and has been a constructive challenge ever since.

The second is the clarity and wisdom that comes from placing the earth at the heart of the coaching relationship. It is the way the concept of 'ego to eco' resonates most clearly with me and is playing a central role in shaping my professional practice and business interests moving forward.

GOSIA: The supervision, as well as the connections and relation which we built as a group, deepened my connection with my purpose and that in turn enabled my 'next steps'. I shifted my coaching practice entirely to support the transformation to sustainable practices and regenerative future, working with individuals, leaders and teams. In my consulting work I now focus on enabling sustainability in manufacturing organizations which is fully aligned with my purpose. For the first time in my life I feel connected with what I do and I truly enjoy it - it feels great!

KATERINA: My presence has improved considerably. To my clients, the enrichment of tools and approaches I would use and the ability to bring more effortlessly to the conversation things that matter, like the environment.

MUSA: Participating in the CCA Supervision Group impacted our company, 'Knowledge Connections', immensely in that it helped us to:

1. Deepen our connections with each other in The Beehive.
2. Shift perspectives about what supervision is and what are the opportunities that can come out of it.
3. Champion each other's work as coaches and facilitators.

Our perspective has been expanded by the ecological lenses through which we see the work we do.

STEPHAN: I have become more courageous in initiating action and care less of what others might think of what I say, write or advocate for. I have also become less afraid of failure, as I no longer try to control the outcomes - I am more comfortable now with the process of coaching and unleashing energy, and I trust that whatever comes out of it is just what needed to happen. More concretely, I am currently influencing a large new UN programme to incorporate a concern for our planet and the needs of future generations.

ALISON: Thank you, Eve and all the group, it was truly a rich inquiry and learning journey we all had. As co supervisors, we found it liberating and so did our supervisees. We enjoyed the permission of 'not knowing'.

CONVENER: As we come back together, we invite you to explore either by yourself or in small groups two questions:

Reflection

CONVENER:
1   How do we enable supervision to be more eco-centric?
2   How do we let the ecology be a resource and partner in doing the supervision?

*

Now let us hear back some of what each group came up with.

- "Stepping into the perspective of different stakeholders. Especially earth / nature, that was really powerful."
- "I really liked the 13th fairy, I'd never come across that before."
- "I found simple presencing and breathing exercises, noticing breath, noticing the air around us. It's all ecology!"
- "The simplicity of that little exercise 'looking out of the window', noticing and seeing what insights that provides to the situation at hand (Eve notes that this brief exercise was used in the experimental supervision group – where we would pause for a couple of minutes, and simply look out of the window, or go briefly into our garden or out of our front door, and take in the sounds, sights and smells around us, take deep breaths, and then return)."
- "Getting out of my own way – I decided to sit with intention before working with a supervisee, consciously inviting the ecology in. I got out of the way. It was beautiful."

We will break for lunch now and this afternoon, we will explore the wider field of coach training and development, of which of course supervision, we believe, plays a crucial part.

Have a good break.

Chapter 13

Eco-Engaged Coach Development and Training

Day Five Afternoon

CONVENER: Welcome back, as we now move from supervision to the wider field of all forms of coach development and training.

When we consider the challenges arising from a western emphasis on individualism and atomistic thinking, rather than on collective and systemic ways of being, we see that coach training is born from this paradigm, and coaching, as it has professionalized, is designed to support the dominant white western privileged hegemony, including the coach training systems and programmes, that now need reimagining.

When we reflect on the work we have done in Day Two, to deepen our understanding of what is happening in relation to our climate and ecology, the journey of awareness we need to travel on Day Three, and the nature of the practice we might craft to support and challenge leaders as they respond on Day Four, we now will explore the coach training that is needed to best enable this practice to emerge.

Let us start this inquiry with a reflection.

Reflection

CONVENER: Please reflect back on your own original coaching training and development programmes you have done since: and write down your responses to the following questions:

- How did it prepare you for the work now required?
- What elements were particularly helpful?
- What elements might get in the way?
- How might every coach, right from the very beginning of their professional journey learn the skills required in the current ecological crisis?

We will explore these reflections in small groups and hear back some of your reflection."

*

Now let us collect up some of your responses:

DOI: 10.4324/9781003153825-13

"My training was very good at giving me a model to work with."

"I was helped to summarise and reflect back, and to listen."

"I had a good understanding of creating a safe space and being non-judgemental."

"We always got to actions."

"I felt that the client was the person opposite me, and that what they wanted was key, so I followed their agenda."

"When there was a coachee and an organisational sponsor involved, we would come to a mutual agreement about what was wanted."

"We could consider other stakeholders, like communities, and the impact on nature."

CONVENER: Thank you for those suggestions. And I am sure you have more you might like to discuss in your groups. But now Eve will reflect on her own journey and share some specific research with coach training organizations and how they are responding.

EVE: As I recall my own training, learning from many generous people for whom I have the hugest respect and admiration, and have reason to be grateful to, I also find myself asking questions. The emphasis was on meeting the wants of the individual client and sometimes of the company as well, via the sponsor. Looking back from the vantage point of today's optics, it feels too narrow a view of the world. I was there to help people be more successful at what they wanted, to achieve their goals, with few caveats. Many assumptions were taken for granted – examples might be that increased profits were good, that growth was good, that reduced costs were good, that people working longer hours were being more productive (without consideration of the impact on families, communities, or well-being). This limited view, is not necessarily right or wrong, it is without thought about the larger context. A lack of awareness at the core, an unconscious way of operating perhaps. Margaret Heffernan talks about wilful blindness, which can be in part unconscious, that 'restricts our view' as we mix with people like us which 'embeds us more snugly inside our affirming thoughts and values' (Heffernan, 2019: 32). Ultimately such blindness can lead to many of the recent crises: from the organizational failures of Wells Fargo to Enron, to the behaviour of Harvey Weinstein and to the UK government in their treatment of a generation of Caribbean immigrants who arrived in the 1950s on the liner HMT Empire Windrush.

Peter Hawkins wrote about possible coaching mindsets from our training we may need to challenge and unlearn (2015: 42–5). Subsequently, he and I added three more (Hawkins and Turner, 2020b and 2020c):

1. The client is the person opposite me
2. I need to consider only their agenda of what they want

3. I leave my experience outside the coaching room
4. Interventions are always questions
5. I should never interrupt
6. Coaching should always end with an action plan
7. Coaching is only about personal development
8. Coaching only happens in the sessions
9. I should not feel their feelings
10. I must never go deep because that is only the realm of counselling and therapy.

These assumptions can severely limit the way we partner the coachee on a journey of collaborative inquiry beyond both our current thinking. For example, I might feel subtle discomfort around the ecologically negative impact the company I am working for is creating and yet I ignore it, because I think it is not part of what I should be doing as a coach.

Peter and I presented these mindsets at a Coaches Rising workshop to several hundred coaches (2020b). Some were in tears feeling somehow that the 'shackles' had been removed. You may not necessarily agree that some of these are limiting mindsets, you may not really have thought about these aspects at all, or you may feel they don't apply to you. Either way, it's important to have the discussion and to experiment in our own practice.

CONVENER: Let's pause for a moment.

Exercise

Take some time in small groups to review this list and consider:

- Which, if any, of these assumptions do you find yourself fully subscribing to?
- Which, if any, of these assumptions feels 'tricky' for you – perhaps you feel you 'ought' to follow this, but find yourself breaking these precepts?
- Which, if any, would you be delighted to let go of?
- Which of these might limit the depth and flow of your relationship with your coachees?
- Which ones could stop you bringing a wider lens that considers community, society and nature as stakeholders?
- What competencies or new rules would you like to add?

You might even like to reframe all of those ten mindsets.

*

EVE: As we learn to work within the context of climate and ecological breakdown, we see that these mindsets will hold coaches and coaching back from the pivotal role that could be played and the contribution we, as coaches might make. How is coach training responding?

As people in our profession are aware, anyone can call themselves a coach or mentor, and training can vary from a weekend course to a Master's programme of several years' duration. There are considerations of what is realistic. However, as we see the damage we have done to our earth, we have to ask what are the responsible changes we need to make as trainers?

As we have seen in those who have offered their stories so far, we live in a world of social inequity, of increasing gaps between rich and poor, of differing health outcomes, and of those who are experiencing the earliest worst impacts of our climate crisis being less likely to have caused them.

How well does coach training prepare us for our shifting world? As mentioned above, coach training grew to serve a particular economic system. How might coach training transform to serve the wider ecology, and within that, the economic and human systems that are being reappraised?

We carried out a small survey to explore the role of coach training in relation to the ecological crisis, and we discovered a breadth of responses. This is not at all surprising, yet we can pull out some threads. We are grateful to the 30 respondents, representing training programmes around the world including, the Americas, Europe and Asia. Some have also provided case studies or gone further than the questionnaire itself that illustrate the way in which some training bodies are engaging with this rapidly emerging context. The responses highlight some areas of discussion, some concerns, some different approaches and opinions, and of opportunities ahead. The diversity of responses mirrors the diversity across the field of coaching practice, and our diversity as humans as we awaken and become aware and explore how to respond. If we repeat this annually, I am wondering what changing patterns we might see?

One of these is that some coach trainers are looking to professional bodies to take a lead, as can be seen in these quotations:

> Q5 What shifts need to happen to enable you to bring climate change and the ecological crisis into your programmes in the future?

A shift in ICF requirements. Once it is mandated by ICF, we will include it. The same respondent explained:

> ...our focus is on meeting ICF requirements; thus we would amend our course if and when ICF includes climate change and the ecological crisis in its requirements.
>
> (Asian coach trainer)

> The professional bodies to take a clearer stand on these issues, that would give permission to coaches to bring this topic to the conversation with the clients.
>
> (Greek coach trainer)

Respondents felt that the need for climate and ecological crisis to fit in with what has gone before, and to feel comfortable, can lead to trainers feeling that tackling some subjects is somehow not allowed. Yet other training organisations, many ICF accredited, have stepped forward and offer deeply systemic and nature embedded training.

Professor Stephen Palmer notes:

> Understandably, there have been concerns raised by training providers about introducing issues relating to climate change during a training session. For training providers, it is important that topics covered on a programme are stated in the course content, objectives and/or learning outcomes and this could include climate change in relation to coaching. This allows potential course learners to decide whether or not to choose the course. It is likely that more courses will include climate change and coaching as a discussion topic on programmes during this decade.
>
> (Palmer, December 2021: Personal correspondence with Eve Turner)

EVE: We will take each survey question in turn.

Q2 What are you currently doing, if anything, in your coaching programmes to address climate change and the ecological crisis? See Table 13.1.

The wide range of responses included some examples of how care for our environment had extended into attending to the wider context of the training:

> The catering is 100 percent vegetarian / vegan and biologically produced. And instead of serving water bottles I offer my participants tap water. And the coffee capsules are 100 percent biodegradable. The tableware is made of bamboo and biodegradable as well.
>
> (trainer, Europe)

EVE: Another European trainer who brought out community work, wrote that they were 'Developing conversations within the community as part of our social impact projects later this year.' In the UK, Edna Murdoch of the Coaching Supervision Academy drew attention to some of the writers their training explored:

> We use teachers such as Thomas Berry, David Abrams and Joanna Macy to enhance awareness of ecology and interconnectedness. This increases

Table 13.1 Current training

Response	Percentage	Number (n=30)
Something	70%	20
Nothing	23.33%	7
This is not relevant	6.67%	2
This is not appropriate	3.33%	1

capacity to hold a bigger frame in supervision conversations. It enables students to notice and highlight systemic elements that may otherwise be ignored...we use Nature's Way by Karyn Prentice, and Catherine Gorham's work – nature as a dynamic co-partner.

Diana Tedoldi of The Nature Coaching Academy in Italy (see case study below on p. 195) intertwines nature into all the training done with seven modules of content from sustainability and climate justice to ecopsychology and somatic practice using yoga and trees:

> The whole approach of my coaching school is focused on expanding my students' awareness about their impact on the social and environmental systems we all belong to. We integrate nature-connection practices as a foundation of our courses...Our coaching programs are directly inspired by the laws observed in nature - we have constant references to environment and biomimetics in our programs.

Other trainers had concerns about the explicit inclusion of the climate and ecological crisis. This included the feeling they may be going against professional ethics. Other respondents did not intend to introduce climate and ecology, or felt it was inappropriate, and explained their reasoning:

> It is not appropriate to push an agenda in a coaching course. Climate change is an undeniably important issue. But in a non-directive coaching programme in which individual beliefs are valued, it is not the place to preach. The way in which the climate change agenda is being forcibly presented can appear politized, anarchistic and anti-capitalist.

Another offered this thought:

> There may be an opportunity here, but this seems quite forced. If the topics of coaching are determined by the client, it feels quite awkward to insist that coaches–and by extension coach training–include specific content on climate change.... I would be open to including this if potential clients begin to talk about wanting/needing this topic, but I'm not 100 percent convinced an advocacy role is appropriate for coach training. Will need to reflect further on this.

There is a breadth in the field, from deeply embedded systemic and nature-based trainings to a desire to see professional bodies and coaching clients create the demand for the inclusion of climate and ecology. Having asked people what they are currently doing, we enquire what their future development decisions might be:

Q4 What plans do you have to bring climate change and the ecological crisis into your programmes in future? See Table 13.2

Table 13.2 Future training

Response	Percentage	Number (n=30)
We already bring this into our programmes	50%	15
We have plans to bring this into our programmes	10%	3
We have no plans to bring this into our programmes	40%	12

Even one year on, as our situation gets increasingly stark, I wonder whether these figures would have shifted. We all have a position, and have choices, actions, power if we choose to use it.

CONVENER: Let's pause for a moment. How would you set up a training programme? What would you want in the content, if anything, related to this field? How might these considerations influence your future development decisions?

EVE: We were then curious about people's awareness of changes at the professional body level. At the time (2021) around half of those taking part in the survey were aware several professional bodies had signed a joint global statement on coaching and climate change. We provided a link to the statement in the questionnaire (www.jgsg.one/joint-global-statement).

Q7 Ten (now eleven) coaching, coaching psychology and supervision professional bodies (AC, AoCS, APAC, APECS, COMENSA, (EASC), EMCC, IAC, ICF, IOC and ISCP) have signed a joint global statement on coaching and climate change. Were you aware of this? See Table 13.3.

We then asked a further question: Q8 Having read this statement, is this something you already discuss with your students, and, if not, is this something you will do in the future? See Table 13.4.

EVE: As we end the sharing of this piece of research, we can see that coach training organizations are no different from any other segment of our world. We have the early adopters who are encouraging the rest of us to take the leap, many curious and explorative, and others who are not yet ready to consider that change is necessary or even how to start approaching it.

Other levers are coming to the fore in addition to nature herself. Since our survey, an update of the Global Code of Ethics, in July 2021, makes further

Table 13.3 Awareness of joint professional body statement

Responses	Percentage	Number (n=30)
Yes	50%	15
No	43.33%	13
Unsure	6.67%	2

Table 13.4 Discussing joint professional statement with students

Responses	*Percentage*	*Number (n=30)*
Yes we do	20%	6
No we do not as yet, but are making plans to do so in the near future	23.33%	7
No we do not, but may do so at some stage	36.67%	11
No we do not, and this is not something we believe is appropriate	6.67%	2
No we do not, and time constraints would preclude us from doing so	3.33%	1
Other (please specify) – e.g. posting up on participant login page, possibly integrating into mission purpose, did in past	10%	3

provision for professional development activities related to the larger context of our lives:

> 3.8 Members will engage in professional development activities that contribute to increased self-awareness in relation to inclusion, diversity, technology, latest developments in changing social and environmental needs.
>
> (www.globalcodeofethics.org)

As well as having their own codes of ethics, many professional bodies are signatories of this shared global code.

CONVENER: Thank you, Eve, that is highly illuminating and thought provoking. Through this and other conversations, we can see that the introduction of climate and ecology is happening at all levels of training, in CPD, and in supervision, but there are many places where it is not happening at all. Sometimes programmes have been up and running for some time, and in other cases trainers are in the foothills of experimentation. And as the number is constantly expanding, we encourage you to investigate what others are doing around the world to develop new approaches. And although this was not part of the research, we might also wonder how are we educating the coachees of tomorrow too?

The research we did, although small in scale, is going to be expanded by Professor Salome Van Collier from South Africa, working with Peter Hawkins, taking forward this research on coach training programmes. Please do share with her or Peter, any interesting new developments you have discovered.

We have come across many examples of ecological and more eco-systemic innovation in training programmes. Diana Tedoldi of The Nature Coaching Academy will now share how she structures her coach training course and aligns this with the requirements for professional accreditation that are already in place.

Welcome Diana.

DIANA: Our coaching training course includes standard coaching training hours (according to the 8 Core Competencies framework set by the International Coaching Federation) and training hours dedicated to eco-somatic coaching. We integrate coaching skills with deep nature connection and eco-systemic awareness.

The program offers different layers of learning so that students can choose how deep they want to dive into coaching with nature in different environments. These include:

- **One-week nature immersion to experience Forest Coaching® and Nature Coaching first-hand:** We dive deep into experiencing eco-somatic awareness, embodied and nature-connected contemplation practices throughout this in-person workshop. The purpose of this part of the training is to:

  - Set a foundation for the coach's presence, deep listening, trust in the other person's potential and self-development capability.
  - Share a common ground of approaches facilitating coaching sessions in partnership with nature and our clients.
  - Explore the synergy between the coach, coachee and the eco-system, exercising our ability to connect with the environment deeply, its intelligence, its unique identity and its life through silence, contemplation and empathy.
  - Build a strong sense of community between participants. The community feeling works as an igniter for the learning process and a gym of collective intelligence, enabling personal motivation and commitment. This is foundational to integrating coaching with biophilic activism (sometimes we feel like lonely pioneers in this field and having a community of like-minded professionals along the journey is vital).

- **Online Eco-somatic training:** where we learn and practice developing an embodied dialogue between inner and outer landscape integrating this approach with the professional coaching framework of the International Coaching Federation. The modules cover:

  - **The Eight Core Coaching** competencies (the model of the International Coaching Federation) and coaching ethics. We integrate foundational coaching skills with nature-connection and eco-somatic approaches to the coaching partnership, with lots of mentoring, practice, lectures.
  - **Somatic and eco-somatic coaching in synergy with nature:** somatic, eco-somatic and deep nature connection practises to experience nature as co-coach during the session.
  - **Emotional intelligence and change mindset:** to empower our coaches in enabling change in the direction our clients' desires.
  - **Eco-psychology, biophilia, deep nature connection and mindfulness with nature:** to empower our clients' ability to connect inner and outer

nature and shift the internal space from which they look at their topics, enabling a broader, holistic, interconnected view.

○ **Climate crisis, climate justice (social and environmental justice), sustainability and regeneration, Earth rights, privilege and oppression**: coaching for personal and collective climate resilience, acknowledging their potential impact on co-creating collective awareness around climate impact, personal and systemic resilience.

○ **Biomimicry (nature-inspired learning) and systems thinking**: to partner with our clients in exploring the systems their part of and the mutual impact, interdependence and potential for change, in synergy with the eco-systemic intelligence.

○ Additional, optional modules (Resource Development for coaches for wish to become specialized in **Forest Coaching® and Nature Coaching**) cover **The Eight Natural Archetypes** (the complete framework behind our approach to coaching and deep nature connection): Desert and Light, The Waters, The Earth, The Tree and The Forest, The Mountain, The Sky (Sun, Moon and Stars), The Volcano, The Great Void). Archetype, transcultural symbology, deep nature connection practices.

We've so far trained 45 coaches in two different cohorts (2019–2021), despite all the difficulties we had to face for in-person workshops due to the spread of the global Covid 19 pandemic. And we have now received ICF accreditation for the training.

The core aspects of our approach focus on nature-connected contemplative practices, the solid community-building focus, and the development of the personal feeling of love for nature or biophilia, which leads on to a desire to be an activist for all living beings' rights.

The 'Biophilia Plan' is one homework we require to get the diploma. Each student must identify daily, weekly and monthly practices to cultivate biophilia in synergy with a buddy chosen among other students. For three months, each student will explore, experiment and share with the buddy the advances in this biophilic self-development plan, writing a report at the end and sharing it with the trainers.

The flowering that we're witnessing in our students' love of nature is, in most cases, intense and moving. We feel that this is a simple practice that significantly impacts our motivation to become Earth's stewards.

CONVENER: Thank you, Diana. And we now invite Jeanine Bailey, who is based in the Middle East and Australia, to share her experiences and learning from a project to train indigenous coaches.

JEANINE: As I share this from where I live, I acknowledge and appreciate I am in the land of the Waddawurrung People who are part of the Kulin Nation, who are the traditional custodians of this amazing land. I pay my respects to their

Elders past, present and emerging and extend our respect to all Aboriginal and Torres Strait Islander people across land, air and sea.

In 2015 I was incredibly grateful to be invited by Rachel Petero to support her vision: Rise 2025.

Rachel is Founder and CEO of Rise 2025 Global. Her goal is to 'empower 100,000 indigenous women and their families to thrive, one coaching conversation at a time, by 2025'. The invitation from Rachel lit up my heart, body, mind and soul. It felt like this was a part of my destiny.

Supporting First Nations people has been a part of my coaching career starting in the Middle East in 2009 supporting Qatari women and men with their work and intentions for themselves and their family, community, country – and the world.

When I met Rachel in Doha, Qatar 2012, she was a woman on a mission to make a positive, powerful difference by mentoring senior Qatari female leaders. It was during the professional coach training Rachel undertook with my business partner (Marie Quigley) and I that Rachel clearly identified her purpose and vision to empower indigenous women.

Rachel identifies herself with 'strong Māori genealogy to her tribal lands and people of Waikato-Tainui, Ngāti Tamaoho, Ngāti Whawhakia, Ngāti Te Ata, and Ngāti Tahinga.' She is inspired by her people - past and present - and particularly by her mother and her grandmothers - who are and were strong community leaders, incredibly resilient and very influential. So much so, one of Rachel's grandmothers played a crucial role in the health and economy of her people, securing land and finance for her tribe to rise and had led the handing over of land in New Zealand back to their original owners. And on return to her home country - Aotearoa - Rachel became the first indigenous credentialed coach with the ICF (International Coaching Federation) New Zealand and began to put her vision into place.

It was also clear to Rachel, the first part of Rise 2025's vision was to train indigenous women to become professional coaches and leaders - and importantly - to rise in influence. Not climb the ladder for the sake of it, but to represent whānau (family and extended family), culture, community and the whenua (land) and sky and sea.

My role was to partner with Rachel to deliver Empower World's ICF Approved coach training programme. And since 2015, when we trained 16 Māori women in the skills of coaching, Rachel and I have run many cohorts, including the first 'all male' cohort in 2019.

The men had witnessed powerful changes for the women who came through Rise 2025 programme - and for their family and communities, and how it was influencing connection to all things living including the whenua (the land). The men wanted to experience it too. And since the first all male cohort, we've delivered mixed cohorts including women and men from Canada, Pacifica and Australia.

This has meant introducing different customs, including words, song, dance, prayer, the Māori karakia, space and protocols such as traditional welcoming ceremonies spoken and 'summonsed' in Māori language and including Acknowledgment of Country which is a welcome protocol in meetings and events recognising Aboriginal people as the First Australians and custodians of their land and seas.

The simple act of sharing the Māori karakia and Acknowledgement of Country on each training day was a powerful reminder of what is truly important to participants and the realisation the programme is much more than a coach training programme. It was a way to connect to heart, mind, soul, Mother Earth and our spiritual, loving and eternal nature together, whilst consciously recognising and respecting the original caretakers of our source of life.

By all participants sharing stories about the whenua they came from, together with their whakapapa (genealogy in Māori), it had the impact of connecting us together and reaffirming our 'core' obligations and purpose. Whakapapa is a line of descent from ancestors to present day and links people to the earth, sky and the origins of the Universe. Whakapapa is also recognised in traditional Māori waiata (songs) sung in schools, communities, ceremonies, gatherings.

Opening protocols, such as karakias (incarnations to ensure favourable outcomes and closing of gatherings), together with beautiful, melodic traditional songs and dance (similar to Pacific Island traditions), as well as, Acknowledgement of Country, sharing whakapapa and more invoke calm, mindfulness and gratitude. They set-up groups for expansive conversations by inviting participants to come into their hearts, body, soul and wise part of their minds. This - together with setting up a psychologically safe place to learn and expand awareness - supported participants to share some of their deepest challenges they had never shared before.

What has been extremely important in the Rise 2025 coach training is to use systemic lenses, asking questions which incorporate the whenua (land), as well as ancestors or future leaders (from the various indigenous cultures) - as a partner - which has been and is incredibly important and powerful to bring these wise voices into conversations.

Never had I felt so grateful in those moments - even though often I didn't understand the full meaning of all the words used. I knew the key words and, importantly, I felt I understood the energy and sentiment behind what was shared.

Whenever we could, Rachel would invite us to stand outside in a circle, barefoot to feel the connection with the whenua (the land) to hear the powerful words of the Karakia - while swaying side to side to feel the connection with the ground with each part of our feel - which would set us up for our day for trust, rapport and growth through asking stretching and challenging questions. It was also a beautiful and fitting way to finish the end of each day - allowing us to leave feeling complete.

I was so moved by these ways of being and the protocols in the Rise 2025 coach training programme that when I volunteered for the role of Pillar

Head, Coaching Excellence for ICF Australasia, I wanted to instigate the Acknowledgment of Country and Karakia for each of our internal meetings and public webinars we offered to members and non-members.

This ritual began (and continues) when I was involved in the creation of Transform 2020 - a month long online programme in November for coaches to support their personal development by offering webinars delivered by both local and international world-class coaches and trainers (including Peter Hawkins, Eve Turner, Alison Whybrow, Josie McLean, Tammy Turner, Marie Quigley, Rachel Petero and other brilliant coaches).

Rachel gifted a recorded and written karakia to ICF Australasia to use for this purpose, and if no-one is available to share the karakia, Rachel's recording was/is used. (It is important to note that gifting a karakia is also a tradition to be respected in Māori culture). So many times people have fed back how moved they are when they hear her words. The opening and closing karakia is as follows:

He Karakia Timatanga (To open a meeting)

Manawa mai te mauri nuku
Manawa mai te mauri rangi
Ko te mauri kai au
He mauri tipua ka pakaru mai te pō
Tau mai te mauri
Haumi e, Hui e, Tāiki e!
Nā Rachel Petero ki ICF Australasia.
Takohatia atu ai nā Whaea Lynda Toki, Ngāti Maniapoto

Embrace the life force of the earth,
embrace the life force of the sky
The life force I have gathered is powerful and shatters all darkness
Come great life force
Join it, gather it, it is done!
From Rachel Petero to ICF Australasia.
Gifted by Whaea Lynda Toki from Ngāti Maniapoto

Karakia Whakamutunga Tawhito (To close a meeting)

Unuhia, unuhia,
Unuhia te uru tapu nui,
Kia wātea, kia māmā te ngākau, te tinana, te wairua i te ara tangata
Koiarā e Rongo whakairia ake ki runga
Haumi e, Hui e, Tāiki e

Draw on, draw on, release us from the mantle of this meeting/class/conference
Draw on the supreme sacredness, so we can stand free,
To clear, to free the heart, the body and the spirit of mankind
Feel free to hang it up, be comfortable
Join it, gather it, it is done!

JEANINE: At the time of writing this, I am coaching a Māori Chief (who I met through the Rise 2025 programme) and I believe we came to this place because of the embracing of culture, protocols, words, energy and the belief we are connected. So when we began our coaching engagement, I consistently checked in with him what he wanted to be put into place which takes into account all that is important to him - which was his *whakapapa* acknowledging and tapping into the wisdom of his ancestors, future generations. We've also included the protocols such as a *karakia* at the opening and closing of each session which sometimes he asks me to say.

I incorporate whakapapa into coaching by asking questions such as:

- What do you believe the land (that is part of whakapapa) would say to you to support you in this situation?
- If your ancestors could share something with you now about this challenge - what would they want you to know?
- When this situation is as you want it to be, what would success be for your future generations in 10 years, 100 years, 1,000 years, ... 1 million years' time?
- As you connect with your whakapapa, what is important to you right now? Or what are you noticing?
- What is important to your *whānau* (family and extended family)?
- What will support Whanaungatanga (forming and maintaining relationships and strengthening ties between kin and communities and beyond)?
- Who do you believe you must be (and commit to do) to ensure *Kaitiakitanga*? (*Kaitiakitanga* means ensuring a sustainable future for all whereby world views and practices maintain and support the wellbeing of the environment and the wellbeing of *Iwi, hapū* and *whānau*).

In 2022 Rachel launched a coach training programme (*Tahi*) incorporating indigenous facilitators, models, processes and approaches. The last cohort I trained was at the end 2021. And although I do feel sadness about not being part of that amazing spiritual place of learning connecting with the people and the land – it's definitely where the programme is meant to be now: to fulfil its inspiring intentions of self-determination, which I trust will support collaboration across cultures, countries and ultimately support lands, seas and sky as part of the collective connection.

I am thankful to ancestors, existing and future leaders, the natural elements that surround us and the spiritual nature within us all - which I learned so much more about through Rise 2025.

Ngā mihi Rachel. You have created a legacy for more than 100,000 *wahine*!

CONVENER: Thank you so much Jeanine for that insight into bringing the earth into training and for those beautiful prayers. And we would like to share one more story, inviting Winfred Nelson, Akua Amoa Okyere-Nyako and Abena Baafi to describe how coaching and mentoring was used to support climate adaptation in Ghana.

WINFRED NELSON, AKUA AMOA OKYERE-NYAKO AND ABENA BAAFI: The Africa Adaptation Programme (AAP), launched a Mentoring and Coaching Initiative (MCI), using a supervised learning-by-doing for a year. The overall goal was to ensure sustenance in human capacity for effective and efficient implementation of climate change programmes.

Thirty-two mentors and mentees were selected who worked in public service including 5 pilot local government assemblies, academia and civil society. Two participants each, representing one experienced (Mentor) and one junior staff member (Mentee) were selected from 14 institutions.

The programme drew from a range of experienced people and organisations including the Lead Climate Negotiator for Ghana (UNFCCC Focal Point) and they came from the following institutions;

- Coaching and Mentoring International, UK.
- Nkum and Associates, Ghana.
- Some Professors at the University of Ghana.
- Presencing Institute, Oxford UK.

The MCI approach included on-the job training, where mentors supervised mentees within each of the 14 participating institutions, as well as off-job training in soft skills (Facilitation, Project Management and Microsoft Project Proposal Development for Fundraising. facilitation, writing and presentations), and technical skills related to climate adaptation and disaster risk reduction.

The training workshops adopted adult and experiential learning methodologies grounded in the Gestalt Organisation and System Development approach. There were also field visits to sea level rise areas and high erosion. These offered opportunities for mentors and mentees to learn practically, making meaning through reflection and developing their own new appreciations of self.

Monitoring and Evaluation mechanisms were also put in place to assess the efficiency, sustainability, continuity and relevance of the MCI.

The MCI received good reviews from all respondents during the evaluation as these reflections make clear:

> 'the mentoring programme has given me new ideas as to how to structure my department to make better use of my staff.'
>
> A mentor 2021

> 'I'm now able to take on more responsibility within the organisation' and lead my colleagues with more confidence. I have honed my communication, presentation and facilitation skills and have better appreciation of the work especially as it relates to climate change.'
>
> A mentee, 2021

'It a was great programme that created a platform for emerging professionals to gain skills and confidence in their field of practice. Network building and the foundation for collaborative research were also fundamental. I have just completed my PhD in a Climate and water.'

A mentee, 2021

'I played an important role to help make climate change in Cities a subject of Interest. (Improving Energy efficiency in Buildings and Urban Heat Reduction in the City landscape through the incorporation of Green Infrastructure).'

A mentee, 2021

The MCI offered a wide range of young people the opportunity to learn practically how to address climate change and other related skills. Replicating this process to cover some more institutions and persons will go a long way to strengthen hands-on capacity in climate change and also ensure sustenance in capacity.

CONVENER: Thank you, Winfred, Akua and Abena.
The research that Eve shared raises several important questions and shows up some of the blockages to trainers and trainings becoming more eco-centric.

Exercise

CONVENER: We would like one group to explore what might be good responses to the following challenges we encountered.

1. The ecological and climate crisis are political agendas and politics should be kept out of coaching and coach training.
2. Coaching is about helping individuals not about wider issues.
3. Ecology is not a coaching core competence.

The other group we would like to focus on the key elements you think should be part of every foundation coach training.
We look forward to what you come back with.

*

CONVENER: Welcome back. Let us hear first from Group One on responses you developed to some of the answers we received.

Group One

CONVENER: Here are some of the responses we generated.

1. **The ecological and climate crisis are political agendas and politics should be kept out of coaching and coach training.**

   a   They are life and human agendas that are already affecting every individual and will increasingly do so.

   b   Every coachee we meet is consciously or unconsciously experiencing some form of eco-anxiety or concern – we find we cannot ignore what is happening for them.

2. **Coaching is about helping individuals not about wider issues.**

   a   We are the ecology and the ecology is us.

   b   No individual is an Island, and we are all part of, and influencing, the wider systems we are nested within, and which are nested within us.

   c   What each and every one of us is doing and not doing will make a difference to whether humanity and many other species survive the next hundred years.

3. **Ecology is not a coaching core competence.**

   a   Yet!

   b   How can you separate an individual from their contexts and the ecology that is flowing through them in every breathe and everything that eat and drink?

   c   Thinking ecologically and eco-systemically is not only an essential competency for being a coach, but for being a human being in today's world.

Group Two

CONVENER: We looked at some of the core ingredients we think should be part of every training, based on the rich examples we have heard today. Thank you all who have shared your rich examples.

1. Start every training session with a connection to the Earth.
2. Close every training session with an opening to the different worlds we are taking our learning into.
3. Teach the Eco-Phase Cycle.
4. Take time for working through the feelings and assumptions we are all bringing to our work, which we worked through on Day Three.
5. Use the exercises in this book.
6. Teach, and then practice, with feedback, the very practical eco-engaged coaching skills we learnt on Day Four.

7. Have a session on how to use supervision to grow your eco-systemic capacity.
8. Provide a list of recommended Supervisors who work eco-systemically.
9. Have a well-chosen short reading list of core readings.

CONVENER: Thank you both groups with some great ideas I am sure we can all use and build on.

As we wind up after your hard work today, we wonder how you might progress the learning from today. Perhaps you might contact your coach training body about something you might like to see in CPD being offered. Or you might like to join with peers to develop your own training group? Or perhaps you are considering what your core values are and how these fit?

On Day Six we will turn to the final stage in the Eco-Cycle. 'Eco-Active' and look at our values and consider ways we can take a more active and broader role in impacting and influencing the responses to climate change and the ecological crisis. We will hear from the professional bodies, consider the codes of ethics, and look at all the people and organisations in our own value-chain that we can influence.

We wish you a pleasant evening and look forward to you being back with us tomorrow.

Chapter 14

Eco-Active

Day Six Morning: Impacting the wider world and developing the coaching profession

CONVENER: Let us start today with two quotes:

> You cannot get through a single day without having an impact on the world around you. What you do makes a difference, and you have to decide what kind of difference you want to make.
>
> (Jane Goodall, 2017)

And now a quote from Paul Polman who transformed Unilever when he was CEO and is a great role model for practical ways of becoming eco-active.

> The solutions to our decades-long global crises – climate change, biodiversity loss, inequality, the racial divided, and poverty among others – lie in empathy and compassion, in systems thinking and in collective action.
>
> (Polman and Winston, 2021: 274)

CONVENER: Today is the penultimate day of our journey together.

Yesterday we looked at how we could fundamentally engage differently in our coaching work, continuing a move from ego-centric to eco-centric coaching, supervision and training. Our intention is to deliver value not just for the coachee (supervisee, trainee), their team, and their organization, but all their stakeholders including the natural ecology, future staff and generations.

Today we will look at being eco-active. Our key inquiry question for today will be:

What can each of us both individually and collectively, do in and beyond our coaching work, in every aspect of our lives, that will make a net-positive difference?

Part of addressing this question involves changing our perspective from a short-term action list orientation to a seven-generation perspective. To get us ready for this work, let's start by doing a beautiful visioning exercise together, that is based on Krznaric (2020: 66–68). Please lead us in this Eve.

DOI: 10.4324/9781003153825-14

EVE: Let's consider, for each of us what does it mean to be a good ancestor? In many indigenous cultures there is a philosophy that the decisions we make today should result in a sustainable world seven generations into the future. So how do we each think of our connection to other generations, that unbroken chain that binds us to those who came before and those who will come after?

So, we'd like to invite you to pause briefly, close your eyes perhaps for a moment, and reflect on what being a good ancestor means to you.

Now we will explore this experientially. Please stand up with space in front of, and behind, you. Close your eyes and think of the world you are grounded in, in the present. Now, take one step backwards, in space and time and imagine someone from a much older generation, who was very important to you when you were a young child, perhaps a grandparent or great-grand-parent, an elder in your community, a great-aunt or uncle. Now, take a second step back and imagine going back in time to when they were a young child, perhaps their fifth or sixth birthday party, sitting with their family, and friends. How are you feeling as you watch them as they start out on their life? What is that person you care about saying? Maybe they're describing their hopes for their life.

We invite you to pause a moment here. Really picture them, feel them, hear their voice (not just the words but the tone, the pitch, the rhythm, their whole embodied essence), observe their movements.

PAUSE

Now, take two steps forward and come back to the present for a moment, to right here, right now. Begin to picture a child or young person that you care about and is important to you now. Perhaps your child or grandchild or great grandchild, or the child, grandchild or great grandchild of someone close to you. Bring to mind their face, their voice, their expressions.

Again, we invite you to pause.

And now, take one step forward in time and imagine them at the age you are now and what they are doing in their life. Now a second step forward in time and space and imagine it is that person's 90th birthday, surrounded by people celebrating with them. You're pleased to see a picture of yourself on their shelf. They are asked to give a speech and that 90-year-old looks at your picture and decides to tell people what they learned from you all those years ago, what you did that inspired them; the lessons you shared… your legacy.

All those years ahead, what will they be saying was the most important thing you did for the generations that followed you? What would you like to be listening in to? Again, take a breath, imagine their voice, their expressions.

PAUSE

And take two steps back to today. What did you hear in that span of time covering several generations? What are you being asked to learn about, do, become now?

What were your thoughts? What is inspiring you? Please open your eyes and write these down for yourself. Now write what you need to learn and develop today to help your capacity to be a good ancestor.

As we sit here in the 2020s, we are all parenting future generations in one form or another. A beautiful book that explores the challenge of *Parenting in a Changing Climate* (Bechard, 2021) provides a wealth of exercises along with stories that are moving, connect on a psychological level and are also told with much humour. One example is how to talk about climate change 'in a moms' group' and avoid the air being 'sucked out of the room'! (Bechard, 2021: 83).

CONVENER: Thank you Eve. To begin to answer these questions, we will soon hear from Zoe Cohen who identifies areas she believes we all need to consider taking action in. Then we are going to work through a number of these. We will:

1. Identify what we need to change to make our business models ecologically net positive business.
2. Examine our complete value chain. First, we will look upstream to all the resources we draw on to make our work possible, and then downstream, at all the people we work with, resources we impact, and how we have a beneficial influence on them.
3. Provide examples of how others are influencing the transformation of our profession.
4. Identify how we can make a difference beyond our own sector and hear examples from others working in communities, through NGO organizations, political campaigning and other forms of eco-activism.

These stages will help us find our unique contribution to the global challenges. They also follow the model outlined by Paul Polman and Andrew Winston (2021) in their chapter: 1+1=11, showing how we can partner with others to greatly multiply our impact in the world.

After doing that exercise considering the generations we have inherited from and our role as good ancestors, we are delighted to invite Zoe to share her simple, yet profound framework, the nine domains of being a climate conscious coach.

ZOE: Thank you so much. I am delighted to share this model that I first conceived in 2019. It really looks across the places that we touch on as citizens and as coaches and what we might start to consider in each.

Self: As coaches, we know that change starts with us as we explored throughout this week. Without awareness and acceptance of the need to change we're not going to get very far. The climate and ecological emergency

Political engagement
Our Practice
Business footprint
Our Ethics
Communiti
Reflection/
Own Life
supervision
Self
Supporting
adaptation

The Climate Conscious Coach:
Nine Domains

© Zoe Cohen shinecoachingandconsultancy.co.uk 2019

Figure 14.1 Nine domains of a climate conscious coach
© Zoe Cohen, 2019

affect everything, everywhere and always; the air, water, soil and food we all – and our clients, their stakeholders and all future generations – rely on.

If we are to become climate (and ecologically) conscious, we must self-educate to play our full, informed part in this chapter of humanity's history, and be worthy ancestors. Self-awareness around this issue is an ongoing process. Everyone has a different starting point. Some people have been informed and concerned for many years, others are just waking up; their awareness stirred by the recent and ongoing citizen movements (e.g. school strikes for climate, Extinction Rebellion among others) and media coverage of climate breakdown events.

As we take responsibility to educate ourselves and our awareness grows, we need to work through the ecological awareness cycle, from denial, grief, anger and blame, to guilt and shame, and finally moving to take responsibility and action (Hawkins and Ryde, 2019; Hawkins and Turner, 2020; See also Day Three, p96). Through this work, we ultimately become part of the movement needed to shift our collective human consciousness. If we don't understand, appreciate and feel the scale and urgency of the emergency we and our children are in, then it is impossible to take proportionate action.

Once we allow ourselves to see and feel the enormity, then we can better calibrate our risk appetite for taking action.

For the purposes of brevity, and in good coaching style, I will walk you through the next seven domains using a series of questions to provoke thought, insight, and action.

Exercise

Please have a pen and paper with you and write down your response in three columns: 1. Thoughts, 2. Insights and 3. Actions.

*

Own life: What do you know about your own household's carbon footprint and ecological footprint? (There are tools you can use such as Carbon Footprint, 2022; and for the carbon and methane output of foods, Ritchie, 2020.) What more can you do to reduce it to as near to net zero as soon as possible? How could your household and family have a sustainability action plan and have a positive ecological and climate impact? What restorative action might you take through how you live, what you buy and consume, the energy and water you use, how you travel, and how your money is invested and used?

Communities: What are the communities (local, business, professional, faith etc.) you are connected to that you could seek to influence? How might you seek to foster greater awareness and accelerate positive change in each of those communities? How might you strengthen your community, building trusting relationships and enabling everyone around you to bring their piece of the puzzle? What circular exchanges might you create or participate in, where you live?

Business footprint: What do you know about your own business's carbon footprint and ecological footprint? What more can you do to reduce it as fast as possible? Do you have and publish your own sustainability plan, and does it have clear targets and measurements? What movements for change is your business aligned with or part of? How might your business become a sponsor of positive impact locally? Do you have a local currency you can use and support? Looking across your supplier network, how local is it?

Political engagement: How aware are you of how the different policies of the political parties in your country have an impact on the climate and environment? How much do you consider these factors in the way that you vote? What local or national activism support to have your voice heard in other ways? Shortly Diana Collett will share an example of the work of the CCA's 'Political pod'.

Our practice: Our work together over the past five days has been about this element of the model: how can we bring the climate and ecological reality into our coaching practice? We might do this in many ways as we have seen from the approaches we have developed and explored together on Days Four and

Five: from voicing our personal values on our website and profile, and in our chemistry conversations, to taking a bigger, wider perspective to our coaching questions to allow in the earth and/or future generations as stakeholders. This is not about putting our 'agenda' out there; it is about raising awareness, which is fundamental to all coaching. And remember, not to do this is also taking a stance – inaction or 'neutrality' is an action.

Our ethics: How does a greater awareness and acknowledgement of the climate and ecological crisis have an impact on the decisions we make as a practitioner, professional and businessperson? How does it affect which clients we chose to work with or the way we choose to develop our coaching careers?

Reflection/supervision: Think back to yesterday's exploration of supervision and coach development, in what ways can we bring the ecological dimension into our own reflective practice? What do we need from our supervisors to develop greater eco-centric practice? If you are a supervisor yourself, how does your practice need to evolve in light of your awareness and action?

Supporting adaptation: This final domain needs a little more introduction, as it is little acknowledged so far in the media or in business. Rapid, deep reduction of our societies' carbon emissions will help to mitigate climate change; adaptation is about accepting what's likely to come and adjusting to it. We are 1.2 degrees above global average heating and we have already caused irreversible impacts such as ice on land melting (IPCC, 2021). We have no choice. We must both mitigate and adapt as if our lives depend on it, because they do. The science indicates that, regardless of what mitigation actions are taken globally in the next few years, the benefit they will bring is avoiding further deterioration decades in the future (while probably also giving us cleaner air, and healthier food and lifestyles in the shorter term). This is primarily because of the lag time between producing CO_2 emissions and their full heating effect on our atmosphere, and because CO_2 remains in the atmosphere for centuries – and in the oceans for even longer. This is why we often use the 'super tanker' analogy to describe turning around global heating.

We are already dealing with deterioration in the present. We need to accept that, for the foreseeable future, extreme weather events, flooding, droughts, heatwaves, typhoons and so on will continue to get bigger, stronger and more frequent. Further, they will have widespread impacts on low-lying communities, small island states, ocean warming and de-oxygenation, food production, forest fires, infrastructure and migration, as well as economics, politics and business.

Coaches have an important role to play, from supporting communities and for-benefit organisations, to working with individuals and groups experiencing 'climate/eco-grief'. Coaches and consultants also have a role in challenging and, helping businesses to adapt their business models and challenging government leaders.

I'm sure you can think of many more instances where coaches and their skills can support the enormous adaptation (change) agenda that is to come.

The task might be gargantuan, but coaches (together with other professionals) are well-placed to bring the humility, courage and will to help us move forward.

CONVENER: Thank you, Zoe, this is a great overview of the multiple domains of our lives, where we can make a positive difference. Once you start to explore the places where each of us connects, the opportunity is endless. We are going to work through these today starting by building on yesterday afternoon and exploring what the coaching professional bodies are doing and how we can influence them and make a difference to the wider professional field.

We will now hear from the Rita Symons, who is going to speak on behalf of the 11 professional bodies who are signatories to the *Joint Global Statement on Climate Change*: (AC, AoCS, APAC, APECS, COMENSA, EASC, EMCC, IAC, ICF, IOC, ISCP).

RITA: In 2019 a number of professional coaches issued a challenge to their professional bodies - what were we doing about the climate and ecological emergency and the global impact this would have on human society, living beings and the planet?

While formal membership organisations are not designed to be nimble in responding to such a challenge, throughout 2019, there were numerous discussions between members from Association of Coaching, Association for Professional Executive Coaching and Supervision, European Mentoring and Coaching Council Global and International Coaching Federation leading to an agreement that the professional bodies would develop and issue a Joint Position Statement on Climate Change. We saw our role as creating a safe space for dialogue, exploration and learning, not prescribing one right course of action.

Inclusion was a priority from the start. The initial organisations invited the Asia Pacific Alliance of Coaches to join the partnership and on 29th May 2020 the Joint Global Statement on Climate Change was issued by these five professional bodies.

Our joint commitment is to:

- Raise awareness and knowledge with our members through sharing information and research.
- Create safe and challenging spaces for coaches, coaching psychologists, mentors and supervisors to reflect on and reconsider their role and their practice.
- Raise awareness and offering support to our clients to enable them to redesign their organisations in response to changing needs and good practices.

- Collaborate to share knowledge between professional bodies and pool resources for free access by all coaches, without the need for membership. We will model the behaviour we seek and always collaborate not compete.
- Develop our thinking and research to further consider how coaching and mentoring can serve to support society's transformation through our client work.
- Work towards carbon neutrality in our events and activities.
- Continue to seek and invite other professional bodies to sign up to this commitment.

We continue to seek other partners in this endeavour and other bodies signed throughout 2020 and 2021 including from psychology: Coaches and Mentors of South Africa (COMENSA), Association of Coaching Supervisors, International Association of Coaching, International Society for Coaching Psychology, Institute of Coaching and the European Association for Supervision and Coaching. We are delighted that affiliation to the Joint Statement has appealed to 11 different coaching/mentoring/supervision and coaching psychology professional bodies and that we now have global coverage. The partnership is considerably strengthened by this diversity.

In November 2020, we held a series of six webinars over a two-week period, each hosted by a different organisation (AC, EMCC, ICF, APECS, APAC, and COMENSA) with each open to all, members and non-members alike, and free of charge. They created space for an open dialogue for participants to discuss any aspect of the Joint Statement. We included Peter Hawkins' Eco-Phase Cycle model as a useful reference to locate their current position in relation to climate change (on the five stages from curious to active). We also asked participants what they wanted from coaching professional bodies in relation to the climate emergency.

In each of the sessions, participants were polled. Between 73 percent and 94 percent of participants felt coaching could make a difference. During the sessions, using action research (Jackson and Cox, 2020: 148) we gathered participant comments and questions regarding whether and how coaching might be used in response to the climate emergency. In many cases, the sessions showed that individuals wanted support to move from personal commitment to action, asking for tools and techniques to help them explore environmental issues in the coaching conversation. One recurrent question across all six webinars was: how do we as practitioners include the climate theme in the coaching conversation in balance with a focus on the client's agenda?

When asked about the role of professional bodies, some participants suggested updating competency frameworks and our Codes of Ethics to make them more explicit about our broader societal responsibility. The Global Code of Ethics was indeed adapted in 2021, making our systemic responsibility much clearer and the ICF code (2021) was also updated.

We embarked on this journey together some time ago, and we now find ourselves in a different environment, in terms of both the urgency and the acceptance of the need for climate action to sustain humanity and the planet. So we are on a journey with an unknown destination! We are committed to further discussion, dialogue and inquiry with members, while also acknowledging the spectrum of views relating to direct involvement by the professional bodies is mixed and ever-changing.

Our priority is to continue on this road with curiosity and compassion and a commitment to impact positively. We will use our joint website to share, raise awareness and signpost and plan to collaboratively hold quarterly round tables to build and sustain energy, creating a space for practitioners to consider ethical action. We remain motivated and sustained by our belief that coaches, mentors, supervisors and coaching psychologists are well placed in terms of their skill sets to be in service to humanity in more and more impactful ways.

CONVENER: I am now going to ask Eve to share the story of starting the Climate Coaching Alliance (CCA) with Josie and Alison. It's a story of how, if we work together, we can all make a big difference to the sector.

EVE: The story of starting the CCA is one of friendship and respect, love and concern, and of collaboration and mutuality. When Zoe Cohen, Linda Aspey and Alison wrote their powerful, open letter to the profession in 2019 they called on 'coaches, coaching psychologists, coaching and coaching psychology professional bodies and our educational establishments to play a vital role in our collective history as a species, for the human and non-human species alike' (Cohen, Aspey and Whybrow, 2019). They described the need to create a network like the *Global Supervisors' Network* which I had set up and led from early 2016 and was free both to join and to attend all events.

So, unsurprisingly, Alison hooked me in! And we invited my friend Josie in Adelaide, south Australia, who was both hugely knowledgeable, and passionate, in the field. The scene was then set, in November 2019. We knew it was important to invite coaches globally and chose a name that mirrored what was happening in psychology through the Climate Psychology Alliance (CPA), who kindly supported us from the start. To have an impact we invited key figures in the field, both professional bodies, and interested individuals, to the first exploratory meeting in December 2019. We are grateful for this widespread support from the start. As I write (summer 2022) the three has become 2000 with geographical, language and subject communities of interest globally. And to ensure that income is no barrier, joining and attending all events is free. You will find that wherever you are on your journey there are like-minded people to travel with you.

We have written a fuller piece about the CCA in Appendix A. You can find this along with encouragement to us all in our work from the CCA's many global communities, responding to four questions from their different perspectives:

1. Standing where you are, how do you see/how are you experiencing the challenges already facing your communities?
2. How does this affect/inform your work as a coach?
3. What do you think the most important challenge is for our profession in attending to this?
4. What is the one encouragement you would like to make to readers?

CONVENER: You might like to think of those questions for a moment, before reading Appendix A.

Eve, perhaps you can share your personal thoughts as you also have a role in one of the signatory professional bodies.

EVE: Thank you. Rita Symons described the work of the 11 signatories earlier. Most professional body executive boards are made up of volunteers voted in by members. So, change is done by consent, listening to and representing diverse views, being involved in dialogue, not top down. And the boards need to approve policy – so the joint statement was an example where all signatories needed board approval to sign.

Here I must be transparent about wearing different hats and how that might influence my perceptions and analysis. At the time of writing, I have recently stepped down from being chair of one of those signatory professional bodies, APECS, since the start of 2020. When people collaborate, it could be easy to be frustrated at the time taken to make these changes, or impatient of the small or large changes to a text requested or required from one of the signatories. But another way of looking at this is that needing approval does mean that no one person or body dominates and that there is accountability back to the membership.

I have always felt collaboration, connection and communication were fundamental to effective leadership, and the joint global statement is a testimony to all three. While there has been collaboration over many years between professional bodies, for example in the *Future of Coaching Collaboration* (FCC, 2022), this has been western dominated. And the joint global statement on climate change is unique in bringing together coaching bodies from around the world.

CONVENER: Let's take a minute to think of the power professional body members have to be eco-active. What ideas come to mind?

Examples might include being able to vote people on or off the boards where applicable, they can let professional bodies know if you don't think they are doing enough or modelling collaborative practice, members can set up working groups, they can post on the bodies' social media pages and so on. And barriers to doing work in this area within our practice can be more imagined than real, as the different codes of ethics might illustrate:

The *Global Code of Ethics* (ten signatories including EMCC, AC, APECS, COMENSA) requires that:

2.8 Members should be guided by their client's interests and at the same time raise awareness and responsibility to safeguard that these interests do not harm those of sponsors, stakeholders, wider society, or the natural environment.

3.3 Members will abide by their respective bodies' statements and policies on inclusion, diversity, social responsibility and climate change.

3.8 Members will engage in professional development activities that contribute to increased self-awareness in relation to inclusion, diversity, technology, latest developments in changing social and environmental needs.

ICF *Code of Ethics* states that:

28 Am aware of my and my clients' impact on society. I adhere to the philosophy of 'doing good,' versus 'avoiding bad'.

Our third speaker on making a difference to the professions is David Lane, whom we heard from on Day One. Today he is going to reflect on the development of ethics in the coaching profession, and how conformity may stop us seeing, hearing and speaking about how coaching can contribute to the ecological and other crises of our time. Eve heard David speak and was struck by his challenge to coaches to stop seeing ethics as conformity, but to appreciate the depth of its complexity and the issues through which it needs to navigate a 'right' path: from economic and environmental to societal and political and to issues of inclusion and diversity.

EVE: David talked about us 'being in the fog and only sort of knowing where we are going' and of our need to be able to live with that tension and find ways to make thought-through decisions rather than seeing codes as the answer, given our complex, international world. I was struck by David's comment that the theories we espoused in our training were dominated by western white males, and the need to see ethics as relational and not as something external to us. David brought in a range and depth of thinking (such as Islamic coaching, Ubuntu coaching, emphasizing the collective too). And he made the case for educating coaches to relational ethics: 'What do we have in common?' 'How can we do what's appropriate?'

DAVID: Thank you, Eve.

Ethics is often presented from the perspective of a limited range of ethical precepts derived from classical philosophy and medical codes of conduct. Developing a code of conduct although not the first action taken by emerging professional bodies in coaching, was certainly an early consideration. However, all the bodies in the field adopted the same approach of defining ethics in terms of an external code to which members were subject.

This is a very culturally bound approach. By not seeing what alternative worldviews might exist elsewhere, they become invisible, which makes it easier to condone by silence. This is certainly the case in the emerging conversations around climate and response. Those who have caused the most damage offer solutions that work for them and ask those most impacted who have contributed little to the crisis to adopt practices bound in a first world mindset. Alternative voices are largely absent and as the COP26 events showed, those marginalised continue to struggle to have their voices heard. This raises the question of what an approach to ethics might include if founded in more diverse cultural contexts and would this provide other ways to view the climate crisis? For example:

- How do you currently make ethical judgements?
- How might cultural context influence those judgements?
- How might alternative ways of viewing ethics influence your practice?

The aim of a conversation needs to open a wider range of ethical perspectives and enable us to begin to ask different questions as we reflect upon practice.

Is a universal code of ethics possible? How do current professional codes in coaching meet or challenge such principles? What alternatives might there be if we take culture into account and given the complexities of a rapidly changing world? We face challenges we can see, but also those we can only imagine and, beyond that, those for which we have not a clue. How might ethics stand up in these different contexts? Diversity is integral to coaching and being a good coach requires working from a diversity perspective because acknowledging and working with difference is about maximising each person's potential. We know from the research that a diverse workforce can improve creativity and innovation but to achieve this also requires better management to prevent conflict or exclusion.

One of the ways in which the world is changing is the emergence of Asia as a dominant power. By 2050 it is estimated that China, not America, will be the largest economy. How then might ideas from the east influence the decisions we make? What eastern concepts of ethics might be usefully considered?

The Western world put an emphasis on individuality in building a confident individual who is self-reliant and self-oriented. This is reflected in the way we define our ethical codes. In Asia and especially China much importance is placed on the collective good of a group or family. Thus, this is much more relational and it's about how one is viewed by others and less about personal pride or ego.

Therefore, from the lenses of an Asian culture, even while ethics in coaching can be viewed as a dyadic relationship of the coach and coachee, it needs to be viewed as bound by duty, a hierarchical view of relationship and the importance of a 'face management'. This is not as a western perspective might think about individual 'face' but a recognition that an individual is embedded in a social network (Yanrong Chang, 2008). We have to think about ethics as a relational and socially constructed process but built across community notions of a wider duty. This provides a broader basis for consideration of bioethics which is more culturally relevant. (Tai and Lin, 2001)

What African concepts of ethics might be usefully considered? What might the world look like from this perspective? Ways to consider transformational coaching in culturally diverse teams have been explored by Stout-Rostron (2019). She argues for an integration of Ubuntu ('I am because we are') and coaching as a way to promote the idea of employers seeing all their people as an integral part of an interconnected system. As Ubuntu recognises, we exist as a community and in belonging - it is a counter to the marginalisation that we currently see. Central to the concept is humaneness. It is a relational concept that governs how we operate in communities and in organisations - thus it is a philosophy of life and a worldview. These beliefs underpin the approach and how the coach operates must reflect these principles. It is not just about learning tools with which to coach but embodying the philosophical stance to community and the world. We see the individual within their relationship to community and the world.

How might a spiritual model influence ethics? Ershad meaning guidance is the Arabic term van Nieuwerburgh and Allaho (2017) use to describe the role of a guide (coach) who develops and grows the individual into his or her desired maximum capacity and degree of accomplishment against a framework of Muslim beliefs about God, self, others and the world. Ershad echo's Islamic ideals of Discovery, Intentions, Pathways and Effort. These principles are used to guide the coaching process.

The Ershad approach to coaching draws upon the teaching in the Qur'an, to create a way to work consistent with its principles rather than starting with western coaching concepts. It asks us to align our lives with the deepest goals and values we find in ourselves - these are considered gifts from God. It is designed to support people on journeys to more aligned and purposeful lives. It can unlock potential, and the pursuit of excellence, based on identifying and pursuing the path that is right for them. There are three components to it - partnership principles, the conversational process and the Alignment Wheel.

Partnership is about building a sense of mutual trust and respect to create the conditions for meaningful and reflective conversation. The conversational process is characterised by patience, humanity, humility and positively. These are all derived from exhortations from the Prophet Muhammad (peace be upon him). There are four phases to the conversation.

1. **Discovery** asks the learner to reflect on self, relationships and their experiences. This enables understanding of current resources and to see their situation from different perspectives.
2. **Intention** asks the learner to turn their attention to the future and to articulate very clearly their intentions.

3. **Pathways** explores ways to move closer to their intentions. They are introduced to the Alignment Wheel which asks them to assess if their pathways are aligned to their values and beliefs. In particular they are asked how a decision might affect relationship with their beliefs, self, life or the universe and significant others. This linkage with values and beliefs is a central feature of the Ershad approach.

4. **Effort** requires the learner to consider what is required to follow the right pathway. In Islam, the right pathway includes care for the environment. There is a rejection of what is seen as environmental colonialism.

The right pathway in Islam is based on teaching its followers that they have to take care of the earth. We act as guardians and are accountable to God for our actions. There are around 200 verses in the Qur'an concerning the environment. Al-Jayyousi (part of the UN Environmental Programme, 2018) argues that Islamic principles of harmony, balance and proportion provide an ethical basis for humans to respect all life and nature.

Of course, other faiths also take a view that we are stewards of the earth. Pope Francis has released a call for Catholics to join the fight through a papal encyclical letter. The Church of England's General Synod has urged world leaders to act. There have been calls from the Hindu, Buddhist and Jewish communities.

Does this mean we can see more informed action to support the environment in other cultures and beliefs? Clearly not, however, the direction of travel is positive and with much of the worlds' population holding religious affiliations this represents an important step.

In conclusion How might this impact on the way we view ethics?

> It behoves us to shift from the individual focus of most dated western individualistic business ideologies and adopt shared worldviews that place greater value on the collective, the community, the organisation, the environment, than the individual.
>
> (Stout-Rostron, 2019, p. 92)

DAVID: If we apply this in coaching the individual leader, this serves the greater good by enhancing each person's journey along the broader path of humanity. It points to a shared more relational view of ethics (Lane, Watts and Corrie, 2016) There are times when the lack of action on key ethical dilemmas such as climate change can overwhelm. The first UN conference on the environment dates back to 1972 and much that was recommended then was ignored. It is more than fifty years since my concern for environmental health emerged (Lane, 1972) and the three calls for action in terms of education in schools, community projects and education of professionals to consider the environment in all they do, remain elusive. Our call in 1994 for action on the role of the built environment as a main contributor the global warming (Lane and Malkin, 1994) was repeated recently (London Build, 2021). Yes, we are acting nowhere fast enough, but many are taking their ethical responsibility for the environment seriously. Many companies and cities have committed to the Race for Zero and those in the built environment have heeded that call with some already at net zero and now going beyond to make a positive contribution to replenish the earth.

However, unless we listen to those currently marginalised and really see ourselves as guardians who live from the earth's income, not its capital, it will not be enough.

CONVENER: Thank you David for those thoughtful insights and challenges. So how does this sit with you? Do you have any questions? Many if not nearly all of us, are members of a professional coaching or mentoring body, so let us practice sector wide collaboration here. We are going to create a joint declaration together from this journey. We are going to do this using a dynamic collective inquiry process.

Exercise

CONVENER: We have six tables each hosted by one of the conveners. Each will table has a different flag and a different seed sentence to complete as follows:

- **Table 1**: The ecological purpose of coaching is to…
- **Table 2**: The ecological values that should underpin all of our work should be…
- **Table 3**: The sustainability plan and objectives to take all activity in our sector to zero carbon emissions by 2030, including conferences, trainings, coaching and supervision work, etc. should be…
- **Table 4**: All coach training including conferences and CPD should include…
- **Table 5**: The ecological ethics that should be in all coaching ethics statements are…
- **Table 6**: Our profession should speak out on… and partner with…

You can move around the tables joining the different discussions and contributing to what is being created on the paper tablecloth. If you are working alone, you can add your comments to what was shared below. As you listen remember this is always work in progress to be built on, it is not right or wrong. The question is whether it is taking us forward as a sector to play a greater beneficial role? Please add and build on what has so far been generated.

Table 1: The ecological purpose of coaching is to...

- Understand the connection between all beings and life on the planet and see them as stakeholders to our work.
- Enable the shift in human consciousness and action necessary to limit the scale of the ecological crisis.
- Heal the split between humans and the 'more than human' world.

Table 2: The ecological values that should underpin all of our work should be...

...inclusion, respect and mutuality, so that the human view does not dominate others, and the impact by humans is seen as a key determinant of what is meant by ethical action.

- Collaboration and partnership.
- Ubuntu.
- Serving the ecological niches that we are part of.
- Being good ancestors.
- Wide-angled empathy and compassion.

Table 3: The sustainability plan and objectives to take all activity in our sector to zero carbon emissions by 2030, including conferences, trainings, coaching and supervision work etc. should be...

- Reducing in person international conferences and trainings which involve lots of airmiles.
- Using locations that have good sustainability policies.
- Using video conferencing to reduce travel.
- Being able to justify the impact of our actions as being worthwhile and exceeding any negative impact.

Table 4: All coach training including conferences and CPD should include...

- Sessions on the ecology and climate crisis.
- A discussion of our role as responsible ancestors and what that would mean for our practice and our relationship with our clients. This will include systemic work including nature as stakeholder.
- Speakers from diverse culture, some speakers under 30, speakers who are or who can represent stakeholders, including critical customers, employees of those we coach, people speaking on behalf of the 'more than human' world.

Table 5: The ecological ethics that should be in all coaching ethics statements are...

- That systemic impact, not just individual client impact, is a key measure of success, and that we proactively need to do good. (This is seen in the most recent ICF and Global Code of Ethics statements).
- That coaches have a responsibility to include exploration unintended consequences and long-term impact of actions that their coaches are planning.
- All coaches having regular supervision and supervisors having eco-systemic training.
- The need to learn from other cultures and traditions, beyond the western white perspectives that dominate coaching.

Table 6: Our profession should speak out on...and partner with...

Speak out on

- Work that is simply focussed on client success, regardless of the impact on society, community, nature etc. It can do this by partnering with other professions, organizational leaders, and by taking a lead for example through its codes of ethics, the trainings it promotes, its own actions (e.g. carbon net zero and beyond).
- The cultures and collective ways of thinking and being, and behaviours, that are holding back the necessary ecological changes.
- How to build greater collaboration across differences and boundaries, necessary to meet the challenges of our times.
- How to develop 'We Q' – collective collaborative intelligence. (Hawkins, 2021 13)

Partner with ...

- Other professional bodies to share ideas and build influence
- Other university departments for joint change projects and cross-fertilization

CONVENER: Thank you, for all your ideas. Can you now each think how you can use some of these ideas we have collectively generated for you to make a contribution to evolving the wider coaching profession? Note down one or two possible next steps.

*

We promised to share an example of political activism and are delighted to welcome Diana Collett to share her experiences and learning, welcome Diana.

DIANA: In December 2020, Gosia and I created a Circle of Interest for the Climate Coaching Alliance about coaching in the political arena. Concerned about divisive political discourse and the lack of action around the globe, we felt coaches could play an important role in helping politicians shift their mindsets. Our first speakers (Wilkins, 2021) explained how many politicians are caught in the difficult space between constituents demanding faster action for a greener future, and the lobbyist jostling for dirty projects. I had visions of coaches supporting politicians to move beyond the constant chatter of conflicting demands to find their own wise decisions for a sustainable future.

In our monthly meetings we provide spaces where passionate coaches can share and create collective wisdom. These meetings attract coaches with a common vision and diverse perspectives. Listening to these perspectives has considerably broadened my views on political engagement. Now I see that using power for the sake of the planet will challenge traditional practices in political decisions-making. Climate-aware coaches can help any public-minded figure embrace a more sustainable mindset, regardless of their position or stance. The aim is to support the development of long-term regenerative public infrastructure, policies, and procedures.

COP26 became the backdrop that incentivised many of our discussions. In the end it both catalysed our greater aspirations and confronted us with harsh political realities:

- Political discourse that reduces arguments to point scoring oversimplifies complex situations and polarises community opinion. At the very time when we need to pull together to make meaningful changes, much of the political rhetoric is divisive.
- To break through this binary discourse, it is helpful to acknowledge that the greenest options are still beyond reach. Most countries can't replace fossil fuels with renewables overnight. We need our public figures to explain change as a pathway to a better future with initial choices that are optimal for now but will lead into the ideal.
- Many people, organisations and businesses are already creating these pathways. They are quietly working on fantastic answers to difficult problems (Tomasdottir, 2021; Hawken 2017).
- But politics always lags behind community change. We may want our politicians to envision the future, but in well-functioning democracies, politicians listen to and mimic the changing views and positions expressed in the community and industry. This type of leadership goes far beyond personal gain from political games.
- We need politicians to actively promote a whole-of-community shift to a greener future. Fortunately, there are some countries already taking this path, such as the nations involved in WeGo (2022). These wellbeing economy governments meet as an alternative to the G7 (Janoo, 2021; WEAll, 2022).

- Wales is an example of a government that has adopted a workable political solution. The Welsh Parliament adopted the Well-being of Future Generations Act in 2015 which incorporates seven Wellbeing Goals. These goals provide a well- considered, pragmatic framework for incorporating complexity into public decision making for the good of all (Davidson, 2021).

Then COP26 confronted us, yet again, with the long-standing ties between politics and money. I discovered how cumbersome the political landscape can be:

- The structural divide between the voices for government, commerce and industry and the voices for community and indigenous needs was impenetrable.
- Crucial messaging about personal responsibility was downplayed by people in power with a messianic belief that future technology will save the world.
- The communities most impacted by climate change are some of the poorest on the planet but they have a lot to teach us about devastation, resilience and regeneration (Phukan, 2021)

Despite the complexity and division surrounding climate issues, here are some simple tips for helping coaches to cut through:

- You don't need to be a climate scientist to support people to find their own connection with nature and the future.
- Halle Tomasdottir, CEO of the B Team (2021), believes coaching can play a critical role in changing mindsets. Coaches can unlock their clients' future thinking mindset by helping them to connect with who they are as people, their relationship to nature and to see the world through the lens of children. Most importantly she advocates a gentle approach that appeals to the heart. Her powerful questions include:

  - Do you think you are doing enough?
  - What more could you do?

- Nadine Andrews, a Scottish social researcher (November 2021), finds that the more connected people feel with the environment, the more they are likely to make changes for the environment. This does not mean spending long hours by clean running streams. Nature is all around us. We constantly breathe it in and adapt to its changes. Coaches can help deepen their clients' connection with Nature by supporting them to reflect on meaningful experiences with nature.
- The topic of climate change often brings up emotions for people (Huntley, 2020). In a non-stop whirlwind of political life these emotions can

influence decisions even when they are barely felt or expressed. Coaches can create spaces where their clients move beyond surface responses to explore their deeper feelings and connect with future actions. Powerful questions that can open up long-term thinking include:

- What legacy do you want to leave for your grandchildren?
- What actions can we all take that will have the greatest good for the most people?

- Coaches can also play a crucial role in shifting the power play. This involves embracing the emergence of new forms of power (Tomasdottir, 2021). Here's where coaches can get creative bringing youth and indigenous voices together with leaders in business and government to discuss big issues, and then supporting everyone to adopt a learner mindset.
- Coaches can help emerging leaders develop skills to build bridges across diversity. This will promote inclusivity whereby people come together with their differences, not in spite of them.

In summary, climate-aware coaching can provide opportunities for people to develop their future focussed public voice with thought and consideration. It has great potential for incorporating environmental truth with power.

CONVENER: Thank you, Diana. This is a great opportunity to pause and reflect. One of the 'light bulb' moments for me was when you were talking about the communities, we are already a part of – we don't have to go far to find places to influence or support people.

Exercise

CONVENER: Either alone or in small groups, please take a few minutes to explore what you could do in the domain that Zoe termed political engagement. These could range from the local to the international; from writing letters or articles to joining or leading campaigns.

*

As we reconvene, it would be good to hear one or two themes from your discussions:

"How little we follow-up with a local or national political representative when we disagree with an action being taken 'in my name'. We discussed why this was: some said, 'lack of time', some, 'lack of interest', while others said 'a feeling of helplessness' or not knowing where to start."

"We said we could talk to our neighbours and friends more, about these areas, and really listen to their views."

"Some of us decided we would change where and how we shop and challenged each other on how much are our choices driven by need rather than want. We decided to do lists of what we could buy less, buy different and ways of reducing waste."

"A number of us are already part of different ecological active groups working with preserving woods, habitats etc. Other were part of campaigning groups both local and national. We all came up with one way we could make a bigger contribution."

CONVENER: Thank you for your offerings. Following on from Zoe's nine domains, we have looked at influencing our professional bodies, expanding our ethical perspective, and influencing the wider political world. After the break we will look at our own vision and values and how we can use these to transform our own life, work, and business.

Chapter 15

Eco-Active

Day Six Afternoon: Eco-Active – developing our vision and values and transforming our own business.

CONVENER: Welcome back. This morning we explored how we can be more *Eco-Active* in the wider worlds of influencing our professional bodies and beyond that to political engagement. This afternoon we are going to come right back to ourselves and look at our values, our vision for the world that we wish for, and how to take this into impactful action that will create the transformation needed.

So let us start with values. James Hillman (1995: xvi–xxii) wrote:

> At the heart of the coming environmental revolution is a change in values, one that derives from a growing appreciation of our dependence on nature. Without it there is no hope. In simple terms, we cannot restore our own health, our sense of well-being, unless we restore the health of the planet.

Many of you will have spent time exploring your own personal core values and indeed, will use these in your practice, whether it's as a coach, a supervisor, a leader or in other ways. At this point, we are going to relook at values that may underpin sustainability or regeneration or a net positive enterprise. And we will be looking at values from a systemic perspective – appreciating the complexity of our values.

First Josie is going to share her work, helping us discover our own deepest values and how we can apply these to our work and use them to monitor our progress.

JOSIE: I will share some of the research I have been undertaking since 2006 that was inspired by the noted systems dynamics scientist, Donella Meadows (Meadows, 1994) and informed by Ronald Heifetz's work on adaptive leadership (Heifetz et al., 2009).

As an aside, I am employing these two terms, sustainability and regeneration, interchangeably because with my living systems worldview they are the same thing. I know that this is not the popular view, but sustainability is an emergent, evolving, and dynamic property of a living system as it attempts to

DOI: 10.4324/9781003153825-15

survive and thrive in connection with the systems around it. In the process, the living system will regenerate or care for the surrounding systems in which it is embedded (or to which it is coupled) because this act supports its own well-being. This shift to caring for surrounding countryside, is echoed in many indigenous cultures. It's a simple idea – shift from exploiting people and planet to caring for both. But it's so difficult for us because the values underpinning the many expressions of exploitation are so deeply embedded in our culture, institutional structures and personal understandings of success and values.

In her seminal conference presentation *Envisioning a sustainable world* Donella Meadows impressed upon us all the importance to envision what we want and went one to emphasize how often vision is missing completely. She goes on to explore envisioning as a muscle we need to exercise more often if we want to usher in a new way of being. Importantly, she also emphasizes that when articulating a vision, we shouldn't feel at all compelled to be able to identify how we will bring that vision into being – implementation is a different thing entirely (Meadows, 1994).

When we observe most political and business leaders, I don't think nearly enough has changed in the intervening nearly 30 years. Although, as the exchange between Zoe and Rob on Day Four (p164) has shown, we are becoming more aware of our crisis of imagination. Envisioning the future we want is an exercise in imagination.

With my colleague, Sam Wells, and inspired by the Meadows' conference paper, we developed a simple, facilitated process, underpinned by a living systems perspective, to help people in communities and organizations envision a sustainable future (Wells and McLean, 2013). At the heart of this process is the question: **How do you *really* want to experience your life and work?**

In answering this question please:

1. Respond in a childlike fashion. Rather than responding with what you think is realistic or reasonable. Respond like a child who knows and demands what they want. Set your inhibitions aside and be vulnerable. Articulate what you really want.

2. Look into, or feel into, your heart, rather than your rational brain. I often use photos or a guided vizualisation, to help people begin the story of how they really want to experience their lives and work. So engage your creative mind and heart – perhaps find a collection of pictures that illustrate your story. Or draw an image or mind map your story.

3. The story is about your experience rather than the tangible form – this is because we are unable to predetermine the tangible form or outcome in the complexity of the future. If for example, you find yourself reflecting on the virtue of a red chaise lounge in your home office; pause and ask yourself why that is important? Is it the vibrant colour that you love or the chance to rest during the workday? How do you really want to *experience* your life and work?

4. Tell the story. It is a story, rather than the usual pithy one-line sentence used in corporate settings. It's not bullet points either – it's a story. A story can convey the complexity of our lives.

Take some time now to articulate your story: how to you really want to experience your life and work?

In a team or organization, I ask people to share their stories in small groups and begin to collect up, in an inclusive manner, the experiences to generate a shared, co-created, story or vision. If you work in a team, you might like to try this. It's really good fun as people share their stories and get to know each other better in the process. I'm reminded by a principle often discussed by Peter Senge: 'It's not what the vision is, it is what it does' (Senge, 2022) that matters. It's not the concrete form of the vision that matters most, it is the fact that a vision is a force for change – a source of motivation and action. The above process of sharing through co-creating has many 'side effects' that are about what the vision does (McLean, 2020):

- Makes our visions responsible.
- Binds the group together to bring the vision into being.
- Motivates us when times get tough.
- Orientates us in complexity.
- Helps us make decisions about what next and monitor our progress as we take iterative steps to bringing our vision into being.

Exercise

JOSIE: Now I'd like to invite you to review your story and identify the important key elements or core messages within it; **your values that are embedded in the story.** Identify the top five-seven values.

*

I have been asking people in different parts of the world and in different contexts (team coaching, corporate culture evolution initiatives, community projects, adaptive initiatives) this question about how they want to experience their lives and work, in slightly different forms, since 2006. Hundreds if not thousands of people. What strikes me most is the strong similarity of their responses. I wonder how the values you identified connect with these categories which often emerge?

- Connection to family, friends, and work colleagues.
- Laughter and fun.
- Learning and creativity.
- Caring, supporting each other in all our diversity – indigenous knowledge.
- Meaning and purpose – contributing to something bigger than the individual.
- Wellbeing – healthy food and environment – healthy children, mental health, physical health – work/life balance – tranquillity.

- Nature – both the intrinsic value of the natural environment and as a contribution to wellbeing.

There are some variations around the edges, but it seems to me, that these are the values of sustainability. It's difficult to prove this assertion, because we would need to apply them and observe over a very long time. However, they are not inconsistent with values in some indigenous cultures – especially the caring for the natural environment including 'all our relations'.

Intellectually, it makes sense that as a species we should know the environment that best sustains us. It should be hard wired into humanity, shouldn't it? This logic is consistent with that of Hamalainen and Saarinen (2007) in their exploration of systems intelligence, where they recognize that we all understand complexity and living systems because we are these types of systems, and we live within them every day of our lives.

One of the things I value most about this approach is that the values are embedded in the vision. They are not made up separate from it as is often done.

The complexity of values

JOSIE: These values you have identified will show up differently or will be expressed differently in different contexts and as you play different roles in your life. They may even be reprioritized in different contexts.

Consider the following as an experiment to recognize the dynamic nature of our values. Choose two values from your list of values embedded in your vision, and consider how they might:

- Be expressed differently in each role below.
- How they may be expressed and/or reprioritized in each context.

1   Imagine yourself as the owner of your business deciding who to work with over the coming year.
2   Imagine yourself as an employee in your business deciding who to work with over the next year.

What did you notice?

My answers for this exercise and the two values I chose – wellbeing, purpose:

1.   As the owner of my business, I take my values and vison into account as I screen potential clients depending upon whether they 'care' for or 'exploit'. What contribution are they making to the planet and community? I also think about paying the wages for everyone – it often propels me into another round of market development!

2. As an employee, I do the client work that I am told to do by the owner. I love the way that most clients are aligned with my sense of purpose. My concern is more for my wellbeing – work hours and nature of the client. How much stress will they add to my load? What do I have to do to look after my stress levels?

Our values are not set in concrete, and we should expect that their expression will vary depending on context. This enables us a flexibility to meet the complexity of the world we live in. (I could say much, much more about viewing values in a systemic manner, but we don't have the time today.)

One last thing, holding your vision and values is an aspirational act. Living them is a process of learning how to live them amongst the values, or beliefs that you have learned from your parents and others in your life who have cared for and loved you. They have passed these beliefs on to you from their parents in some cases – and so your personal values reflect the cultural values of your ancestors. You are also trying to live these aspirational values in a system that is underpinned by other values and a different vision. Renegotiating these beliefs, or values takes time and is sometimes deeply emotional. Reprioritizing values is at the heart of adaptive work and leading adaptive change. (Heifetz et al., 2009).

Before we leave the vision and values you have identified for living your life and embedding into your work, let's address a big question. **How will you monitor your progress?** My answer to this question is likely to raise some resistance because we have been raised in a system or culture that in underpinned by, and privileges, the mechanistic paradigm (see p128).

The following draws on thoughts previously published (Wells and McLean, 2003; McLean, 2020). Upfront I want to say that my intention is to persuade you to let go of SMART goals and embrace either a combination of quantitative goals and qualitative indicators or monitor your success via responsible qualitative data. My argument follows several points of logic:

1. If we acknowledge that we are living and working in a complex system, then prediction itself over time and space is increasingly unrealistic.
2. If we can't predict the future, what process do we use?
3. Learning your way forward in an iterative fashion, using your vision and values as your guide.

Our complex world of life and work is increasingly uncertain. If the Covid-19 pandemic has taught us nothing else, it has highlighted the natural uncertainty of our interdependent world. We now see the inter-relatedness in ways that were previously as if hidden behind a veil. The veil is now dropped. As the pandemic progresses, we have become increasing adept at quick short-term changes as necessary to all types of systems – education, work, health, transport, food, political, technological and so on. We now see how each of these is entwinned in our lives and entwinned within each of us and between us. Predicting what will happen tomorrow or next month, is no longer taken for granted. We have come face to face with the reality the future is uncertain. SMART goals, Specific, Measurable, Achievable, Realistic and Time-bound require us to predict into the future. Which we can't. Or if we think we can, we will likely cause great stress for ourselves as the distance increases between what we predicted and what happened.

There is a truth that you can control a small part of the world, but it is very small – self limitingly small. If you want to play big, you need to accept that your vision, holding complexity in its story, will guide you more helpfully than a set of singular (reductionist) predetermined goals. You need the flexibility to respond or adapt as new information and events emerge. Hold your vision lightly and let it flex to accommodate other people and circumstances.

Consider your day-to-day life for a moment. How many of you have shifted jobs, changed houses, chosen a partner, decided to have children (or not, but have them anyway!)? How do you do this without SMART goals or organizational KPIs? Consider this possible process. You think about what you'd like your life to be like. You look around (observe), decide what is happening (interpret), make decisions and take some action (experiment). At some point you do an audit to decide if your action has brought you closer to your desired future (your vision) or not. You take corrective action accordingly. As a model this process may look like Figure 15.1. (This model is adapted from Heifetz, 2009).

JOSIE: Indicators of success are often qualitative in nature. For example, I have one indicator for the workshops I deliver. It is the sound of laughter. If I am not hearing laughter, I know I am not yet being successful. The laughter is an indicator. I can't measure it - it's more of a vibe. It's a sign of people being relaxed, engaged, and taking a few risks to learn together. So sometimes it has a nervous quality to it too. So it's more than decibels, because I could just bring a comedian in for decibels of laughter – there is a quality to the 'right' degree of laughter that I know when I hear it.

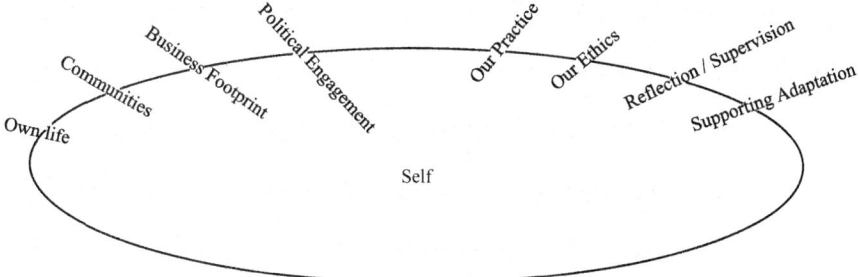

Figure 15.1 Vision as a cradle for experiments to learn your way forward (McLean, 2020: 72)

You could spend time identifying indicators of success for you and your business. And I used to do this with organizations and teams but I do it less often now. Because the pre-identified indicators have a tendency to act like goals for people who are culturally predisposed to setting goals. An indicator is just that. It's not a goal to develop actions to achieve. It's an outcome, a physical expression of your vision and its embedded values, that emerges from your interactions in the system. It seems much simpler and more pragmatic to employ the vision and values directly.

Vision and values as your indicators of success

JOSIE: Your vision and the values that you previously identified as being most important to focus upon at this time, can help you monitor your progress by stepping back and observing. Ask yourself: As a result of what I did, am I closer to what I envisioned or further away? Am I seeing the expression of each of the values more fully? Or less fully? These are subjective questions and answers that you will gain data and information about – some from within your body. Do you feel more joyful when working with these particular people, or less? You will be able to collect some quantitative data to help you answer your questions too – but you will not be able to measure everything that is important. In a group a more responsible, yet still subjective, response is possible. The degree of responsibility increases with the number of perspectives (Meadows, 1994, 1998). So how can you gain other perspectives on your vision and the lived expression of the values that you say are important?

Of course you can change your vision at any time you want to. And you can reprioritize values or change them completely based upon what you have learned.

There is a wonderful fluidity in this process that recognises the aliveness of the systems we live and work within. We can learn to dance within these systems in ways that stress us less. In ways that enable opportunity spotting to enliven possibilities, rather than closing down possibilities in the quest for achieving a predetermined goal that was probably out of date the moment that the ink dried.

My vision and values audit

JOSIE: 1. Keep your vision lively: review the written story of your vision. In what ways might you alter this given what you know now?

2. Complete the following table (Table 15.1) recognizing that the values that we separate out in a table are living in expression together all the time – they are not separate from each other.

JOSIE: In closing this section, please let me say that there is a lot more to be said and I refer you to my book (2020) if you are interested to explore this topic

Table 15.1 Vision and values audit

Value	Example of its lived expression	Am I experiencing more or less of this?	My evidence (quantitative and qualitative)
#1			
#2			
#3			
#4			
#5			

further. But, most importantly, for our world to change we need a new way of being. New ways of being include new ways of thinking, deciding to act and monitoring our progress (Snorf and Baye, 2010) if we are going to usher in new ways. Old ways of monitoring progress, privileging goals and measurements, are a part of the problem and they need to evolve as a part of the solution. The process shared above is not THE way – it is one way that we might evolve our understanding of a new way of being.

Working systemically requires a vigilance about the influence of old structures (tangible and in our heads) that hold us within old paradigms.

CONVENER: Thank you, Josie. Bronnie Ware wrote a best-selling book *The top five regrets of the dying* (Ware, 2011) that reflected her work as a palliative carer in Australia. The first regret she identified is:

> I wish I'd had the courage to live a life true to myself, not the life others expected of me.

(2011: 34)

Exercise

CONVENER: Perhaps now is a good time to pause and reflect on your courage levels. The courage to move beyond the expectations of our past, or the people around us. The courage to experiment with ways of moving beyond our ego wants and cravings, to living from our source and deeper purpose and ways of being that create a treble benefit for ourselves, our human community and the 'more than human' world.

*

As many of us know, there can be a big difference between our 'espoused values' and our 'values in action' and the road to hell is paved with good intention. As Peter often says action plans are devised by the left hemisphere neo-cortex, but change is always embodied. To address this, we have asked Andy Miller to explore moving values from our heads to our hearts and bodies and from espoused to enacted.

ANDY: It takes 'good grief' to supercharge my climate action plan? Whaaat?

Carbon cutting is just like coaching – it happens best when we get out of our heads and into our hearts. Now that you are transforming your business, you can amplify your efforts by approaching carbon and sustainability from a loving heart-centred space, full of gratitude. My work with companies in Vancouver, Canada, illustrates that cutting carbon is largely a matter of the heart: of aligning and priming and 'living' our values and emotions to spur action. Sadly, it seems the best values are laid to waste without vibrant and vulnerable emotional underpinning.

Emotions inform our values. I've found that without leaning-in to big emotions like grief and fear, biospheric value-activation suffers, and our ability to cut carbon is limited. Many of us take a conceptual and cognitive approach to cutting carbon. We rely on facts and data, we study carbon literacy, we think of carbon reduction from a knowledge-based perspective. Don't get me wrong, climate literacy, facts, and data are important, as we discovered on Day Two (pp Witzany 55–66). But just as factual logic doesn't work to persuade people to take climate action, or believe that Earth is in crisis, climate action plans that are inspired or informed only by facts and data and information miss the mark. By now it's common knowledge that companies that engage in fact-based climate education programs, to inform and guide their climate action plans, struggle to meet their climate and carbon cutting goals, and often fall short!

I've found that deep emotion and value activation is the crucial missing link in supercharging a climate action plan, enabling cuts of 50 percent or more (Miller, 2022). Without it cuts of 10–20 percent are the norm (Climatesmart, 2022). Why? Most of the carbon reduction comes from behaviour change, not technology, and behaviour change doesn't happen from the head. Often, we get stuck thinking carbon cutting is mostly a technology fix, and we spend our time thinking about our carbon action plan conceptually and cognitively. Well surprise, surprise, we can't cut much carbon with our head talking at us at the speed of light. As Greta (Thunberg) says....blah blah blah. We need to swan dive into our heart, ooze love, and befriend our grief, fear, among other big hairy emotions.

And as you well-know, behaviour change is complex stuff, driven by emotional contributing factors, childhood experience, and so much more. The trick is to conscientiously activate and use emotions, like grief, sadness, fear and anger, through vivid imagery, to inform our values. Then, by linking our biospheric-values to Earth-connected behaviours, our climate action plans become less intimidating and more attainable. The key is using imagery to engage emotion, commit to values, and activate behaviour. Some of us find this challenging.

Cutting carbon by upwards of 50 percent, something we'll all have to do fairly quickly, is a tall order requiring significant re-alignment of lifestyle. The quickest and easiest way to get there? Prime your emotions and values. Here's how I do it.

First, watch the HOPI First Nation-inspired Qatsi film trilogy Koyaanisqatsi (Life out of balance), Powaqqatsi (Life in transformation) Naqoyqatsi (Life as war), Samsara, Baraka, One Earth, or a similar **visual tone poem** film, devoid of narration, and full of music and vivid imagery (see details for all these films in the references). This will get you in the mood for a deep check-in with your emotions and values. You can find the films divided into short segments, so you don't need to watch them in their entirety! I show my clients a 20-minute compilation of bits of all these film-tone-poems. Then, with our hearts wide open from vivid imagery, we do emotion and values check-ins, which I refer to as priming.

I learned a helpful tool from a CCA colleague. Please, take the values you developed just now with Josie and for each one identify and connect deeply with a personal role model, who enacts this value. Then look at their behaviour and actions as a guidepost for your own.

Having done that now create a daily practice, a simple set of activities, to reinforce those behaviours, and map out your journey, step by step. This process helped me and my clients, and could help you, build courage and confidence to bring your future story to life.

Now, for one or two of the values you would like to enact more, identify a cue, like drinking a cup of coffee/tea, that triggers you to think about those feelings, values, and Earth connection. Then the magic will happen.

With core Earth-centred emotions and values being triggered with intention every time you pour a cup of coffee or tea, people end up thinking A LOT about Earth, and feelings, and the importance of values. Then, writing and adhering to a climate action plan becomes second nature.

The secondary beauty of priming your values and emotions in relation to your carbon emissions, is that you feel proud, and start talking vulnerably to friends and neighbours about your actions, and how important climate is to you, and your grief, and how you found gratitude on the other side, and you become more inspiring and attractive than before. Rather than having climate shut down a conversation, when you speak of climate from your heart, rich with emotional and value clarity, it opens up conversations based on vulnerability, trust and intimacy. And you end up experiencing the radical hope of Joanna Macy. Or develop the capacity to hold seemingly diametrically opposed emotions like hope and despair simultaneously.

CONVENER: Thank you, Andy. I really like the use of the imagery of films to help immerse people in an experience of life and I love the way you share how to trigger behaviour changes too – attaching it to something we do regularly.

We now want to go further on the path from vision, to values, to behaviours, to practical commitments. As mentioned before in what Zoe shared, we want to help you examine your life and businesses from an *eco-active* perspective. We are going to each look critically, looking into the mirror in **examining our own business impact**, and holding up the mirror to each other's business.

PETER: We are going to start by looking at global human goals remembering back to Day Two about the key science indicators of the ecological and climate crisis and identifying what this tells us the global human family need to achieve together. Then we can look at our own goals and how we can each make our own unique contribution to what must be a total human effort.

The United Nations announced their Sustainable Development Goals (SDGs) for the next 15 years, in 2015. They were intended to 'provide a shared blueprint for peace and prosperity for people and the planet, now and into the future' (United Nations, 2022). There is both a great deal of acceptance of, and critique of, the SDG goals. We are going to use them because they have a usefulness but as we do, let's remember the systemic perspective we have been developing and recognize that these are goals that are underpinned by assumptions and political compromises that reflect the inadequacies of the very system we are trying to transform.

- The words sustainable and development together have been criticized by many. By combining the two words we assume that sustainability and economic development as prescribed by the reigning neoliberal economic model is possible.
- The previous commentary by Josie raised the topic of goals being a reductionist approach (that also failed to deliver the intended result through the previous set of Millennium Goals).
- The goals do not show fully the underlying dynamics that connect the separate areas such as poverty and climate change.
- The tendency for people to choose a goal to work with – rather than recognizing the need to work with them all. Combined, they are an integrated vision.

Acknowledging these and other shortcomings, let's work with the 17 SDGs for a little while as we each reflect upon our own businesses.

Sustainable Development Goals

Goal 1. End poverty in all its forms everywhere.
Goal 2. End hunger, achieve food security and improved nutrition and promote sustainable agriculture.
Goal 3. Ensure healthy lives and promote well-being for all at all ages.

Goal 4. Ensure inclusive and equitable quality education and promote lifelong learning opportunities for all.

Goal 5. Achieve gender equality and empower all women and girls.

Goal 6. Ensure availability and sustainable management of water and sanitation for all.

Goal 7. Ensure access to affordable, reliable, sustainable and modern energy for all.

Goal 8. Promote sustained, inclusive and sustainable economic growth, full and productive employment and decent work for all.

Goal 9. Build resilient infrastructure, promote inclusive and sustainable industrialization and foster innovation.

Goal 10. Reduce inequality within and among countries.

Goal 11. Make cities and human settlements inclusive, safe, resilient and sustainable.

Goal 12. Ensure sustainable consumption and production patterns.

Goal 13. Take urgent action to combat climate change and its impacts.

Goal 14. Conserve and sustainably use the oceans, seas and marine resources for sustainable development.

Goal 15. Protect, restore and promote sustainable use of terrestrial ecosystems, sustainably manage forests, combat desertification, and halt and reverse land degradation and halt biodiversity loss.

Goal 16. Promote peaceful and inclusive societies for sustainable development, provide access to justice for all and build effective, accountable and inclusive institutions at all levels.

Goal 17. Strengthen the means of implementation and revitalize the global partnership for sustainable development.

PETER: Look at this list and please hold the whole for a moment. Can you imagine your business contributing to all these requirements? It is a huge endeavour! And just because it is huge and difficult, we tend to choose one or two goals that we can contribute to. Transformation is difficult work and so let's not short change it. Put a tick against which of these you can easily contribute to. When I did this exercise previously, I was amazed at how many I could either make a direct contribution to, or an indirect contribution, by whom I support and work with. Look through the list and identify with a cross, those that are problematic for you. Now think harder, where are the connections to your business? What do you need to consider that you do not presently consider? Then look at what you could do to increase your contribution to all the SDGs.

This is by no means simple. It is complex and the answers will be different for each of us. We need to think harder to be responsible for the transformation we require.

Spend some personal reflection time and make some notes.

Now let us move from looking through the binoculars at what others are doing or not doing and look in a magnifying glass at what we are currently doing. This is something that we as the Conveners are still working through for ourselves. Tim Cook, Apple CEO, said: 'The stakes are high, and failure is not an option… if you have not developed a plan [on climate and sustainability], you have failed at your job'. (Quoted in Polman and Winston, 2021: 213). This is an example where we are going to set objectives and targets – in the words of Paul Polman: 'if some of your targets don't make you uncomfortable, you are not pushing hard enough' (Polman and Winston, 2021: 105–106). We cannot criticize big corporations for not having ambitious sustainability and net zero plans, if we have not done this for our own businesses.

Exercise

PETER: Please open your laptop or phone and click on your own website, LinkedIn or Facebook page for your business.

Now complete the following questionnaire:

1   Does your business have a clear purpose statement of how it intends to make a positive difference in the world?
2   Does your business have a clear sustainability plan, with strategies, targets and progress?
3   How could the sustainability plan be improved in the light of this workshop journey?
4   What percentage of your website and social media promotes what you do and what you have achieved and what percentage why you do the work and who and what it is in service of?
5   What percentage of your social media (blogs, tweets, LinkedIn articles) is promoting you and what percentage is enabling others to make a difference?
6   What percentage of resources you mention are you selling for gain and what percentage are you giving away or promoting other people's resources?
7   Reading through your website, write down which of your values are being clearly demonstrated. Against each write examples that the website shows of these being enacted in practice.
8   Is the website, LinkedIn profile, etc., clear about the criteria of who you work with, who you associate and partner with and who you do not?
9   Is it clear how you work with others to deliver benefits for your customers, suppliers, employees, organizations, their stakeholders, communities, future generations and the ecology?
10   How do you show and demonstrate your commitment to diversity and inclusion, both in what you say and the associates and organizations you partner with?
11   In the light of the above answers what would be your top three to five commitments to changing what you do and what you articulate on the web and other social platforms?

a.

b.

c.

d.

PETER: In this case these do need to be SMART which used to stand for Specific, Measurable, Achievable, Realistic and Timely. Polman argues that the A needs to be changed to Audacious and Accountable and challenges us: if we can think we can achieve them, then are we being as ambitious as the world needs us to be? The R needs to be Results orientated, not the goals you want to and think you can achieve, but what the world and the science is telling you is necessary (remember back to Day Two). Science informed goals are becoming the standard expected in corporate organizations globally today – let's step up to that standard too.

*

Don't worry if you have not managed to finish every question, or have got stuck, because it is an ongoing process of learning, and we can invite help from others.

Paul Polman (Polman and Winston, 2021: 128) invites us 'To find knowledgeable critics and invite them in', and modelled this at Unilever, by inviting in NGOs like Oxfam, Greenpeace, and Forum for the Future, to critique what they were doing around the world. We all need critical friends who will hold up the mirror for us, so look around and chose someone who you think will be the most challenging and supportive person for you and your business.

Give them the links to your website, LinkedIn, Facebook and any other social media for you and your business and ask them to fill in the same questionnaire as you have just done. At the same time, you will perform the same audit for their business. Each write down your top three to five recommendations for each other's business (i.e. responses to question 11).

Let's go even deeper now with some peer coaching and consultancy to refine your actions. Please find a quiet space as a pair, real or virtual, and spend two hours together – an hour focussed on each person and business in turn. Please be fearlessly compassionate with each other stepping fully into the role as a compassionate friend. Work through how you have both answered the questions on the business in question, and end by agreeing the three to five actions and targets that will best support its transformation into a business that is net positive – giving more than it takes from the world around you.

CONVENER: Our work never happens in isolation. Every business is dependent on many resources and suppliers that enable our work to happen.

We invite you to list as many of those resources and suppliers that support your business as possible…

*

Participants shared the following:

Energy for heating and lighting and air-conditioning where we work.
I.T. equipment and platforms – computers, web-designers, website hosting, apps, Zoom, LinkedIn, Facebook, Calendly, Microsoft teams, Slack, Asana, CRM platform.
Materials: Stationery, printing, business cards, brochures, videos.
Finance: accountants, advisors, banks, pension funds, insurance – both professional insurance, building insurance, car and travel insurance, life insurance.
Travel: car, public transport, air flights.
Administration: people I employ to help me with my admin.
Outreach: LinkedIn, places where I blog, post a podcast, Twitter, publishers, YouTube, Vimeo, video editing.
Training centres and universities: That I use to run or attend training programmes, and conferences, their energy suppliers and caterers.
Clothes and accessories: I buy for work and to impress!
Professional development: attending courses and conferences
Others:

PETER: Wow! What a lot of resources, so many of us use!

We have looked at how we can develop our business and many coaches I work with are now registering their business as a *B Corp*, to both have their business audited for how it is net positive, but also to show publicly their support for this approach. We can extend our *eco-active* responsibility even further by how we influence our value chain. Let me give one example.

Many years ago, I woke up to the fact that I, like many others in the very privileged world of middle-class people with long-term employment, living in western countries, probably had more influence through my pension scheme, than through the ballot box. Also, that very few of us take responsibility for what others are doing with our money – banks, pension funds, insurance, etc.

The year 2020 saw a major tipping point. In 2019 only 14 percent of European pension schemes were taking climate impact into consideration in their investments, which quadrupled jumping to 54 percent in 2020 (Institutional Asset Manager, 2020). Also 2020 saw the third largest pension fund in the USA (New York State Pension Fund) divest from coal mining, and 'committing to sell its investments in any oil, gas, oil-services and pipeline companies that do not have clear plans to abandon the fossil fuel business by 2025'. In 2020 they reduced their investments from US$12 Billion to US$2.6 Billion in fuel-related companies and committed to selling their shares in all companies that contribute to global warming by 2040 (Barnard, 2020).

Ireland was the first country to take a lead and disinvest completely from fossil fuels (Lahn, 2018). Other countries have been slow to follow its lead, but in 2019, Norway's Sovereign Wealth Fund, which has been built on oil and gas revenues – announced significant disinvestment in fossil fuels. 2020 also saw a further dramatic slide in the Exxon share price, which in December 2020, was trading at just over US$35 a share compared to just over US$65 at the end of 2019, which analysts see as more than the Covid 19 effect and in a big part attributable to their climate impact.

When I did this exercise, I felt proud that I had been actively ensuring that my pension fund was not invested in coal, oil and gas, and was only in so called ethical investments. But then I realized we should not stop at disinvesting from the many companies whose ecological costs can now be shown to be greater than their profits (Cohen, 2020). I had to go further. I had to start to ask those who managed my pensions and investments to show me each quarter what beneficial social and ecological value my pension money has created through what it has been invested within.

We can also apply this to energy and catering, for example, by asking questions of conference organizers. There is always more we can all do to make a difference everywhere we go.

Exercise

PETER: Now please write a table like the one below (Table 15.2).

- Fill in column one, all the resources and suppliers from the list above (p240), that you use to do your work.
- In column two, please put a tick against the ones you are *eco-informed* about – meaning that you know the ecological impact (as much as is humanly possible) - and a cross against the ones you are not informed about.
- In column three, list what you are doing against each item to influence the reduction of this resource's negative ecological impact and in a different colour what you are doing to increase its positive impact in the world.
- Now in column four, what you can do to increase your influence.

PETER: When you have completed this exercise, find a partner and share your tables. Help each other come up with three to five specific agreements and actions on how you can increase your influence based on your 4th column. Add these to your previous list of three to five outwards facing business actions.

*

CONVENER: We sometimes feel we don't have a voice, but Peter's story shows that we can each play a role. And I find that encouraging. We are going to end today with an overall reflective exercise that may help explore *eco-active* in another and complementary fashion. Alison will share her model of **Regenerative Coaching** and help us use this model to reflect on our own work.

Table 15.2 Beneficial social and ecological value assessment

Resources and Suppliers	Aware of their ecological impact	I actively influence their impact by…	I could increase my beneficial influence here by

ALISON: On Day One (p40), we shared Paul Hawken's 12 practical questions (Hawken, 2021) around whether our contribution was making a positive difference to contributing to a regenerative future. Paul Hawken talks about regeneration as a default mode of life – all living organisms are regenerating. You are right now as you engage with us here. The cells in your body are regenerating. So, his primary question to us is how can we get an alignment of humanity's way of life with biology and work with nature's natural regenerative force? (Commonwealth Club of California, 2021)

I have created a circle that allows a more nuanced assessment. The 12 segments each represent one of Hawken's questions (figure 15.2).

Imagine a scale running from the centre of the circle to the outer perimeter, the centre point of the scale being -10, the mid-point 0, and the outer perimeter +10.

Reflecting on your own choices you might ask the following questions:

1. To what extent does your life, coaching practice, coaching business restore land, create life, heal the future, reduce poverty, prevent disease, etc.? Perhaps score each dimension on the -10 to +10 scale.

Having answered the first question, then ask:

2. How do you achieve these things in your life, practice, business … create life, heal the future, regenerate, provide workers with dignity, decrease global warming, restore land…

3. From today's workshop which of these areas could you actively increase the score on and how?

Finally, the final question:

4. What insights has this exercise generated for you?

Perhaps you might test out this circle, explore what insights emerge and how it assists you make sense of significant decisions in relation to a flourishing future and what needs to change for that future to emerge.

CONVENER: Thank you, Alison.

We are going to end today with a quotation from the great inspirational leader Archbishop Desmond Tutu, who died in late 2021:

If you are neutral in situations of injustice, you have chosen the side of the oppressor.

And so we come to the end of our sixth day together. We hope that you have enjoyed the deep dive into looking at the intentions you hold for your life, business, and its impact socially and upon the more than human world. Exploring ways of being more *eco-active* in your life and work. Awareness

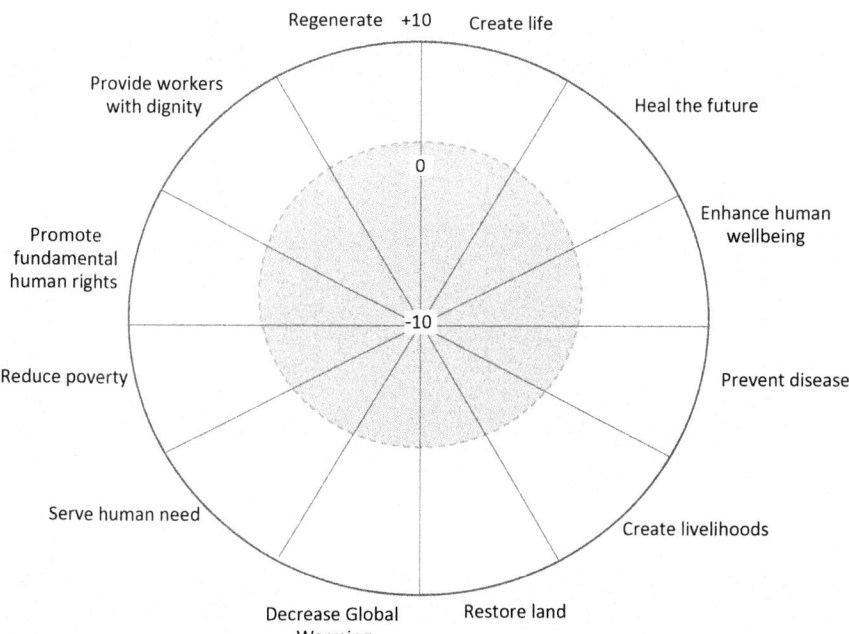

Figure 15.2 Segments representing Hawken's questions

and responsibility are at the heart of this exercise and conversation with others. Interesting that these two qualities are also at the heart of coaching as we practice it.

Tomorrow we will draw threads together and bring our week to a close. Thank you.

Chapter 16

Integrating our Learning

Final Day Morning

CONVENER: Welcome to our final day together and once again we will start with a beautiful short poem, with many thanks to Jenny for allowing us to use it:

Beyond the darkness, a new world comes

Beyond the far horizon, the sun will return again,
Though darkness be with us now, its shadows will not remain.
These dark times will depart at the coming of the sun,
For always light follows darkness, somewhere the dawn's begun.

The spinning of the Earth brings to us both darkness and light,
As soon as the darkness begins, tomorrow is in sight.
It may be a while yet before its light and warm once more,
But it is coming, for it has never failed to come before.

Upon the hint of dawns first light, we can begin to see,
The world tomorrow is bringing to us, to you and to me.
It will not be yesterday, for that was a darkness ago,
And what is coming after the dawn, we have yet to know.

The dark night may have been a struggle, so hard to get through,
That the world coming with this dawn, is so different and new.
Perhaps the darkness changed many things, that we are now afraid,
But in this new dawn, love can be found, and a future remade.

Jenny Maryl (2020) (Jennymaryl.uk)

CONVENER: Today, we will finish earlier with just a shorter morning and afternoon session. In the morning we will:

- Integrate the complete journey not as steps, but as an on-going cycle we constantly need to go round.
- Centre ourselves in our own purpose, values, and being.
- Learn from indigenous traditions and hear from two young African coaches.
- Explore the journey from ego to eco-coaching.

DOI: 10.4324/9781003153825-16

This afternoon we will:

- Discover ways of resourcing ourselves and our work – so we work from and are supported by 'Source'.
- Explore how we will each take our work in this area forward.
- Make our farewells.

And so, as we come to our last day, we can reflect that in the previous six days we have journeyed together, we have been right round the Eco-Phase Cycle, but we have not come to a place of completion, as this is a cycle not a linear journey. We will all need to continue to cycle many times around and through all these stages, in the hope that each time we will be able to do so in a deeper, more inclusive and challenging way. But we do not, and cannot do this journey alone, and today we will explore how we can ask for help from fellow travellers and from the 'more than human' world.

PETER: So let us now turn to an eco-systemic spiritual practice called: *Opening the Seven Levels.*

For a number of years, I have had the privilege of being an inter-faith spiritual celebrant, facilitating weddings, child blessings, funerals and other rites of passage. In more recent years I have trained other spiritual celebrants in this important work. One of the core practices happens before the ceremony and is for the celebrant to prepare themselves through the practice of opening to seven levels of awareness.

Every coaching session is, in some way, a rite of passage, so this is a practice we can do as coaches before each coaching meeting. (Hawkins, 2021b: 308–9)

Picture the individual you are about to coach, listen to their voice and sense their presence in your imagination. Bring to mind the journey they are on, their life and context.

Now please travel with me on a guided meditation, as you picture one of your coaching clients, an individual or team.

The first level is to open to the individual or individuals, and to picture them with love and compassion.

Then to refocus on the relational connections. What do you know about the coachee's family, about their team and their relationship to their boss and their peers? What is the quality of these relational flows?

Thirdly, open to the wider community of immediate stakeholders, the customers of their work, the suppliers and support systems of their work, the investors and funders of the work and the communities in which the organization operates.

We then move our focus to those who are often ignored. The not yet and unreached customers and organizational partners, future employees, and wider future generations that will be affected by what we do today, the organization's '13th fairy', the stakeholder they may be ignoring at their peril.

At the fifth level, the attention then moves to the whole interconnected human family, all 7.9 billion of us that share this planet. How is this person contributing to our wider connected human world? Are they giving more than they are receiving? Are they and their work moving us to a more inclusive and equitable world?

And then we focus on the more-than-human world of all the sentient beings that surround us, and the elements that support and flow through us in the air we breathe, the food and drink we consume, the non-renewable resources we consume. What are the connections and flows between this person and their ecology?

Finally, we open the door to the mystery of oneness and inter-being – that which connects everything, beyond time and space, beyond words, and certainly beyond our own limited comprehension.

Now please come back down through the levels, sensing how the wider systemic levels, live in and flow through, the more immediate and local levels.

Exercise

Personal reflection: Use the space below to record and allow the emergence of your experience of that visualization. Then, if you are able, share what you became aware of as you participated in the visualization. Give each other 15 minutes to share your emerging awareness at each of the seven levels.

*

PETER: Welcome back.

What we know from the experience of many coaches and practitioners who have experimented with mindfulness practices like this one, is that when you open to some new awareness within you, even though you never mention it, the client starts talking to that same level, as though they had only been awaiting your readiness.

We can also learn from those who have maintained their connection with earth and wider eco-systems much better than those of us who live in the industrialized, urbanized and transient, so-called 'developed world', which we might also call 'The world of the indigenous orphans.'

> We are all indigenous orphans, migrants on the road, separated from our sense of place and belonging. We have lost, not only our sense of rooted-ness in a landscape, but the feeling of being held in time by the rhythm of seasonal celebrations, rituals of the year, and life transitions ceremonies.
>
> (Hawkins, 2019a: 64)

Most of us, and indeed nearly all of us, in the white privileged world of the European diaspora, are indigenous orphans. We have lost our connection with our deep home in nature and the wider ecology. We are no longer rooted in place, community, both human and more-than-human, and in the geology, topography of landscape, climate and seasons.

So we are now going to spend some time exploring the theme of how we as **indigenous orphans – find our way home**.

Chief Seattle, First Nations Elder, Leader and Teacher in North America, said in 1855:

> Will you teach your children what we have taught our children?
> That the Earth is our mother?
> What befalls the earth befalls all sons of the Earth.
> This we know: the Earth does not belong to man, man belongs to the earth.
> All things are connected like the blood that unites us all.
> Man did not weave the web of life, he is merely a strand in it.
> Whatever he does to the web, he does to himself.

PETER: Alison, will you please share your thoughts on what we can learn from indigenous wisdom traditions.

ALISON: Thank you, Peter. Yes, we urgently need to find our way back home, remembering that home is a way of being and living rather than a place. For guides we can be helped by those whose indigenous connections are still vibrant and alive. Writers and teachers such as Robin Wall Kimmerer (2016), Anita Sanchez (2017), Melanie Goodchild (2021) and many others.

Robin Wall Kimmerer (2016: 9) challenges and up-ends our scientific humancentric world view, which place humans at the pinnacle of creation.

> Within Native ways of knowing, human people are often referred to as 'the younger brothers of Creation.' We say that humans have the least experience with how to live and thus the most to learn—we must look to our teachers among the other species for guidance. Their wisdom is apparent in the way that they live. They teach us by example. They've been on the earth far longer than we have been and have had time to figure things out.

She encourages us to develop deep humility and learn from our elder brothers and sisters in creation. Kimmerer goes on to say (2016: 222):

> The land is the real teacher. All we need as students is mindfulness.

Anita Sanchez gives us indigenous orphans hope as she points out that at root, we are all indigenous. She tells us that we all have access to 'elder wisdom' and healing 'because ultimately, at the beginning, we all came from indigenous tribes, even if we have lost our story of those origins.' (Sanchez, 2017: 15). All of us have come from the Earth, are made up of elements of Earth, are sustained by what flows and grows on Earth, are intimately connected with all other sentient beings and elements that we share this one Earth with, and eventually we will all return to, and become the earth of Earth.

Sanchez (2017: 5) defines 'indigenous' very inclusively:

A truly indigenous person is one who has intimate connection with Mother Earth and who embraces all human beings in order to get along with them.

She also teaches us about the role of elders, and as you listen to her attributes of elders, please ask yourself which of these capacities you think it is important for coaches and mentors to develop and ask yourself which you need to most develop?

Elders are people who are steeped in the traditions or the passed-down knowledge of a community or tribe.
People who have taken on the role of healers, cultural leaders, and spiritual teachers.
An Elder can be truly funny…a twinkle in their eye…
The innocence of a child and the deep wisdom of the ancient.
They know magic exists and are playful with it.
They know how to balance things, creating harmony and connection.
They hold the understanding that every thought and action has an impact that either lends support to or destroys our interconnected life and community.
They share the message that the honor of one is the honor of all; and the hurt of one is the hurt of all.
They understand we are all connected

(Sanchez, 2017: 16–18)

For Sanchez, elders are the carriers of the four sacred gifts that were visioned by a Mohican Elder in the Rocky Mountains in 1996 and then brought into life in a ceremony with:

twenty-seven indigenous elders from the four colors and directions – Elders from the North American tribes representing the red and south direction, a Buddhist elder from Tibet representing the yellow and east direction, a Sami elder from Finland representing the white and north direction, and two elders from African tribes representing the black and west direction.

The four gifts they collectively brought and shared were:

the power to forgive the unforgiveable…
the power to heal…
the power of unity. The power to come together…
the power of hope. The ability to dream, to see wellness and the powers to attain it.

(Sanchez, 2017: 3)

The importance of all humanity finding their way back *home*, to an Earth-connected way of being, has now become critical and urgent.

The re-emergence of the felt experience of belonging in the larger consciousness (our collective interdependence and evolution), not only on our own lives but in our connections to all things, requires action from all of us, you and me. …. This wisdom is necessary for humanity to move forward and evolve.

(Sanchez, 2017: 16–17)

Exercise

ALISON: So let us pause and experientially explore these four gifts. As I read out the questions, I invite you to write your responses, as you consider each gift.

The first sacred gift is the power to forgive the unforgiveable…

Who and what do you need to forgive? Write down the name of someone in your life that you are still angry with. Right down the wrong or injustice you feel they did to you. Now picture that person as a teaching that had been sent to you – what was this encounter asking you to learn? Can you try replacing the complaint or blame with a thank you?

The second gift is the power to heal…

What needs healing or making whole inside you, in order that you can better heal others?

Notice what is disconnected, that needs connecting. Write these down. Then write what you need to do to bring about that healing.

Now write down who in your life and wider contexts most needs your healing energy and how you will bring it to them.

The third gift is the power of unity. The power to come together…

We all have much more power in our capacity to convene than we realise.

Write down the most important challenge and purpose that you could convene people, in your network and beyond, to come together and address.

And the fourth gift is the power of hope. The ability to dream, to see wellness and the powers to attain it

What new dream, or vision has been arising in you through these days we have been together?

How can you take that vision, dream and purpose back home with you to your communities? Write your vision and dream down and then write down who you will share it with when you return home.

As we take a break you might like to go for a walk and contemplate your answers or walk with a friend and explore them together.

*

CONVENER: Welcome back. Before we continue the final part of our journey, we would like to invite two people to share their stories of trying to keep that connection to Earth. They demonstrate two approaches to the challenges and opportunities brought about by climate change. The first is from Denis Opio, a young coach, mentor and climate advocate in Uganda.

DENIS: For many thousands of years generations have been inheriting nature as it was left to them. The current generation seems not to be minding or perhaps mindful of what state of nature they will pass over to future generations that will follow them. Let me put it this way, families tend to put their children first in their plans and actions, valuing their responsibilities to them. Parents will consider how their children will be left after them, that's why they will educate, feed, care, love and ensure their children's good health. Being a loved grandparent is the best gift you can get from your children, as the lineage continues. The question is which parent would want their children to lack natural oxygen and move with cylinders on their backs just because they couldn't protect the current nature in the names of modernization, technology advancement and industrialization.

I am a youth living in the middle of nowhere surrounded by the tears of innocent nature that see the end of their life in every second that passes by, due to the toxic gases from human money-making investments that give little back to nature. The unpaid care that nature provides to humanity is no longer recognized and appreciated by the world today and because of this we have seen constant sufferings of all kinds like floods and prolonged droughts, among others. The current trend of life has paved the way for difficulties that we are not prepared for and that will last for more billion years to come. Humans can be hugely creative, make changes in technology and rapidly innovate. But why does this need to be done at the expense of nature?

Climate change is real, and it affects you 360 degrees, it doesn't know whether you are rich or poor, rural or urban, educated or uneducated, tall or short, black or white. And it is not just the ecology and fellow beings that suffer. When the reality of climate change bites, life on earth can become a nightmare for some humans, and over time more and more people will be affected. A peasant farmer in my village would definitely throw blame on God or some bad omen affecting the community for little or no rainfall to the only source of living he has available to him: agriculture. He or she will not know that there is something bigger than that to blame. The eco-systems are being destroyed every second that passes, through human actions: bush burning, deforestation, degrading rivers and swamps. For us our religious practice is Christianity, and The Bible tell us that in the beginning God created heaven and earth and put life on Earth and He was pleased with his work. On the last day of his creation, he created man and gave him the mandate to look after his creation. But because of unsatisfied needs of man, he has destroyed what was created before. Any human action that leads to breaking of the eco-system of nature will have immeasurable impact on life.

There is a need to change people's attitude, mindset, and belief in the way they do things and to having nature first, at heart, mind and soul, before taking any action towards nature. Be nature first before you think of destroy-ing it, this will make you think of securing your life first. If we all take up this responsibility step by step, future generations will appreciate their inheritance from us. The fresh brains of young people are not being involved, yet they are the solutions of tomorrow. Therefore, involvement of young people or youth in any action needs to be considered and their voices respected, because they are the sustainability model for our legacy on earth.

CONVENER: Thank you so much Denis for your impassioned plea on behalf of all beings. Now we hear from Reshma Khan, in Kenya, whose work has had a profound impact on how she practices, both personally and professionally.

RESHMA: I see the whole world as a connected WHOLE. The earth has continued to show us the importance of recognizing that we are a WHOLE organism. Just like the Earth has a connected eco-system, so do we as individuals and humans, and there are different levels to this:

MY LIFE HAS BOTH A PERSONAL AND PROFESSIONAL SIDE - what goes on in my home also affects how I show up at work. Very much in line with the Co-Active model (which is part of my core model of coaching). In my coaching I am very cognizant of this WHOLENESS and from the very beginning, my chemistry sessions with the thinker (coachee), include 'wheel of life' reflections that include both the personal and professional lives.

AS A WHOLE HUMAN BEING WE ARE AN ECO-SYSTEM - Often our bodies tell us something is going on. I am a highly intuitive person and use this as a coach, particularly with thinkers who may be feeling certain dis-ease or feelings in their bodies. Just like when the earth is on fire, flooded, or when the rains don't come on time or at all, and we know this is a microcosm of something bigger, our bodies go through similar patterns and so I work with the client to understand this.

The Power of the Earth to Inspire, Illuminate and Heal. Nature IS a cure for our distracted minds, if we choose to listen. I often recommend to my clients some reflection time in nature between sessions – since Covid 19 emerged, I haven't met people in person otherwise I would also do sessions with local cli-ents walking in the forest coaching. There is a great book I read in the last year – The Nature Fix, (Williams, 2017) that helped land some of this for me.

My own grounding as a coach - Often with really busy days it is important to ground before meeting a client, and after. Sometimes, prior to meeting a client, spending some time in my essentially zero-waste kitchen, where we make our own yoghurt, or I make my own chemical-free, earth friendly body lotion, helps me slow down, observe fermentation or creation, and gets me in a space where I can be completely present with my client. Often, after a session, particularly with a client who may have gone to depths that may affect me emotionally, I walk into the garden, take off my shoes, stand on the grass and ask Mother Earth to take all that I have held for the client, let it pass through me with all the trauma it elicits, and go into the earth.

By the way, when I say 'garden' I mean a little slice of whatever forest is left in Nairobi, where I live - I am very blessed to live in a rented home with my husband that is full of indigenous trees, animals, and peace, next to a river. We grow our own food wherever possible and have a permaculture gardener who stays with us (today's dinner is most likely a famous family recipe for stew made with cassava that we picked over the weekend, along with some parsley, chillies, and other herbs to add in from whatever I pick in a few hours).

I would love to do more with coaching from a zero-waste mindset and would love to explore what that could look like. I am thinking through signature 'Journeys' of coaching, where people will come on local, low carbon trips, to indigenous wisdom areas, learn from locals, and also explore their leadership as we go – and so that's my vision!

CONVENER: With Reshma and Denis's ideas in mind, this afternoon we will explore how we resource ourselves for the work we all need to be doing. However, before we start on that, there are, and always will be, more questions, that have arisen, but let us address one or two which seem important to enable us to complete this journey.

One person has asked: 'You have said much about how we include the ecology in coaching, but is this not still ego-coaching with a wider focus?'

Another asked: 'How do we truly move beyond ego-coaching?'

Great and important questions. Josie and Peter, can you start to address these.

PETER: We leave ego-coaching behind when there isn't an 'I' doing the coaching, and a 'you' being coached. When the coaching goes beyond the short-term individual wants or needs of the coachee. When we are listening through the immediate, through the local, through the internal and external stakeholders, and beyond through the 'more than human' world, to the life that flows through us and surrounds us. Where we are both in service of what life is requiring of us.

When in the words of the Katha Upanishad:

> What is within us is also without. What is without is also within. One who sees difference between what is within and what is without goes evermore from death to death.

(as quoted in Ravindra 2014: 51)

Where there are at least two seekers and two learners in the coaching room, both learning, unlearning, and discovering together.

> Ravindra (2014:26) also for me, captures the essence of egoism, which he says emerges when we: 'think that we are the maker of ourselves and of our destiny'. And he tells us the Sanskrit word for ego is ahamkara which literally means I am the doer.
>
> No path from ego to eco, that I have found, is easy or comforting or self-advancing. Perhaps to move beyond ego-coaching is to work at discovering how both coach and coachee can be supporting each other on this path through how they engage together.
>
> But that does not mean we need to be saints or sages. Just constantly alert to when our ego or the coachee's ego, starts to take over and dominate. At these moments it is important not to become judgemental of oneself or of the other but interested in what this rising up of our egos is telling us about what needs attending to, and then gently to have the courage to call our collaborative inquiry back to its bigger purpose.
>
> Another question that has been asked is: 'where do I start?'
>
> Firstly, you already have begun by coming on this journey. Also, the only place you can start is with yourself. Each day, each moment, each new meeting, and each thought and feeling that arises. Each is a new start.

JOSIE: The path you are pointing to, Peter, is a deep and lengthy spiritual path that some may feel is beyond them. I know that I have felt challenged by your extraordinary path.

The pathway from ego to eco, also represents a different way of perceiving the world. A fundamentally different way of seeing, thinking and doing in the world. But I am not sure we can jettison an older way completely – I see it more of an expansion of bandwidth (to use an IT metaphor). And it may be helpful to make some connections back to our previous days of work that we have already done together.

I want to refer us back to the dialogue we shared about a systemic way of perceiving the complexity of our world that is a series of nested subsystems within systems. This was not a beginning of my path, but a catalyst to deepening my understanding of how the world works.

If I track this part of my journey, it began inside my head. In hindsight, it was initially very ego driven because I was trying to understand how to design systems that would deliver specific, predetermined outcomes. (Maybe you can also hear an echo of the Newtonian or mechanistic worldview in my intention?) As I practiced perceiving my work within different organizations and communities, through this lens of living systems, I became more and more aware of my inability to deliver specific outcomes. I became more aware of needing to invite others into the work of change – leading it themselves. Practice is very important – we need to practice.

I also became more aware of needing to hold my own solutions or vision for outcomes, very lightly. I began to recognize that I could 'pervert' the outcomes and diminish the energy that people brought to the work by being too focussed on what I wanted. I began to experience what Elisabet Sahtouris taught us about a mature sustainable eco-system, 'self-interest is negotiated, which leads to compromises and cooperation or intelligent dynamic harmony' (Sahtouris, 2005). These compromises included my need to compromise. Why would I do that? Because my self-interest is dependent upon the health of the whole – Covid 19 has surely taught us all that now. For me, caring for the health of the whole – human and more than human is the logical outcome of a systemic perspective.

A systemic perspective is also a window into the spiritual for me. Recognizing the layers of interconnectedness and inter-being. Caring is another word for loving. Viewing the world, the people, myself, through a lens of loving.

I am reminded of my friend Alexander Caillet, a wonderful coach living in Boston, who shared a fabulous question with coaches in 2003 at an ICF conference in Sydney 'what would love do now?'. Maybe, the extension of our own pathways to strengthen our eco-coaching, can be aided by the practice of asking this simple question: What would love do now to optimize the health of the whole?'

This short story is just part of my story – you have one of your own. A pathway or a dance – we are all in this process of life.

CONVENER: Thank you, Josie and Peter. Our invitation to you, over our lunch break, is to spend time in reflection, to create your story of moving from ego to eco-coaching and then if you want to share it with a colleague, noticing how your pathway from ego to eco is already alive in you. Appreciate your stories and your pathways because it expresses itself differently in each one of us.

Chapter 17

Resourcing Ourselves for the Future Journey and Farewell

Final Day Afternoon

CONVENER: Welcome back. I hope you enjoyed exchanging stories of your journey from Ego to Eco Coaching.

Another crucial question that a few of you mentioned in the break and reflects what many said in our research questionnaire is 'How do we support ourselves for the journey ahead and the great work we all need to do?' It is that we will now turn and ask Peter to talk about.

PETER: Resourcing ourselves sustainably and working from 'Source'. We can only do this important work in and for the wider world if we are ourselves are sustainably resourced – if we are working from renewable energy and if we are working from source rather than from effort. (Hawkins, 2019a)

Exercise

So let us take off our shoes and socks and all stand up and form a circle with some movement space between us. Put your feet about shoulder width apart and relax your knees, so you have a little bounce in your body. Feel you connection with the Earth beneath you. Imagine you have roots that go through the floor and any floors beneath that, deep down into the Earth.

Cup your hands and imagine you are drawing energy up from the Earth, like fresh water from a very deep well. Bring this energy up to your heart. Still breathing in, allow your hands to reverse with fingertips touching, as you push that energy up from your heart above your head, until you are touching the sky as high above your head as you can.

Breathe out, through your mouth, allowing your hands and arms to float down, reaching out either side of you.

Repeat this practice two more times, and now, as you breathe out, think of something that makes you smile or makes your heart sing.

Working in pairs, decide who is A and who is B. If you cannot find a partner to do this with, you can still do this by yourself and possibly even record yourself doing it.

DOI: 10.4324/9781003153825-17

For the pairs can A now spend 2 minutes constantly repeating and completing the sentence:

'When I work from effort I......'

As you do this take up the posture, stance, facial look and voice quality that characterizes how you are when working from effort.

*

Can A now spend 2 minutes constantly repeating and completing the sentence:

'When I work from source I......'

As you do this take up the posture, stance, facial look and voice quality that characterizes how you are when working from source.

Now A now spend 2 minutes constantly repeating and completing the sentence:

'I could work more from source by.........'

As you do this increase the breathing, posture, stance, facial look and voice quality that characterizes how you are when working from source.

Thank you. Now swap over and B your time starts now. Please complete each sentence.

*

Let us hear back from you what you discovered about working from source:

'I felt so much more alive.'
'I noticed I breathed deep and fuller.'
'I started to smile and as I did so, my eyes opened wider.'
'I uncrossed by legs and could feel the connection with the floor.'
'My whole heart area opened.'

'It really helped when my partner kept repeating the start of the seed sentences in such a non-judgemental way.'

'I felt really supported by my partner – as I changed, I could see them changing.'

PETER: Thank you. We have started to look at being resourced from within, through our breath, our connection with the Earth beneath us and the sky above us.

Now I would like to invite you to get a glass of water and a piece of fruit you enjoy.

Please take the glass of water and look at it, smell it, notice what a miracle and blessing it is. There are so many planets in the universe and very few have this life-giving water. Feel the gratitude for this water, that falls as rain, as manna from the heavens.

As we look at the water, Alison is going to read the traditional thanksgiving to water of the indigenous Haudenosaunee people of America (quoted in Kimmerer 2016: 311):

> We give thanks to all the Waters of the world. We are grateful that the waters are still here and doing their duty of sustaining life on Mother Earth. Water is life, quenching our thirst and providing us with strength, making the plants grow and sustaining us all. Let us gather our minds together and with one mind, we send greetings and thanks to the Waters.

Now take a small sip of the water. Hold it in your mouth and savour the taste. The taste carries the history of the journey that water has taken. The rocks and soil and rivers it has flowed through. Taste them. Then let it trickle down the back of your throat and see how far you can sense the flow of that water through your body.

Then take another sip and see if you can follow the flow even further this time.

*

Let us now turn to the fruit. Hold it up to the light, Look at its shape its colour. Feel its texture. Smell its perfume. Notice where it was once joined to its parent tree. Its belly button where once there was an umbilical cord through which it was fed and nurtured. Imagine that tree if you can. The soil it is fed and nurtured by. The wind blowing through its leaves, the rain watering it. Its flowering attracting the bees and insects to come and pollinate it. The pollinated fruit forming from the flowers.

Try opening the fruit and looking and feeling the different qualities of its inside. Find its seeds, waiting to be carried by wind, or birds, or animals to create offspring of the tree. Hold those seeds and feel the future potential lying patiently and expectantly within.

Now look back at the fruit and see if you can imagine those that shopped for this fruit, the shop keepers who arranged it for sale. Those that transported it to the shops. Those that packed it for shipping or transportation. Then the fruit pickers with their baskets working in the sun and rain. The farmers who planted and pruned and tended the trees. The bees and insects that pollinated it. The rain and sun that nurtured it. The soil it grew in. The trillions of small microbial beings and the fungi that created the soil. The stars that died that released the carbon that formed the biosphere of our living Earth. All that contributed to the life that produced this fruit and brought it here into your hands.

Find how your heart wants to respond to this awareness.

Now taste the fruit, just one small morsel. Hold it in your mouth and savour its unique taste and smell.

Slowly chew it and feel how it changes as you break it down. Again, as it moves down through your body see how far you can still sense it on its journey.

*

I am reminded of the lovely line by the spiritual teacher Ram Dass, who spent many years studying and training in India and who said: 'we are all walking each other back home'. Home is not a fixed place in time, but the place where we feel connected to a deeper source, and where we feel greatly resourced, by all that we are part of. So let us practice that now.

Let us start by completing the following seed sentences:

Home for me is where…….
What I need to connect with more in my life is…
What is disconnected in the world that I can help connect is…
The biggest contribution I am being called to make in the world is…
What in nature can most help me, teach me and support me in doing this work is…

How I can bring this into my practice as a coach, mentor, supervisor, trainer, leader is…

Now take what you have written and find a partner and together go for a walk, in nature. If you are reading this on your own, perhaps you can arrange a phone call with a colleague to undertake this exercise together? Adjust the time accordingly.

Each person will have an hour to work through their emerging answers. The other will become the coach and will help them take each of their emerging sentences and experiment with taking these emerging responses, into connection with the wider world around them. Try rehearsing how they will enact this in their lives, right now.

Here is an example. I have written: 'Home for me is where I garden.' So, my coach would ask me how would you garden right here, wherever we are on the walk? How will your gardening change when you get back home? Don't tell me, show me.

Then they might ask: 'Show me how you will start your coaching and supervision sessions differently as a gardener – imagine you are about to coach me right now.'

I have written I need to connect more with singing and music. So, my coach might say – what are you being called to sing right now? Let the singing emerge through you and share it with me and the Earth around us.

Let me assure you and warn you, I am not a good singer! But I hope these examples give you a flavour of how we can move the coaching from the cognitive and verbal 'talking about things', into embodied expression, and doing in the moment.

Have fun experimenting and trying out new ways of being.

Importantly we do not know all the answers now. As we have experienced, we need humility as a guiding quality. Humility opens us to the not knowing. Humility opens us to the notion of listening deeply to each other and Earth. Humility opens us to working closely with others – human and more than human - in genuine collaboration. Humility is needed to learn our way forward into a new, old way, of living and working on this planet.

*

CONVENER: Welcome back from your explorations together, I hope they brought some of your intentions into embodied lived reality in an enjoyable and playful way. Josie is now going to help all of us explore: **What needs more resources and what no longer serves us.**

JOSIE: Before we close, it would be remiss of us not to consider those things we do or think, those things we have already 'birthed' that need more attention from us to mature, along with those things that are no longer serving us and the health of the whole. It is easy to add more and more things to do to our to do lists without pruning them! And our lists become so long that we don't have the space for the new to birth, develop and grow. Treat the outcomes of this exercise as an experiment; try the changes you reveal to yourself and see what happens.

This exercise is different from, but influenced by, Liberating Structures (Lipmanowicz and McCandless, 2014) and the Three Horizons (Sharpe et al., 2016).

As we work through this exercise, I am going to use the pronouns you and we together. You may be called to do something, and so we are all called. There is no 'you' in an interbeing world, so I am trying to reflect that – perhaps in a clumsy way – consider it my/our experiment.

I invite you/us to begin with a blank piece of paper and fold it (or draw lines on it) to divide it into three columns. Within the context of your/our emerging vision story with Alison and Peter, please consider each question, one column at a time.

Column 1: The activities you/we undertake that sap your/our energy

The life-affirming world is full of life affirming energy. So let's begin to make some space by identifying all the activities you/we do that sap your/our energy. Maybe stand up for a moment and assume the physical stance of being drained of your/our energy. What does it look like and feel like? What activities contribute to this state?

Having listed them, decide if someone else (who is not drained by them) can do them for you/us or if you/we can just drop them. If you/we have to do these

Table 17.1 Activities designing our way forward

Activities that sap energy	Activities that are old patterns that do not serve the whole	Seeds of the future

activities, how can you/we time them so that they are bookended by activities that energize you/us, and so do not deplete you/us so much? Maybe some of these activities are merely a distraction from what you/we are being called to do now?

Column 2: The activities that you/we undertake that represent patterns of the past that are not going to serve the health of the whole into the future

Make a list and please do not overwhelm your/ourself! Remember, that because everything is interdependent, changing just one thing will influence changes in many places. In a living system, it is also true that some changes have greater influence than others. Remember the famous 'butterfly effect'; a butterfly flaps its wings on one side of the planet and causes a hurricane on the other side of the planet (or not because it is unpredictable). So small changes can have big effects.

Once you have made your list, choose just one thing that you/we might let go of either now or within the next two weeks. Choose the thing that you/we intuitively believe will give you the 'biggest bang for your buck'. Or to word it more formally, feel into, and choose the item that you believe represents the highest degree of leverage for you.

Column 3: What are you/we already doing that represents the seeds of the new ways of being you/us have visualized here today?

Often the future already exists in the present, like a tendril from the future or a seed in the present. What experiments or innovations are you/we already employing?

Having listed them, in what way might these, or even just one, require more resources of energy, time or attention from you/us?

This is a process that you/we could employ on a regular, say quarterly, basis. Gradually altering the patterns that exist in your/our life and work - but always connect to, and emerging from, your/our vision of the patterning that is required for future generations.

CONVENER: Thank you/us.

It is time to end our journey together, to pack our bags, both physically and metaphorically, with all the riches we have received. With Josie's help we have chosen: what we need to leave behind and compost from the past; what we need to hold on to; and what new learnings and seeds of new ways of being we will carry in our bodies and our being.

Eve: So here we are, at the end of our workshop. What a journey it has been. I can feel my sadness at this part of the journey ending, and yet my joy knowing that we will still be walking this path together, each imperfectly, but each determined to make a difference. On Day 1 in our Welcome letter, we shared the poem by Drew Dellinger, Hieroglyphic Stairway. The words 'What did you do, once you knew' now come to me in dreams, going for a walk, or sitting by my PC, texting on my phone. I can't 'unknow'. That sits with me, drives me, keeps me alert. I wonder how you are feeling now, as you stayed with us, reading and joining in the many exercises. We thank you so much for your time and your contributions.

Let's acknowledge that some of us may be very knowledgeable about the impact we are having on our shared home, may have the science at our fingertips, while others of us may worry that we don't 'know enough' or have concern about not overstepping the mark between reflecting what is there – our interconnected world – and dictating an agenda. To me what really matters is that we take care of each other, and that embraces all beings, not just human beings, as we have seen together how inextricably we are all intertwined. As Sanchez says we can be a life-giving connection to others:

> We have the capacity and the opportunity, by virtue of simply being alive, to be conscious men and conscious women, aware that we are a precious part of a larger whole. We can choose to become aware, awake and curious, to act as participants in the development of ourselves and in support of the harmonious development of our communities.
>
> (Sanchez, 2017: 25)

In the end I think this book has been about love, for each other, for animals, plants, rivers, seas, the mountains, for the human and 'more than human' world that makes up our extraordinary planet. Through love we know we can achieve anything. So, we echo the role that Joanna Macy and Chris Johnstone invite us to play as participants in Active Hope (2012), turning to our hearts and therefore our actions to contribute to the healing needed to sustain our beautiful world through the profession we all embrace: coaching, supervision, mentoring, leadership. Our work is both an intimate dialogue between two or more people and an expression of the wider, systemic world that we are part of.

Please do reach out to us and share your thinking, your feelings, and your ideas. And we are going to end with each of the Conveners sharing a wish for this community and a wish for the world.

ALISON: That each of us can reach out, gain that sense of deep interconnectedness with our fellow humans and the more than human world. That we can allow our vulnerability and 'not knowing' a place. That we stop looking to fix what's 'out there', and see the joy, abundance and possibility of living from our own hearts, creating space within to hold ourselves, each other and our communities as we travel forward together. That we wake up to all that is required of us to heal ourselves and our world with love and compassion.

EVE: The deepest gift we can offer is our love and acceptance, our ability to listen, truly hear and to avoid, as much as we possibly can, judgement; to understand that we are full of assumptions, from our culture, our backgrounds, our religions, our social class and so on; to bring humility; to dialogue; to bring other perspectives. My wish is that we simply open ourselves up to all the possibilities and at times responsibilities, this brings. For us to thrive (and for me us is all beings in the world) we need to think generations into the future and let our wishes for those generations – whether animals, plants, insects, humans or other beings - guide our actions now.

JOSIE: My wish for this community is that we remain connected as we collectively learn how to expand and reshape coaching and more. Coaching as a way of working with people has such an important role at this moment in history. Our interactions with individuals, teams, groups, organizations and communities, have the power to liberate our innate creativity and imagination to unleash new possibilities. Creative experiments to learn what may work, are our birth right as an expression of nature and Earth. May your energy be a positive creative force for the benefit of future generations.

PETER: That as humans we all help each other to wake up to what we have done and what we are doing to this beautiful planet that gives us so much. That in all our work, whatever form it takes, we focus on healing the destructive split between the human and 'more than human' world and learn the humility to listen deeply to what life and the wider world are teaching us. Blessings be on all you do, and be, in the next phase of your journey.

CONVENER: Fare well, go well and make well.
   All will be well, and all manner of things will be well.
   Thank you for joining us on this journey and thank you for all you will help happen to heal our world, on your future journey.

Glossary

Anthropocene   The National Geographic Society defines *The Anthropocene Epoch* as "an unofficial unit of geologic time, used to describe the most recent period in Earth's history when human activity started to have a significant impact on the planet's climate and eco-systems" (online, 2022). While officially we are in the Holocene, which began a relatively short time ago after the last major ice age, increasingly there is talk of the *Anthropocene*, with many suggesting it started around 1950. The common factor is seen as the impact of human activity on carbon and methane in Earth's atmosphere. The *Anthropocene* could be the sixth mass extinction our planet has witnessed and the first to be brought about by just one species, homo sapiens. What we do may influence its course through our choices going forward, while understanding that with living systems we will not have the final say.

Co-evolution   This is the understanding from scientists such as Bateson (1972), Lyn Margulis (1998), and Freya Mathews (2019) that all evolution is co-evolution. That we do not just adapt to our environmental niches, we are in a dynamic relationship with them, which means that as they change us, we also change them. This dynamic is related to the ideas of inter-being, inter-dependence and is relevant to living systems (also known as complex adaptive systems).

Deep ecology   *Deep ecology* was the term coined by Arne Naess (1987) to distinguish it from 'shallow ecology' which was a study of our eco-systems as if they were separate from ourselves, displaying a human-centric concern focused on how nature can be best sustained for the benefit of *Homo Sapiens*. Deep ecology recognizes the intrinsic value of all living beings and views humans as just one particular strand in the web of life (Capra and Luisi, 2016: 12). Deep ecology argues that the only way homo sapiens can reverse the trajectory from being the destroyers of sustainable ecology to becoming positive contributors to ecological health is by embracing a fundamental shift in consciousness - from seeing nature as

DOI: 10.4324/9781003153825-18

outside ourselves, and the environment as something that surrounds us - to realizing that nature and the environment are also part of us, and that we are an indivisible part of the wider ecosystem (Hawkins and Turner 2020: 122).

Degrowth   The study of how we can decrease human consumption, particularly from the *Global North* (see below), to lessen the unsustainable burden on the Earth's regenerative capacity. It is estimated that those in the *Global North* need to reduce their consumption in all aspects of their lives by 50 percent. This has huge implications for the economic system as we have come to know it. The circular economy is one way of decoupling economic growth from the ever-increasing use resources. *Degrowth* is also a mindset shift away from the addiction of the white western world to continuous economic growth and the false belief that more is better (Hawkins and Ryde, 2020: 46–50).

Eco-literacy   Ecological literacy is a term that was first coined by David Orr and furthered by Fritjof Capra during the 1990s and is the capability to understand the behaviour of natural systems and act in accordance with that understanding. Their belief is that eco-literacy will be fundamental to forming sustainable societies.

Ecology   *Ecology* is a term used for the discipline that studies living organisms and how they interact with each other and the environment in which they exist. The term has been extended to other types of systems too, for example social ecology, where the living systems are people, and the study is of their relationships with each other and the institutions they live and work within. At the heart of understanding an ecology is the focus upon relationships.

Eco-systemic awareness   Drawing on both *systemic awareness* and *deep ecology* (see above), *eco-systemic awareness* recognizes that:

> The environment is not something that is external to the human species, it is within us. The rain that falls on the earth flows through my body, the radiation of the sun that warms the earth, penetrates my skin and changes the chemical balance within me, the air of my ecological niche is in constant interchange with the air within me with every breath I take. There is a constant flow between me as a system and the world in which I live and breathe and the boundary between the two is an artifice, a conceptual boundary necessary for our thinking, but that is never stable of fixed in nature. This is the principle of co-abiding.
>
> (Hawkins, 2015b)

Eco-systemic literacy   This is the capability to combine eco-literacy (see above) with systemic awareness (see below), which creates an understanding of how ecologies contain, and are contained within, nested ecological systems. Thus, every ecology is impacted by the ecological subsystems within it and by the wider ecologies it is nested within.

Global North and Global South	The concept of *Global North* and *Global South* (or *North–South divide*) is used to describe a grouping of countries along socioeconomic and political characteristics. The *Global South* is a term often used to identify lower-income countries on one side of the so-called divide, the other side being the countries of the *Global North* with richer countries who industrialized earlier and built their wealth on exploitation of both poorer countries and of the ecology. The term does not refer to a geographical south hemisphere, as much of the *Global South* is geographically within the Northern Hemisphere.
Greenhouse effect, climate change and climate crisis	In 1824 a French scientist, Joseph Fourier, revealed the phenomena of what he termed the greenhouse effect. His work was built upon in 1896 by Swedish chemist, Svante Arrhenius, who determined that the burning of coal was implicated in the increasing concentration of carbon dioxide as a gas that contributed to the greenhouse effect – the way in which the atmosphere altered to release less heat from Earth to escape into space. At the time, this discovery was seen as a good thing as it was associated with saving Earth from future ice ages and with increases in crop yields. (www.historyextra.com/period/modern/climate-change-warnings-history). It has taken many years of scientific research to confirm that the greenhouse effect, or climate change as it subsequently became known, is the result of human activity. Today, the United Nations describes climate change as the defining crisis of our time, happening even more quickly than we feared. Its devastating consequences are felt globally in different ways in different locations: rising temperatures fuelling environmental degradation, natural disasters from floods to droughts, weather extremes, food and water insecurity, economic disruption, conflict, and terrorism. And accompanying this, sea levels are rising, the Arctic is melting, coral reefs are dying, oceans are acidifying, and forests are burning. As the amount of carbon in the atmosphere increases and temperatures rise, more parts of the world are being exposed to increasing frequency of extreme weather events that may also produce mass migration of people as they realize that they can no longer live in their homelands.
Greenwashing	*Greenwashing* is a term that was coined in the 1980s. It occurs when an organization, usually a business, deliberately and deceitfully misinforms its customers about the environmental and social costs of its products. Deceitful marketing spin and even entire marketing campaigns are employed to assuage the public that the business is sincere in its attempts to ensure positive sustainability impacts. It aims to cover-up and move attention away from the ecological damage and costs their work, practices, supply chain and products

are externalizing. An external cost is one that is imposed on others or the natural environment as a result of a transaction that they are not directly involved in, and not taken into direct account in the organization's profit and loss statement.

Inter-being This concept has roots in many spiritual traditions, indigenous religions and philosophies such as Baruch Spinoza's notion of 'ethic of interconnectedness'. It can be understood through Bateson's (1972) writing about 'the seamless web', Thich Nhat Hanh, the Vietnamese Buddhist monk, writing on 'inter-being', or Martin Luther King's teaching of how we are all part of an "inescapable network of mutuality, tied in a single garment of destiny". (King, 1964)

Kincentric ecology This term was first used by Enrique Salmon (2020: 1332) who wrote: "Indigenous people view themselves and nature as part of an extended ecological family that shares ancestry and origins. It is an awareness that life in any environment is viable only when humans view that life surrounding them as kin. The kin or relatives, include all the natural elements of an eco-system". This view resonates with a deep ecology perspective.

Newtonian Atomistic thinking The modern scientific paradigm, while having it roots in scientists like Galileo and Francis Bacon, is often seen as originating with Sir Isaac Newton, who promoted the scientific method as we know it today. It is a process that privileges objectivity and quantitative measures. This worldview also uses what is known as a reductionist approach to understanding the functioning of the world and universe, through examination of smaller and smaller parts. It looks at life as being constructed from separate building blocks and studies parts rather than relationships between the parts. Where relationships are examined, it assumes a direct and liner causation.

Sustainable Development Goals The eight Millennium Development Goals concluded in 2016 and were superseded by the 2030 Agenda for Sustainable Development that was adopted by all United Nations Member States in 2015. The agenda provides a shared blueprint for peace and prosperity for people and the planet, now and into the future. At its heart are the 17 *Sustainable Development Goals* (SDGs), which are an urgent call for action by all countries, both developed and developing, in a global partnership. The goals recognize the interdependency between different social and physical contributing factors to sustainability and specifically include action to address climate change (SDG 13). For example, the SDGs highlight that ending poverty and other deprivations, improving health and decreasing social injustice are as important as more direct actions to reduce carbon emissions, deforestation, and ocean acidity. (https://sdgs.un.org/goals)

Systemic awareness   This moves beyond systems thinking to recognize first that we cannot observe a system objectively as we affect and are affected by any system we engage with. An important distinction is that the observer perceives herself to be a part of the system. Secondly, *systemic awareness* recognizes that every system has different systems within it and is nested in many levels of larger systems. All these nested systems are in constant dynamic relationship with each other, resulting in a constantly changing and moving world. Also, the dynamics of the larger systemic levels are nested in all the smaller systemic levels nested within it. A human being is part of a family, community, team, organization, culture, species, Earth, solar system, universe, and all these flow through and live in and through the individual human. Thus, we need to be aware of the dynamic dance up and down the systemic levels.

Systems thinking   Some trace the origins of *systems thinking* all the way back to Heraclitus who Ison (2008: 144) quotes as observing that 'You cannot step into the same river twice, for fresh waters are ever flowing in upon you'. Throughout the 20th century a growing number of disciplines began to experience difficulties with the reductionist approach because they were losing sight of the whole. Scientists from disciplines, including biology, mathematics, physiology, economics, philosophy, and sociology, began to recognize common patterns and so an inter-disciplinary conversation began at the Macy conferences in the 1940s and 1950s, and developed into the General Systems Theory project in the 1960s. Systems thinking is, however, a broad church. There are different types of systems, for example, simple, complicated, and complex and a distinction can made between 'hard' or 'soft' systems. Different types of systems behave differently. Most commonly, systems thinking is assumed to be associated with simple systems (e.g., a temperature thermostat in your home or office) which are mechanical (hard) and goal seeking. These types of systems have control engineered into them. Importantly a common perspective in systems thinking is often that of objectifying the system as something separate from the observer.

White Western Privilege   This refers to countries ruled by people and ideas that came from Europe and the European diaspora. Besides white people who live in Western Europe, it also includes the white people who live in the USA, Canada, Australia, New Zealand, South Africa, South America, and other regions and countries. It recognizes that these people have and still do benefit from their ethnicity, owing to colonization, slavery, industrialization, and over-extraction of the ecological wealth of the world. (Ryde, 2019)

Appendix: Climate Coaching Alliance

The why and the what

The Climate Coaching Alliance (CCA) is variously described as an enterprise, organization, community, learning laboratory and movement. Whatever it is, it is not a traditional hierarchy. It has been deliberately nurtured since its inception in November 2019 employing an understanding of the behaviours of living systems. These understandings form the principles for the development and operation of CCA. As co-founders, we have always been clear that although we would provide some structure and co-ordination, it would be as little as possible. The CCA also provides all services free of any monetary consideration and as inclusively as we can.

The central DNA for CCA was seeded from two inaugural meetings of a small group of senior coaches in December 2019. We shared our concern about the climate and ecological crisis and the fact that coaching as a profession did not seem to be responding, with enough urgency. We asked the participants what their vision for a responsive coaching profession might be and as they shared, one by one, we all felt the energy for change rising (McLean, 2017). As the alliance has developed, we have consistently responded to requests for direction with questions and requests to act from the member's own intrinsic motivation, to follow the energy. The mental models of the traditional hierarchical structures are difficult to overcome but the participative nature of our culture is strong and growing.

A powerful example of the power of emergent intrinsic motivation have been four special events – global 24-hour conversations, March and October 2020, and Festivals in March 2021 and 2022. Initially in support of Coaching at Work's Climate Coaching Action Day, which started in March 2020 and are now annual, ours have grown, so that in March 2021 there were 100 conversations as members offered their own contributions to the day and connected with each other internationally to deliver workshops together. It also highlights the generosity of so many CCA volunteers. Since then there has been a proliferation of communities being set up by members, by language, geography, and interest. It was a clear example of the inherent creativity within all nature, including we humans, and a demonstration of the whole being greater than the sum of the parts. And in March

2022 the CCA 6-day Global Festival again had some 100 events worldwide. Membership stood (as at August 2022) at more than 2,000.

Emergence is an amazing and surprising quality of a living system, representing the collective response of a system to changes within its environment or within itself as a living system with boundaries. Systemic change emerges from the system, and emergence poses what is perhaps THE challenge for all leaders now. Emergence does not thrive within conditions of control which is the dominant management style in traditional hierarchies. Within the CCA we are also learning in an action research style, about how to 'lead' organizations founded on the principles of living systems.

As we continue to develop and nurture, rather than build the CCA, we are conscious of working with members' energy, liberating it as much as possible. To do this, we engage as often as possible. For example, although a core co-ordinating group developed a strategic update late in 2021, it was also worked on at monthly members' meetings. Our purpose, operating principles and values are the DNA from which our movement grows and breathes. As the CCA pursues its purpose, we hope to have influenced enough coaches and professional bodies globally by 2025 to no longer be necessary within the system.

The CCA is open to all and welcomes everyone, wherever members are on their journey or their background – coaches, supervisors, and psychologists are joined by scientists and sustainability practitioners to explore change. We are all learning together. Find out more and join us at www.climatecoachingalliance.org. Membership and attendance at events will remain free so that income is not a barrier.

The CCA's communities

We also wanted to share the CCA's breadth and depth by reflecting views from some of our communities worldwide (www.climatecoachingalliance.org/local-communities). So, we invited them to consider these four questions (see Chapter 14). The responses vary from one voice to multiple voices, and some done in dialogue. The questions were:

1 Standing where you are, how do you see/how are you experiencing the challenges already facing your communities?
2 How does this affect/inform your work as a coach?
3 What do you think the most important challenge is for our profession in attending to this?
4 What is the one encouragement you would like to make to readers?

CCA Asia (India community) Rashmi Shetty and Jaya Bhateja

I The challenges already facing our communities

The challenges through the Covid 19 pandemic are immense, with uncertainties in the market, the fear is palpable. The risk in expansions in the different sectors are apparent. While trust was the first factor that was affected, the pressure points of the leadership were retention and loyalty. This increased pressure hurts the mental health and well-being of the leaders. However, the challenges have a flip side to it too. Online provider communities and home entrepreneurs have thrived. Many women have quit corporate careers to follow their passion in cooking. Practical start-ups are slowly but surely increasing! However, the planet /climate is still not a stakeholder in the decision making.

In the past five years, climate change concerns are inviting attention from all corners of the globe. While at one level there is a lot of damage done, I also see a lot of awareness, action, initiative, concern and policy change taking place. Be it ground, water or air, there is awareness, interest and initiatives to deal with the huge challenges. We are quite in trouble with pollution levels all-time high in northern part of India affecting children, pregnant women and elders more adversely.

A few years back I saw an 'oxygen parlour' in Greece, and I laughed, saying 'what's the point of having an oxygen parlour?', because the idea was so alien to me. As the pandemic has continued with polluted air, fresh and clean oxygen has become a luxury that even money cannot buy.

Covid 19 has impacted progress towards the 2030 Agenda for Sustainable Development in Asia and the Pacific region. This includes the WHO's SDG goal 13 on taking urgent action to combat climate change and its impacts.

2 How this affects/informs our work as coaches

Coaching managers are facing a difficult time with attrition. With budgets in companies slashing, a few preferred personal coaching while many others did not want it. Mental well-being became a prime concern. The second Covid 19 wave put a lot of personal pressures too on many leaders.

This helps me connect the dots systemically and informs my curiosity and awareness about my client. I feel responsible to create awareness about climate change for my clients during my coaching engagements. I am constantly thinking of tools, perspectives, and resources to be integrated or introduced to clients.

Stress levels contributed by pandemic, lack of social interaction/movement / long duration of working from home is impacting everyone's well-being. In such a situation, I feel personal responsibility to take extra care of my health and well-being to be able to combat the situation for myself and my clients. Clients are looking for more support than ever and each time meditation, music and breathing exercises bring a lot of relief to them.

3. The most important challenges for our profession

- The most important challenge for our profession attending to this is to integrate personal development with climate change by making climate change a priority conversation in boardrooms.
- Connecting climate change with personal responsibility and accountability.

- The reduction in assignments had forced many coaches to drastically cut their charges which affected the market.
- Many others not credentialed, masked mentoring as coaching and gave away advice rather than helped the clients with an inner reflection.

The last two years established coaching as a profession with the number of coaches increasing. However, what should have helped in the visibility of coaches ironically confused the market.

4 An encouragement to others

The climate is now a definite stakeholder in business decisions. Every leader/ individual HAS a role to play in the way the priority is given to the planet. Climate crisis is real. The extremes of temperatures and the loss as a result of floods and natural disasters are proof. If we do not wake up now, we will be out of our grace period. We need to consciously live a life that has each one coexisting with Nature. As humans we are the only ones in the pyramid who are irresponsible. However, the last two years has shown us who is the BOSS!

I would also like to say that what we are doing today has an impact for our generations and hence choosing our actions are important to create a safe environment. Let's try changing what we are struggling through today and leave a better world behind. The first step to every change is awareness and second is action. Taking personal responsibility and accountability for controllable actions can help us make a drop of a difference and if everyone tries a bit from their side, things can change. With country, global, local, and community level initiatives we can save some damage. Start today, start small.

Look for opportunities to contribute at your workplace and home, conduct sessions for younger generations, write about possible solutions are some small and manageable actions; something is better than nothing.

CCA Canada from Janet Mrenica, co-convenor of the CCA-Canada monthly Circle

1. The challenges already facing our communities

Maintaining a continued presence by members during a revolving pandemic when there are what is perceived to be increasing pressures in our lives has negative impacts.

2. How this affects/informs our work as coaches

This has informed the work of CCA Canada member coaches to prioritize learnings around grief, loss, and sorrow, build their competency in this regard, and to come to meetings where demo coaching is found.

3. The most important challenges for our profession

To dive into one's personal emotions and do the work required to hold space for climate anxiety, fear, sorrow, and loss. Without this work, we are not prepared to meet the wall of climate anxiety that is in front of us.

4. An encouragement to others

Clients come to us with their topic, their inquiry, their story. If coaching as a profession takes the world view that all is related, as coaches we can bring the world lenses to the attention of our client.

This supports and promotes their wellbeing. There are tools available for coaches to support clients to take a world system view in working through their topic. Ask for this suite of knowledge!

CCA French-speaking: Andra Morosi and Roselyne Lécuyer in dialogue captured by Fiona Chandler

In our roles as co-leaders of the CCA French speaking community, we agree on three key challenges that perhaps might resonate with others.

One is acknowledging that there is a great willingness of our members to be active and share ideas and initiate actions. Yet the level of engagement is varied – some can offer more time and others less – the challenge is to capture a more balanced and sustainable involvement. The voluntary nature of working with CCA FR, and our true passion for playing our part in the field of climate action can pull us away from our coaching livelihoods and might lead to a potential neglect of other aspects of our roles as entrepreneurs – loss of income or in some cases, putting off aspects of our personal lives. Besides, passion for a cause can sometimes have a blinding effect and move us slightly astray from the right posture and detachment inherent to our profession. We recognize the challenge to grow this voluntary community, and with that can be clear on the expectations. It helps to remember to be explicit in how we engage and acknowledge the give and take in our work with the CCA community – the ethics of engagement as we might call it.

As coaches, being a part of CCA FR has informed us in different ways. We have gained a greater knowledge and understanding of the systemic nature of climate change and sustainability. This has allowed us to add a dimension to our coaching and our professional contribution in building an environmentally sustainable society. However, there are times as a coach, when we are engaged with clients who might be in 'climate denial'. It is sometimes a difficult line to walk and raises the issue of our legitimacy of this in our coaching role.

In overcoming this, the words that come to mind are congruence and alignment. This is a need to tolerate and be at peace with this rather than taking sides and feeling that there is something that we need to bring from our value system to 'waken' the client. If the topic emerges, it is not to be avoided, and based on our clients' response, we can decide on whether to explore it further. We need to ensure a level of neutrality and not allow our judgements to get in the way of our coaching.

Finally, some encouraging thoughts to share. First, is the value in staying in balance with our role as a coach. There is no doubt that the knowledge of the issues that we share in CCA FR adds legitimacy to our coaching relationships, as and when it is appropriate. Second, we have a great sense of the sharing that goes on in CCA FR – we are not alone in our challenges. We feel less lonely and know there are others in our community facing these same dilemmas. We are stronger thanks to our sense of belonging for being part of this community. What we get is what we can offer and share.

CCA Japan – via co-leader Aya Usui

Table 1

Questions	Cory McGowan	Adrian Gen Tsukamoto	Tensei Yoshida	Aya Usui
1. The challenges already facing our communities	In the rural Japan town I live in, the tourism industry is being hit very hard. From extreme weather events to instability caused by the virus, it is making what was already a challenging industry very challenging to thrive in. This has the further impact of forcing the community to reconsider whether having its main identity as a tourism destination is sustainable or not.	There are more people visiting and now living in my community due to higher temperatures experienced during the summer months and more recently due to the pandemic. The demand for home ownership and construction has really increased over the past 2–3 years. Houses being built are utilizing more concrete and removing trees from their properties to make way for water-thirsty lawns and flower beds.	I believe that in Japan, there are very few touch points for citizens to understand the scientific facts about the climate crisis, compared to other developed countries. I believe that this is due to a number of political issues. Therefore, I think the major problem is that citizens' awareness has not been aroused.	In Japan, where I live, the population is aging and declining, but there is still a tendency to cut through mountains, forcibly build new buildings, and construct high-rise condominiums. Citizens' awareness of climate change and sustainability is below the level where action is needed at the individual level. The level of awareness varies from person to person, so I feel it is important to establish some standard level.

Questions	Cory McGowan	Adrian Gen Tsukamoto	Tensei Yoshida	Aya Usui
2. How this affects/ informs our work as coaches	One of my main missions is to bring clients to the local area and to partner with local service providers so that they also get to benefit from the income I get from my clients.	I am experiencing a slight shift in levels of interest towards social and environmental success metrics.	I feel that there is a significant gap between agreement that coaches have a mission to help evoke awareness among people and organizations, and that coaches themselves have a responsibility to commit to the climate crisis first.	As suggested in ICF Core Competency #7 (ICF, 2022) 'Evoke Awareness', coaches need to think about the question of what sustainability means to them, and then ask the same question to the leaders involved to stimulate.
3. The most important challenges for our profession.	No matter how far removed our clients are from where we are working, we have to consider our part in supporting our local community. While it is a great advantage to be able to coach virtually so easily now, we are still an important part of a local community ecosystem.	The traditional short-term indicators of economic success still overshadow the social and environmental metrics of success.	I think it is how we as a coaching industry, in terms of our professional ethics and competencies, position addressing the climate crisis as an essential knowledge and competency for the coaching profession. This means that we need industry-wide consensus and systematic action.	In addressing this issue, I think it is important to respect the values of each coach in Japan, while at the same time building a standard of thinking for the coaching industry to some extent and raising awareness among many leaders.

Questions	Cory McGowan	Adrian Gen Tsukamoto	Tensei Yoshida	Aya Usui
4. An encouragement to others.	Japan is a place of incredible resilience, and this is not unique to the people here, although is a powerful part of the culture. We have the technical capability to be more connected across the globe than we ever have been, so sharing lessons of resilience from places like Japan is one way we can all overcome challenges together.	Market leaders in certain industries are beginning to expand their criteria for success. In doing so, I am beginning to see how they are applying pressures within their value chain to encourage other suppliers to adhere to new standards (i.e., greener logistics).	Let's work together so that we can say to the next generation and the one after that, 'We have done all we can for your generation.'	In the wake of Covid 19, I feel that the world is shifting from the traditional values of winning and taking to the values of 'giving'. I believe that with more humility, love, and respect, the impossible will become possible if people work together worldwide.

CCA Oceania: Heather-Jane Gray

1. The challenges already facing our communities

Physical: bush fires, drought, plagues of mice and locusts, massive storms, plus an increase in tremors, earthquakes, and floods.

Economic: adverse impact on rich-poor divide, social degeneration, impact on farming communities, loss of livelihoods, etc.

Political: polarization, frustration with in-action, increased lobbying… 'disenchantment' with political leaders (- were we ever 'enchanted'?! Is 'political leadership' a non-sequitur?!)

Technological: Increase in the use of solar, wind, electric vehicles, and large-scale innovation in the private sector, impacting sustainability across many industries, etc.

2. How this affects/informs our work as coaches

- 'It makes it easier to have more meaningful and action-oriented conversations about climate change as it is now finally in main-stream media.

- Having more personal/ emotional conversations with people experiencing the overwhelm of the truth/ impact.
- Climate change also coming up more often in coaching supervision.

3. The most important challenges for our profession

There is a need for further education – for instance, our SG Climate Action Team (CAT) have or are still completing a variety of relevant Continuing Professional Development courses.

These give coaches more confidence for integrating their understanding about the human impact – including complexity, systems thinking, agile growth mind-sets, reflection, and sustainability into their practice.

4. An encouragement to others

We are in this together – our CCA community can become a movement in support of the wider large-scale change/movement. Many of us work with leaders of large organizations across the private, public, and not-for-profit sectors and can have a huge impact. It is also worth working with individuals, small and medium-sized enterprises, and local community groups – this grassroots approach is powerful and unifying, especially in the wake of the pandemic. Coaches have the necessary transferable skills to hold the space for essential conversations to take place.

We can then build on 'reality-based optimism' for action wherever possible.

CCA Spanish speaking – Elsa Valdivielso-Martínez (Spain), CCA-Hispano

I. The challenges already facing our communities

In Spain, luckily or not, the current ecological crisis can be more or less hidden with all the Western commodities that have been developed (heating, air conditioning, etc.). However, there are still challenges that cannot pass unseen, such as air pollution, excess of rubbish everywhere, plastics and microplastics, extreme weather conditions, loss of biodiversity and loss of natural spaces.

I experience these challenges with a mix of emotions, almost as if I was grieving the loss of our planetary well-being. At times, I feel sad. I can notice how things have been changing in my community and how it has become very industrialized. Where before there were unpolluted and untouched areas, now it has all been dominated by humans.

At other times, I feel I am part of the problem and I know what I should be doing to reduce, for instance, my carbon dioxide emissions to zero. However, it feels too much of an effort or loss to give up everything known to me thus far. And the reaction I experience when I face these challenges is through anger, frustration, and a sense of powerlessness.

2. How this affects/informs our work as coaches

This whole situation is very present in my coaching philosophy, and it informs my practice, mainly my values. As a coach, I have been recently reminding myself of my values and I have been faced with an ethical dilemma to not get stuck in the business-as-usual loop and to not help maintain the unsustainable status quo.

This situation keeps me grounded and keeps reminding me why I started to coach and, although challenging, it is helping me make decisions to make sure that my coaching has a positive impact. I want to help people transition into a more sustainable way of being and to return to the roots.

3. The most important challenges for our profession

The understanding that a shift in mentality is required. It is currently more of an ethical dilemma or moral responsibility, rather than the actual impact of the crisis. It is believed, by some coaches, that the ecological crisis topic cannot be imposed to our clients, as this is the coach's agenda. However, I do not believe that the current crisis is a political subject or an agenda. It is a reality that is affecting all of us as we *are* nature.

I think that coaching needs to keep redefining itself and developing its ethical code of conduct. Nowadays, coaching feels to be 'everything and nothing'. What is coaching serving? What are our ethical boundaries as coaches?

4. An encouragement to others

I would love to say: you are not alone. We are a lot of coaches sharing the same values and concerns and CCA represents a big part of this community. On many occasions I doubt myself, and I question whether I really want to keep my values so alive and present in my everyday coaching, as at times it can become exhausting. However, in every CCA monthly meeting that I attend, it reminds me to not give up and it reconnects me with my values. This helps me keep going and to be in peace with the planet, others, and myself.

CCA USA: Alexandra Arnold, Andy Miller, Anne-Marie Brest, Christopher Eng, David Drake, Don Maruska, Janet Mrenica, Meredith Bell, Phil Metzler, Tamara Staton, Lennie Noiles and Mayme Doumbia

The community broke into two groups and considered questions 1 and 2 together and 3 and 4 together with overlaps:

1. The challenges already facing our communities

In just the past few years, human communities in the US have already seen a vast range of climate impacts, from unprecedented fires throughout the western US (and the secondary impact of smoke lasting for months during the summer and fall), as well as wild *derecho* windstorms and tornados, flooding, and severe storms throughout the country. The toll on animal and plant communities around us is far from understood, but in addition to daily news of impacts on the land and waters around us, it is easy to see with our own eyes vast swaths of forests burned or struggling from drought and insect infestations.

2 How this affects/informs our work as coaches

Coaches can play an important role in expanding the conversation. There is an urgent need to normalize climate conversations and have more of them in every realm of community, from conversations between parents and their children, to conversations at a national level. As coaches, we can shift clients from their personal agenda to a larger agenda that holds the life of the planet in its centre. Some coaches see their role as helping clients open to new possibilities or see the systems they are connected to; others see their role as setting a 'table' for the conversation and inviting the client, when they are ready, to sit at the table; others see that by helping clients shift themselves to a more sustaining self, so they can expand that perspective to a more sustainable world.

Being authentic and vulnerable, especially with our grief, fear, and other heavy emotions, is essential. Many coaches stressed that it is essential to acknowledge the grief, sorrow and loss that are part of the conversation, in order to move beyond resistance. To encourage clients to see that it is acceptable to have negative emotion around the climate crisis. One coach described this as realizing that people are 'frozen' and helping them to 'thaw'. Others explained that our coaching already involves having the client notice their way of 'being in the world' and 'being in the world' today involves dealing with the climate crisis.

3 and 4 The most important challenges for our profession. And an encouragement to others

And yet, coaches stressed that the conversation actually has potential to be joyful and fun! Feeling helpless does not mean you are helpless. One coach explained, "I always strive to bring my clients to a place of vulnerability because regardless of who they are, this typically results in joy, gratitude and understanding on the other side; vulnerability breeds vulnerability and it always leads to a richer, valuable actionable dialogue". Another coach described, "Again it's about acknowledging and providing a space that it's ok to feel the emotions: grief, anxiety, helplessness, passivity. I strive to help people talk about it and take the first steps in bringing it into their lives and maybe start to do something about it. The conversations don't always have to be about doomsday". For parents speaking with children, likewise, the solution seems to be finding the opportunity and space to be authentic in order to bring forth an open channel to start a genuine dialogue.

We also want to encourage readers to consider the larger structures and systems in their efforts and desires to make a difference towards climate change. There are some fantastic systemic changes that are on the brink of possibility including key ways to bring emissions down. Getting involved with one of those efforts that aligns with your mission and beliefs would be greatly beneficial to both participant and the larger effort and world.

Nature is a powerful guide. Nature is the best teacher: regenerative, cyclical, resilient, and available at any moment. One coach describes, "as a nature-based coach, nature provides or leads the discussion, and it always informs the conversation and dialogue. I live in a place where climate change is at the forefront of our lives; it's right there every day so I let my clients discover and be led by nature".

CCA Yorkshire: Jeremy J Lewis and Jan Brause – Joint Facilitators of the Yorkshire Climate Coaching Café

1. The challenges already facing our communities

- Profit versus sustainability, many organizations are not embracing the fact that becoming more sustainable will ultimately benefit them financially.
- Recognizing the complexity of the climate crisis and the interconnectedness between for example, social justice, diversity, and the climate.
- Senior leaders often not really understanding the need to respond to the climate crisis.

2. How this affects/informs our work as coaches

- Own our responsibility/duty to raise these issues with our clients and combine inquiry with advocacy.
- We can help leaders recognize the value of embracing climate and sustainability issues encouraging our clients to consider the planet and humankind as a stakeholder.
- Hold a space for exploration, being curious together, daring to raise the issue of the climate crisis in our work, calling out what we notice.

3. The most important challenges for our profession

- Creating a more systemic approach to our work as coaches and differentiating between rhetoric and reality by making sure we have good science and reliable data in support of our work.

4. An encouragement to others

- Start small, start local, and encourage dialogue – change starts with conversation and understanding each other's perspective so that we avoid getting into polarization and making others 'wrong'.

References

2 degrees Institute (2022). Global CO2 levels. www.co2levels.org (Accessed 8 August 2022).

ABC Foreign Correspondent (2021). *Dead on arrival*. Available at: https://www.abc.net.au/foreign/dead-on-arrival/13515760 (Accessed 8 August 2022).

Abrams, D. (1996). *The spell of the sensuous*. New York, NY: Random House.

Ahmed, N. (2020). Green economic growth is an article of 'faith' devoid of scientific evidence. Available at: https://medium.com/insurge-intelligence/green-economic-growth-is-an-article-of-faith-devoid-of-scientific-evidence-5e63c4c0bb5e (Accessed 8 August 2022).

Alave, K. (2011). Philippines ranks third on climate change vulnerability list. Available at: https://globalnation.inquirer.net/14987/philippines-ranks-third-on-climate-change-vulnerability-list (Accessed 8 August 2022).

Albrecht, G. (2005). 'Solastalgia' a new concept in health and identity. *PAN (3)*: 41–45. Available at: https://www.academia.edu/21377260/Solastalgia_A_New_Concept_in_Health_and_Identity. (Accessed 8 August 2022)

Allen, K.E. (2019). *Leading from the roots*. New York, NY: Morgan James Publishing.

Amidon, E. (2012). *The open path: Recognizing non-dual awareness*. Boulder, CO: Sentient Publications.

Andrews, N. (2021). *Emotional connection to climate change attitudes in Scotland*, Talk at CCA Political Pod meeting, 4 November 2021.

Angelou, M (2014). 19 inspirational Maya Angelou quotes. Available at: https://www.biography.com/news/maya-angelou-quotes (Accessed 8 August 2022).

Aspey, L. (2022). With the earth in mind. Available at: www.aspey.com/with-the-earth-in-mind (Accessed 8 August 2022).

Bachkirova, T., Jackson, P., and Clutterbuck, D. (2021). *Coaching and mentoring supervision: theory and practice* (eds). Maidenhead: Open University Press.

Banerjee, N., Cuchman Jr, J.H., Hasemyer, D., and Song, L. (2015). Exxon: the road not taken. Available at: https://insideclimatenews.org/content/Exxon-The-Road-Not-Taken (Accessed 8 Augusst 2022).

Baraka (1992). Documentary film. Ron Fricke and Mark Magidson Films. USA.

Berto R.(2014). The role of nature in coping with psycho-physiological stress: a literature review on restorativeness. *Behavioural Sciences* (Basel); *4(4)*: 394–409.

Barnard, A. (2020). New York's $226 billion pension fund is dropping fossil fuel stocks. *New York Times*, 9 December. Available at: www.nytimes.com/2020/12/09/nyregion/new-york-pension-fossil-fuels.html (Accessed 8 August 2022).

Bar-On Y.M., Phillips R. and Milo, R. (June 2018). The biomass distribution on earth. *Proceedings of the National Academy of Sciences of the United States of America (PNAS), 115(25)*: 6506–6511. Available at: https://doi.org/10.1073/pnas.1711842115 (Accessed 8 August 2022).

Barks, C. (2010). *Rumi: The big red book*. London: Harper Collins.

Bateson, G. (1972). *Steps to an ecology of mind*. New York, NY: Ballantine Books.

Bateson, G. (1979). *Mind and nature: A necessary unity*. New York, NY: Dutton.

Bateson, G. (1991). *A sacred unity: further steps to an ecology of mind*. New York, NY: Harper Collins.

Bateson, G. and Bateson, M.C. (1987). *Angel's fear: an investigation into the meaning of the sacred*. London: Rider.

Bateson, N. (2021). *What is systems change?* Regenerative Confluence conference. 11 May 2021. Virtual conference.

Bateson, N. and Hawkins, P. (2016). Working and living systemically a dialogue between Nora Bateson Director of the Bateson Institute Stockholm and Professor Peter Hawkins. (online) Available at: https://www.renewalassociates.co.uk/resources/videos/ (Accessed 8 August 2022).

BBC (2021a). Sea level rises, other changes irreversible, UN warns. 9 August 2021. Available at: www.bbc.co.uk/news/live/science-environment-58142632 (Accessed 8 August 2022.

BBC (2021b). Air pollution: Coroner calls for law change after Ella Adoo-Kissi-Debrah's death. 21 April 2021. Available at: www.bbc.co.uk/news/uk-england-london-56801794 (Accessed 8 August 2022).

Bcorporation (2022). About B Corp certification. Measuring a company's entire social and environmental impact. Available at: www.bcorporation.net/en-us/certification (Accessed 8 August 2022).

Bechard, E. (2021). *Parenting in a changing climate: Tools for cultivating resilience, taking action, and practicing hope in the face of climate change*. New York, NY: Citrine Publishing.

Bendell, J. and Read, R. (Eds) (2021). *Deep adaptation: navigating the realities of climate chaos*. Cambridge: Polity Press.

Berry, J.W. (1997). Immigration, Acculturation, and Adaptation. *Applied Psychology: An International Review (45)*: 5–34.

Berry, T. (2006a). Nature and humans. interview with Caroline Webb. Available at: www.youtube.com/watch?v=pWdo2vpr-Rs (Accessed 8 August 2022).

Berry, T. (2006b). *Evening thoughts: reflecting on earth as sacred community*. San Francisco, CA: Sierra Club Books.

Berry, W. (1983). *Standing by words*. San Francisco, CA: North Point Press.

Berry, W. (2015). *Our only world: ten essays*. San Francisco, CA: Counter Point.

Birch, J. (2022) (Ed). *Coaching supervision groups: resourcing practitioners*. Abingdon: Routledge.

Blake, W. (1950). Auguries of Innocence. *Poets of the English language*. New York, NY: Viking Press.

Borst, A. (2021). *Life at zero distance*. CreateSpace Independent Publishing Platform.

Bradshaw, C. (2021) Underestimating the challenges of avoiding a ghastly future. *Frontiers in conservation science*, 13 January. Available at: www.frontiersin.org/article/10.3389/fcosc.2020.615419 (Accessed 8 August 2022).

Breines, J.G. and Chen, S. (2012). Self-compassion increases self-improvement motivation, *Personality and Social Psychology Bulletin, 38(9)*: 1133–1143.

Brewer, M.B. (1999). The psychology of prejudice: Ingroup love and outgroup hate? *Journal of social issues, 55(3)*: 429–444.

Brick, C. and van der Linden, S. (2018). Yawning at the apocalypse. *Psychologist, 31(9)*, 30–41. doi:10.17863/CAM.33639.

Brock, V.G. (2012). *Sourcebook of coaching history.* CreateSpace Independent Publishing Platform.

Brundtland, B. (1987). *Our common future.* Milton Keynes: Open University Press.

Buhner, S.H. (2002). *The lost language of plants.* Chelsea, VT: Chelsea Green Publishing.

Buhner, S.H. (2014). *Plant intelligence and the imaginal realm.* Rochester, VE: Bear & Company.

Capra, F. and Luisi, P. L. (2016). *The systems view of life: A unifying vision.* Cambridge: Cambridge University Press.

Carbon Footprint (2022). Carbon Calculator. Available at: www.carbonfootprint.com/ca lculator.aspx (Accessed 8 August 2022)

Carroll, M. and Shaw, E. (2013). *Ethical maturity in the helping professions making difficult life and work decisions.* London: Jessica Kingsley.

Castelloe, M.S. (2019). Coming to terms with eco-anxiety. *Psychology Today.* [online] Available at: https://www.psychologytoday.com/gb/blog/the-me-in-we/201801/coming-terms-ecoanxiety (Accessed 8 August 2022).

Carson, R. (1962). *Silent spring.* Boston, MA: Houghton-Mifflin.

Chaskalson, M. and McMordie, M. (2017). *Mindfulness for coaches.* Abingdon: Routledge.

Cheng, S.K.K. (1990). Understanding the culture and behaviour of east Asians - A Confucian perspective. *Australian and New Zealand Journal of Psychiatry, 24(4)*: 510–515. doi:10.3109/00048679009062907.

Chi, X., Kohler, T.A., Lenton, TM, Jens-Christian, S., and Marten, S. (2020). Future of the humans climate niche. *Proceedings of the National Academy of Sciences, 117(21)*.

CILT (2021) *2021 Making the journey to zero carbon. Focus January 2021.* London: CILT.

Clay, J. (2021). Breaking boundaries: The science of our planet. Directed by Jon Clay. [Documentary]. USA. Netflix.

Climate Coaching Alliance (2022). *Welcome to the Climate Coaching Alliance.* Available at: www.climatecoachingalliance.org (Accessed 8 August 2022).

Climate Change Commission (2022). *Climate change impacts.* [online] NICCDIES. Available at: https://niccdies.climate.gov.ph/climate-change-impacts (Accessed 8 August 2022).

Climate Psychology Alliance (2022). *Facing difficult truths.* Available at: www.climatep sychologyalliance.org (Accessed 8 August 2022).

Climatesmart (2022). *Good math makes for great momentum.* Available at: https://clima tesmartbusiness.com/our-data (Accessed 8 August 2022).

Clutterbuck, D. and Megginson, D. (2011). Coach maturity: An emerging concept in Wildflower L. And Brennan, D. (eds). *The handbook of knowledge-based coaching: from theory to practice.* Hoboken, NJ: John Wiley & Sons.

Cohen, R. (2020). *Impact: Reshaping capitalism to drive real change.* London: Ebury Press.

Cohen, Z., Aspey, L., and Whybrow, A. (2019). *An open letter to coaches, the coaching & coaching psychology professional bodies and coach educators.* LinkedIn, 23 July. Available at: www.linkedin.com/pulse/open-letter-coaches-coaching-psychology-professional-bodies-zoe-cohen-1e/ (Accessed 8 August 2022).

Collins, R. (1988). The micro contribution to macro sociology. *Sociological theory, 6(2)*: 242–253.

Commonwealth Club of California (2021). *Regeneration with Paul Hawken*. 3 November. Available at: www.commonwealthclub.org/events/2021-11-03/regeneration-paul-hawken (Accessed 8 August 2022).

Cosier, S. (2019). The world needs topsoil to grow 95% of its food – but it's rapidly disappearing. *The Guardian*, 30 May. Available at: www.theguardian.com/us-news/2019/may/30/topsoil-farming-agriculture-food-toxic-america (Accessed 8 August 2022).

Covey, S. (2020). *The seven habits of highly effective people*. London: Simon & Schuster.

Dasgupta, P. (2021). *The economics of biodiversity: the Dasgupta review*. London: HM Treasury.

David, S., Clutterbuck, D., and Megginson, D. (2013). *Beyond goals – effective strategies for coaching and mentoring*. Farnham: Gower Publishing Limited.

Davidson, J. (2021). #future gen: lessons from a small country, CCA Political Pod Event, 1 September.

de Haan, E. and Gannon, J. (2017). The coaching relationship. In T. Bachkirova, G. Spence, and D. Drake (eds.), *The Sage Handbook of Coaching*. Sage Publications.

d'Henin, L. and Mundle-Garratt, C. (2022). Building relationships with politicians, CCA Political Pod Event, 3 February.

de la Sablonnière, R. (2017). Toward a psychology of social change: A typology of social change. *Frontiers in psychology, 8(397)*. doi:10.3389/fpsyg.2017.00397.

Deep Time Walk (2022). Walk. Experience. Act. Available at: www.deeptimewalk.org (Accessed 22 January 2022).

Deloitte (2020) *The Deloitte Global Millennial Survey 2020*. Available at: www2.deloitte.com/content/dam/Deloitte/global/Documents/About-Deloitte/deloitte-2020-millennial-survey.pdf (Accessed 8 August 2022).

Dellinger, D. and Bioneers (2022). *Hieroglyphic Stairway*. Available at: www.youtube.com/watch?v=fjc4rmJdA3k (Accessed 8 August 2022).

DiGirolamo, J.A. and Tkach, J.T. (2019). An exploration of managers and leaders using coaching skills. *Consulting Psychology Journal: Practice and Research, 71(3)*: 195–218. http://dx.doi.org/10.1037/cpb0000138.

Dryden, H. and Duncan, D. (2021). *Climate regulating ocean plants and animals are being destroyed by toxic chemicals and plastics, accelerating our path towards ocean pH 7.95 in 25 years which will devastate humanity*. Last revised: 11 March 2022. Available at: https://papers.ssrn.com/sol3/papers.cfm?abstract_id=3860950 (Accessed 8 August 2022).

Dunphy, D., Griffiths, A., and Benn, S. (2007). *Organizational change for corporate sustainability: A guide for leaders and change agents of the future*, 2nd ed. Abingdon: Routledge.

Earth Stations (2021). *Earth stations - A resource guide for planning a Deep Time Walk*. Available at: www.deeptimewalk.org/kit (Accessed 8 August 2022).

Edelman (2022). *Edelman trust barometer 2022*. Available at: www.edelman.com/sites/g/files/aatuss191/files/2022-01/2022%20Edelman%20Trust%20Barometer%20FINAL_Jan25.pdf (Accessed 8 August 2022).

Edmondson, A.C. (2019). *The fearless organization*. Hoboken, NJ: Wiley.

Einzig, H. (2017). *The future of coaching: vision, leadership and responsibility in a transforming world*. Abingdon: Routledge.

Elkington, J. (2018). 25 years ago I coined the phrase "Triple Bottom Line". Here's why it's time to rethink it. *Harvard Business Review* [online]. 25 June. Available at: https://hbr.org/

2018/06/25-years-ago-i-coined-the-phrase-triple-bottom-line-heres-why-im-giving-up-on-it (Accessed 8 August 2022).

Emery, M., Hubbell, K., and Miles-Polka, B. (2011). A field guide to community coaching.

Fenton, J. (2019). 'Four's a crowd'? Making sense of neoliberalism, ethical stress, moral courage and resilience. In *Ethics and social welfare, 14(1)*: 6–20. Available at: www.tandfonline.com/doi/full/10.1080/17496535.2019.1675738.

Figueres, C. and Rivett-Carnac, T. (2020). *The future we choose – surviving the climate crisis*. London: Manilla Press.

Fleming, P. and Macy, J. (2007). Guidelines for a council of all beings workshop. In Seed, J. Macy, J., Fleming, P. and Naess, A. *Thinking like a mountain – towards a council of all beings*. Gabriola Island, BC: New Catalyst Books.

Flores, H. (2018). Climate change vulnerability: Philippines ranks 3rd. *The Philippine Star*, 21 March [online]. Available at: www.philstar.com/headlines/2018/03/21/1798866/climate-change-vulnerability-philippines-ranks-3rd (Accessed 8 August 2022).

Forde, C. (2022). *Bright new world: How to make a happy planet*. London: Welbeck Editions.

Fredrickson, B. et al (2008). Open Hearts Build Lives, *Journal of personality and social psychology, 95(5)*: 1045–1062.

Future of Coaching Collaboration (2022). Available at: https://futureofcoaching.org (Accessed 8 August 2022).

Germer, C.K. (2009). *The mindful path to self-compassion*. New York, NY: Guildford Press.

Gift Based Coaching (2022). *Open Hearted Presence*. Available at: https://giftbasedcoaching.org (Accessed 8 August 2022).

Gillespie, E. (2021). *The Omerata of consultancy*. 17 March [online]. Available at: https://edgillespie.medium.com/the-omertà-of-consultancy-bf26b116e3e0 (Accessed 8 August 2022).

Gillespie, S. (2020). *Climate crisis and consciousness – Re-imagining our world and ourselves*. Abingdon: Routledge.

Global Agriculture (2022). *Soil fertility and erosion*. Available at: www.globalagriculture.org/report-topics/soil-fertility-and-erosion.html (Accessed 8 August 2022).

Global Code of Ethics (2021). *The global code of ethics. For coaches, mentors and supervisors*. Available at: www.globalcodeofethics.org (Accessed 8 August 2022).

Global Footprint Network (2022). *Ecological footprint*. Available at: www.footprintnetwork.org/our-work/ecological-footprint (Accessed 8 August 2022).

Global Monitoring Laboratory (2022). *Recent daily average Mauna Loa CO_2*. Available at: https://gml.noaa.gov/ccgg/trends/monthly.html (Accessed 8 August 2022).

Goodall, J. (2017). *Jane Goodall's vision for the future*. [online] 29 June. Available at: www.ecowatch.com/jane-goodall-video-2450327473.html (Accessed 8 August 2022)

Goodchild, M. (2021). Relational systems thinking: That's how change is going to come, from our earth mother. *Journal of awareness-based systems change, 1(1)*: 75–103.

Gore, T. (2021). *Carbon inequality in 2030 –Oxfam report*. [online] 5 November. Available at: www.oxfam.org/en/research/carbon-inequality-2030 (Accessed 8 August 2022).

Gorham C. (2022). Nature as dynamic co-partner in group supervision. In J. Birch (Ed.) *Coaching Supervision Groups - resourcing practitioners*. Abingdon: Routledge.

Grant, A.M. and Cavanagh, M.J. (2004). Toward a profession of coaching: Sixty-five years of progress and challenges for the future. *International Journal of Evidence Based Coaching and Mentoring, 2(1)*: 1–16.

Graves, C. (1974). Human nature prepares for a momentous leap. *The Futurist, April*.

Gray, D.E., Garvey, B., and Lane, D.A. (2016) *A critical introduction to coaching and mentoring*. London: Sage Publications.

Ho, D.Y.F. (1994). Face dynamics: From conceptualization to measurement. In S. Ting-Toomey (Ed.), *The challenge of facework: Cross-cultural and interpersonal issues*. 269–286. Albany, NY: SUNY Press.

Hall, S. (2015). Exxon knew about climate change almost 40 years ago. *Scientific American*. [online] 26 October. Available at: www.scientificamerican.com/article/exxon-knew-about-climate-change-almost-40-years-ago (Accessed 8 August 2022).

Hamalainen, R.P. and Saarinen, E. (2007). *Systems intelligence in leadership and everyday life*. [online] June 2007. Systems Analysis Laboratory, Helsinki University of Technology, Finland. Available at: https://sal.aalto.fi/publications/pdf-files/systemsintelligence2007. pdf. (Accessed 8 August 2022).

Harari, Y.N. (2022). The Surprisingly Low Price Tag on Preventing Climate Disaster. *Time* [online] 18 January. Available at: https://time.com/6132395/two-percent-climate-solution (Accessed 8 August 2022).

Harding. S. (2021). Deep ecology and the Deep Time Walk. In *Holding space: a guide to organising and facilitating a Deep Time Walk* Available at https://drive.google.com/file/d/19bJSTYRQLnz7rj3Ps0esNcJWRe-GtSjD/view. Accessed 8 August 2022.

Hawken, P. (2017). *Drawdown: The most comprehensive plan ever proposed to reverse global warming*. London: Penguin.

Hawkin, P. (2021). *Regeneration: ending the climate crisis in one generation*. New York, NY: Penguin Random House.

Hawkins, P. (1985). Humanistic psychotherapy supervision: a conceptual framework. *Self and Society: Journal of Humanistic Psychology, 13*(2): 69–79.

Hawkins, P. (2004). A centennial tribute to Gregory Bateson 1904–1980 and his influence on the fields of organizational development and action research. *Action Research, 2*(4): 409–423.

Hawkins, P. (2005). *The wise fool's guide to leadership*. London: O Books.

Hawkins, P. (2006). *Coaching supervision, in J. Passmore (ed.) Excellence in coaching*. London: Kogan Page.

Hawkins, P. (2011). *Leadership team coaching: developing collective transformational leadership*. 1st ed. London: Kogan Page.

Hawkins, P. (2011b). Expanding emotional, ethical and cognitive capacity in supervision. In Passmore, J. *Supervision in coaching*. London: Kogan Page.

Hawkins, P. (2012). *Creating a coaching culture*. Maidenhead: McGraw Hill.

Hawkins, P. (2015). Cracking the shell. *Coaching at Work, 10*(2): 42–45.

Hawkins, P. (2017a), *Tomorrow's leadership and the necessary revolution in today's leadership development*. Henley Business School. [online] Available at: https://www.henley.fi/2017/08/16/catch-up-with-our-latest-research-tomorrows-leadership-and-the-necessary-revolution-in-todays-leadership-development (Accessed8 August 2022).

Hawkins P. (2017b) The necessary revolution in humanistic psychology. In House, R. and Kalisch, D. (Eds). *The future of humanistic psychology*. London: Routledge.

Hawkins, P. (2018). *A systemic primer*. Renewal Associates.

Hawkins, P. (2019a). Resourcing – the neglected third leg of supervision. In Turner, E. and Palmer, S. (Eds), *The heart of coaching supervision - working with reflection and self-care*. Abingdon: Routledge.

Hawkins, P. (2019b). *Systemic organizational learning and the coevolution of organizational culture. Chapter 10 in Örtenblad, A.R. (ed) The handbook on the learning organization.* Oxford: Oxford University Press.

Hawkins, P. (2020a). *We need to move beyond the high-performing teams.* [online blog] 5 July. Available at: www.renewalassociates.co.uk/2020/07/we-need-to-move-beyond-high-p erforming-teams (Accessed on 8 August 2022).

Hawkins, P. (2020b). *We are all in this together: Coronavirus, climate change, collaboration and consciousness change.* [online blog] 20 September. Available at: www.renewalassocia tes.co.uk/2020/09/we-are-all-in-this-together-corona-virus-climate-crisis-collaboration-and-consciousness-change-2 (Accessed on 8 August 2022).

Hawkins, P. (2020c). *Let the wider ecology do the coaching.* [online blog] 19 November. Available at: www.renewalassociates.co.uk/2020/11/let-the-wider-ecology-do-the-coa ching (Accessed 8 August 2022).

Hawkins, P. (2021). *Leadership team coaching: developing collective transformational leadership.* 4th ed. London: Kogan Page.

Hawkins, P. (2021b). *Ecology mental health and eco-spirituality. Chapter 25: 299–310 in Aris, S., Garraway, H. and Gilber, H. (eds) Spirituality, mental health and wellbeing handbook.* Shoreham-by-Sea: Pavilion.

Hawkins, P. (2022). *Leadership team coaching in practice: case studies on creating highly effective teams.* 3rd ed. London: Kogan Page.

Hawkins, P., Allan, J., and Turner, E. (2021). Supervision: widening the lens and perspective - the art of reflective practice. In Smith, W.A., Boniwell, I., and Green, S. (Eds). *Positive psychology coaching in the workplace.* Cham: Springer.

Hawkins, P. and Hutchins, G. (2020). How we step up to the world's challenges. [online] Available at: www.renewalassociates.co.uk/resources/videos (Accessed 8 August 2022).

Hawkins, P. and McMahon, A. (2020). *Supervision in the helping professions.* 5th ed. London: Open University Press/McGraw Hill.

Hawkins, P. and Ryde, J. (2019). *Integrative psychotherapy: a relational, systemic and ecological approach.* London: Jessica Kingsley.

Hawkins, P. and Schwenk, G. (2006). *Coaching supervision.* London: Chartered Institute of Personnel and Development.

Hawkins, P. and Schwenk, G. (2021). The seven-eyed model of coaching supervision. In Bachkirova, T., Jackson, P., and Clutterbuck, D. (Eds). *Coaching and mentoring supervision – theory and practice.* 2nd ed. Maidenhead: Open University Press.

Hawkins, P. and Shohet R. (2012). *Supervision in the helping professions.* 4th ed. London: Open University Press/McGraw Hill.

Hawkins, P. and Smith, N. (2013). *Coaching, mentoring and organizational consultancy – supervision, skills and development.* 2nd ed. Maidenhead: Open University Press.

Hawkins, P. and Turner, E. (2017). The rise of coaching supervision 2006–2014. In *Coaching: An International Journal of Theory, Research and Practice,* 10(2): 102–114. doi:10.1080/17521882.2016.1266002.

Hawkins, P. and Turner, E. (2020a). *Systemic Coaching – delivering value beyond the individual.* Abingdon: Routledge.

Hawkins, P. and Turner, E. (2020b). The necessary revolution in coaching – towards eco-systemic coaching. In *Coaches Rising Coaching Summit.*

Hawkins, P. and Turner, E. (2020c). The necessary revolution in coaching: from ego-centric to eco-centric coaching and supervision. Presentation to APECS, virtual, 23 November.

Hawkins, P., Turner, E., and Passmore, J. (2019). *The manifesto for supervision*. Henley-on-Thames: Association for Coaching, and The Centre for Coaching, Henley Business School.

Hawkins, P., Turner, E., and Roell, C. (2022). *Coaches response to the climate crisis: a research report*. In press.

Hayhoe, K. (2019). A conversation with Katherine Hayhoe. November. Available at: www.youtube.com/watch?v=8YuYbep_lAA (Accessed 8 August 2022).

Heffernan, M. (2014). *Talk at the University of Bath to launch her book The Bigger Prize*.

Heffernan, M. (2011). *Wilful Blindness*. 1st ed. London: Simon & Schuster.

Heffernan, M. (2019). *Wilful Blindness*. 2nd ed. London: Simon & Schuster.

Heifetz, R., Grashow, A., and Linsky, M. (2009) *The practice of adaptive leadership: tools and tactics for changing your organisation and the world*. Boston, MA: Harvard Business Press.

Herrington, G. (2021). Update to limits of growth: Comparing the world model with empirical data, *Journal of Industrial Ecology*, 25(2021): 614–626. Available at: https://advisory.kpmg.us/articles/2021/limits-to-growth.html (Accessed 8 August 2022).

Hickel, J. (2020). Quantifying national responsibility for climate breakdown: an equality-based attribution approach for carbon dioxide emissions in excess of the planetary boundary. *The Lancet Planetary Health*, 49): 399–404. [online] 1 September. Available at: www.thelancet.com/journals/lanplh/article/PIIS2542-5196(20)30196-0/fulltext (Accessed 8 August 2022).

Hickford, A. and Blainey, S. (2021). *A radical new way to align intervention in transport, digital and energy infrastructure is needed in order to meet the goal of achieving net zero carbon emissions from transport by 2050. Making the journey to zero carbon*. London: CILT Focus.

Hillman, J. (1975). *Loose ends*. Zurich: Spring Publications.

Hillman, J. (1995). A psyche the size of the world: a psychological foreward. In T. Roszak, M.E. Gomes, and A.D. Kammer (Eds), *Ecopsychology: Restoring the earth healing the mind*. Berkeley, CA: Counterpoint.

Hodge, A. (2013). Creative supervision. Talk delivered at Coaching at Work conference, 2 July.

Holding Space (2021). Holding space: a guide to organising and facilitating a Deep Time Walk. Available at: www.deeptimewalk.org/kit (Accessed 8 August 2022).

hooks, bell and Angelou, M. (1998). Angelou, *Shambhala Sun*. [online] January. Available at: www.hartford-hwp.com/archives/45a/249.html (Accessed 8 August 2022).

Hopkins, R. (2021). *What is to what if: unleashing the power of imagination to create the future we want*. Chelsea, VT: Chelsea Green Publishing.

Horn, J. (2020). On motivation, active hope and sustainability leadership: A master's dissertation reflection, Pagbubuo. Available at: https://jenhorn.substack.com/p/motivation-of-sustainability-leaders (Accessed 8 August 2022).

Horn, J. and Wehrmeyer, W. (2020). Developing a framework for understanding the personal motivations of sustainability leaders. *Journal of Management for Global Sustainability*, [online] 8(2). Available at: https://journals.ateneo.edu/index.php/jmgs/article/view/3398/3210 and https://archium.ateneo.edu/leadership-and-strategy-faculty-pubs/15 (Accessed 8 August 2022).

Howard-Vyse, A. (2020). *Creating circles of active hope, Pt 1*. Medium.com, [online] 26 April. Available at: https://medium.com/@alicehv/creating-circles-of-active-hope-pt-i-e5dc2dae21bb (Accessed 8 August 2022).

Huntley, R. (2020). *How to talk about climate change in a way that makes a difference. Sydney.* Murdoch Press.

Hutchins, G. and Storm, L. (2019). *Regenerative leadership: The DNA of life-affirming 21st century organizations.* Wordzworth Publishing.

Ilango, A.V. and Kalyan, S. (2015). *Coming home to earth: space, line, form.* Chennai: Fooniferse Arts; Bengaluru: Sublime Galleria.

Institutional Asset Manager (2020). *Percentage of European pension funds taking climate risks into account quadruples in 12 months.* Available at: https://www.institutiona-lassetmanager.co.uk/2020/08/26/288965/percentage-european-pension-funds-taking-climate-risks-account-quadruples-12 (Accessed 8 August 2022).

ICF (2021). *Code of Ethics.* Available at: https://coachingfederation.org/ethics/code-of-ethics (Accessed 8 August 2022).

International Coaching Federation (2022). *ICF Core competencies.* Available at: https://coa chingfederation.org/core-competencies (Accessed 8 August 2022).

Intergovernmental Panel on Climate Change. Available at: www.ipcc.ch (Accessed 8 August 2022).

International Panel on Climate Change (2021). *Climate change 2021, The physical science basis, summary for policymakers. WGI contribution to the sixth assessment report of the IPCC.* Available at: www.ipcc.ch/report/ar6/wg1/downloads/report/IPCC_AR6_WGI_SPM_final.pdf (Accessed 8 August 2022)

International Panel on Climate Change (2022). Climate change: a threat to human wellbeing and health of the planet. Taking action now can secure our future. *IPCC Newsroom.* 28 February. Available at www.ipcc.ch/2022/02/28/pr-wgii-ar6 (Accessed 8 August 2022).

Iordanou, I., Hawley, R., and Iordanou, C. (2017). *Values and ethics in coaching.* London and Thousand Oaks, CA: Sage Publications.

Jackson, P. and Cox, E. (2020). *Doing coaching research.* London: Sage Publications.

Janoo, A. (2021). Why we need economic systems change to combat climate change, Climate reality project virtual training, October 2021.

Jayne, T. (2020). Earth connection: Exploring our human relationship with the Earth. *Consciousness, spirituality and transpersonal psychology,* 1: 49–61.

Jinpa, T. (2015). *A fearless heart.* London: Piatkus.

Joint Global Statement on Climate Change (2020). *Joint global statement on climate change from the professional bodies for coaching, coaching psychology, mentoring and supervision.* Available at: www.jgsg.one/joint-global-statement (Accessed 8 August 2022).

Johnstone, C. (2021). Active Hope practices to inspire positive change. AC Podcast https://www.associationforcoaching.com/page/CITCC_podcast_series (Accessed 26th September 2022)

Jordan, M. (2015). *Nature and therapy.* Hove: Routledge.

Just, T. (2010). *Postwar: A history of Europe since 1945.* London: Vintage Books.

Kahn, M.S. (2014). *Coaching on the axis.* London: Karnac.

Kahn, O. (2021). *When looking at the causes of climate change, supply chain organisations are far from blameless. Making the journey to zero carbon.* London: CILT Focus.

Khan, H.I. (1972). *The Sufi message volume X1.* London: Barrow and Jenkins.

Kalyan, S. (2022). *The porous self: Is there really a dividing line between human and nature?* Talk given at Climate Coaching Alliance Festival, 6 March. Available at: www.climatecoachingalliance.org/event/2022-global-festival-the-porous-self (Accessed 8 August 2022).

Kalyan, S. (2019). Creativity as intrinsic ecological consciousness in Foster, R., Mäkelä, J. and Martusewicz, R.A. (eds). *Art, ecojustice, and education: intersecting theories and practices.* 152–165. New York: Routledge.

Kalyan, S. (2020a). Bird and Line: Tracing Indian aesthetics to find an 'Ecosophy' of personal art practice. *The Trumpeter: Journal of Ecosophy.* 36(1). Available at: http://trump eter.athabascau.ca/index.php/trumpet/article/view/1574 (Accessed 8 August 2022).

Kalyan, S. (2020b). A porous consciousness in and as artistic practice: Re-engaging with classical Indian philosophy and aesthetics as a living tradition. *Journal for Artistic Research.* Available at: https://doi.org/10.22501/jar.614184 (Accessed 8 August 2022).

Karpman, S. (2022). *The Karpman drama triangle.* Available from https://karpmandrama triangle.com/ (Accessed 8 August 2022).

Kashtan, M. (2014). *Reweaving our human fabric: working together to create a non violent future.* Oakland, CA: The Fearless Heart.

Kasotia, P. (2022). The health effects of global warming: developing countries are the most vulnerable. *UN Chronicle.* [online]. Available at: www.un.org/en/chronicle/article/hea lth-effects-global-warming-developing-countries-are-most-vulnerable (Accessed 8 August 2022).

Kegan, R. (1994) *In over our heads.* Cambridge, MA: Harvard University Press.

Kegan, R. and Lahey, L. (2009). *Immunity to change: how to overcome it and unlock the potential in yourself and your organization.* Cambridge, MA: Harvard Business Review Press.

Kimmerer, R. (2017). Speaking of nature - Finding language that affirms our kinship with the natural world. *Orion.* [online]. Available at: www.orionmagazine.org/article/spea king-of-nature/ (Accessed 8 August 2022).

Kimmerer, R. (2020). *Braiding sweetgrass. indigenous wisdom, scientific knowledge and the teachings of plants.* London: Penguin Books.

King, M.L. (1964). *Methodist Student Leadership Conference address.* Lincoln, Nebraska. Available at: www.americanrhetoric.com/speeches/mlkmethodistyouthconference.htm (Accessed 8 August 2022).

Klein, N. (2014). *This changes everything–capitalism v the climate.* London: Allen Lane.

Kline, N. (2020). *The promise that changes everything–I won't interrupt you.* London: Penguin Life.

Kline, N. (1999). *Time to think –listening to ignite the human mind.* London: Cassell Illustrated

Kline, N. (2015). *More time to think – the power of independent thinking.* 2nd edition. London: Cassell Illustrated.

Korten, D.C. (2007). *The great turning: from empire to earth community.* Bloomfield, CT: Kumarian Press and San Francisco, CA: Berrett-Koehler Publishers.

Koestler, A. (1976). *The ghost in the machine.* (2nd paperback edition). Picador, London.

Knight, S. (2009). *NLP at work: The essence of excellence* (3rd. ed.). Nicholas Brealey Publishing.

Kolbert, E. (2015). *The Sixth Extinction.* London: Bloomsbury Publishing.

Koyaanisqatsi. (1982). Motion Picture. *Institute For Regional Education.* New Mexico, USA. Qatsi Productions. USA. American Zoetrope. Francis Ford Coppola. George Lucas. San Francisco California. USA.

Krznaric, R. (2020). *The good ancestor: how to think long term in a short-term world.* London: W.H. Allen & Co.

Kuhn, L. and Whybrow, A. (2019). Coaching at the edge of chaos: a complexity informed approach to coaching psychology. In Palmer, S. and Whybrow, A (Eds). *A Handbook of Coaching Psychology: A Guide for Practitioners.* (2nd edition). Abingdon: Routledge.

Kurio, J. and Reason, P. (2021). voicing rivers through ontopoetics: a co-operative inquiry. (with Jacqueline Kurio). *River research and applications, special issue: voicing rivers.* London: Wiley.

Lahn, G. (2018). *What Ireland's fossil fuel divestment bill means for global climate action.* Available at: www.chathamhouse.org/2018/07/what-irelands-fossil-fuel-divestment-bill-means-global-climate-action (Accessed 8 August 2022).

Laloux, F. (2014). *Reinventing organisations: a guide to creating organisations inspired by the next stage of human consciousness.* Brussels: Nelson Parker.

Lane, D.A., Watts, M., and Corrie, S. (2016). *Supervision in the psychological professions: Building your own personalised model.* London: Open University Press.

Lane, D.A. (1972). Education in environmental health. *Community Health, 4,* 149–156.

Lane, D.A. and Malkin, J. (1994). *Global warming and the built environment: the challenge, in Samuels, R. and Prasad, D.K. (Eds) Global warming and the built environment.* London: Spon.

Latour, B. (2017). *Facing Gaia: eight lectures on the new climatic regime.* Cambridge, UK: Polity Press.

Lawrence, P. and Moore, A. (2019). *Coaching in three dimensions.* Abingdon: Routledge.

Lenton, T. et al. (2019). Climate tipping points — too risky to bet against. *Nature.* [online] 27 November. Available at: www.nature.com/articles/d41586-019-03595-0 (Accessed 8 August 2022).

Lipmanowicz, H. and McCandless, K. (2014). *The surprising power of liberating structures: simple rules to unleash a culture of innovation.* Liberating Structures Press.

London Build (2021). *London Build 2021 - The UK's biggest festival of construction.* Available at: www.londonbuildexpo.com (Accessed 8 August 2022).

Loving Kindness Practice (2022). *Loving kindness resonance: guided meditation by Mark McCordie.* Available at: https://insighttimer.com/markmcmordie/guided-meditations/lkm-loving-kindness-meditation (Accessed 8 August 2022).

Lumber, R., Richardson, M., and Sheffield, D. (2017). *Beyond knowing nature: Contact,* emotion, compassion, meaning, and beauty are pathways to nature connection. PLoS ONE, [online] 9 May. Available at: https://doi.org/10.1371/journal.pone.0177186 (Accessed 8 August 2022).

Lushwala, A. (2012). *The time of the black jaguar: An offering of indigenous wisdom for the continuity of life on earth.* Ribera, NM: createspace.com.

Mace, M. L. and Mahler, W.R. (1958). On-the-job coaching. In H.F. Merrill and W.R. Mahler (Eds), *Developing executive skills.* American Management Association. 99–110.

Macy, J. (2022). *The council of all beings.* Available at: www.rainforestinfo.org.au/deep-eco/Joanna%20Macy.htm (Accessed 8 August 2022)

Macy, J. (2009). *The great turning.* [online] 29 June. Available at: www.ecoliteracy.org/article/great-turning (Accessed 8 August 2022).

Macy, J. (2014) *Joanna Macy and the great turning, with filmmaker Chris Landry.* Documentary. [online] 8 April. Available at: https://vimeo.com/ondemand/greatturning (Accessed on 8 August 2022).

Macy, J. and Johnstone, C. (2012). *Active Hope - How to face the mess we're in without going crazy.* Novato, CA: New World Library.

Manos, J. (2007). *Ghetto plainsman.* Fort Worth, TX: Temba House Press.

Margulis, L. (1998). *Symbiotic planet: A new look at evolution*. New York: Basic Books.

Maryl, J. (2022). Beyond the darkness, a new world comes. (poem). *Journey to Light* collection. Available at: https://jennymaryl.uk/2021/02/06/beyond-the-darkness-a-new-world-comes-poem (Accessed 8 August 2022).

Martin, L., White, M., Hunt, A., Richardson, M., Pahl, S., and Burt, J. (2020). Nature contact, nature connectedness and associations with health, wellbeing and pro-environmental behaviours. [online] 18 January. *Journal of Environmental Psychology, 68 (April)*, 1–12. Available at: https://www.mauritian-wildlife.org/mwf-files/files/files/Martin%20et%20al%202020%20contact%20NC%20health%20wellbeing%20and%20PEB.pdf (Accessed 8 August 2022)

Mathews, F. (2017). Panpsychism. In Oppy, G. and Trakakis, N.N. (Eds). *Interreligious philosophical dialogues, Volume 1*. London: Routledge.

Mathews, F. (2019) *Living Cosmos Panpsychism. In W. Seager (Ed) The Routledge handbook of panpsychism*, London: Routledge.

Mathews, F. (2021) *The ecological self*. London: Routledge.

Maturana, H.R. and Varela, F.G. (1987). *The tree of knowledge*. Boston, MA: Shambhala Publications.

McGrath, M. (2018). *Sir David Attenborough: Climate change 'our greatest threat'*. [online] 3 December. Available at: www.bbc.co.uk/news/science-environment-46398057 (Accessed 8 August 2022).

McGrath, M. (2021). Climate change: IPCC report is 'code red for humanity'. [online] 9 August. Available at: www.bbc.co.uk/news/science-environment-58130705 (Accessed 25 March 2022).

McLean, J. (2013). Hermeneutic circles: gaining a collective understanding of what was learned, one by one. *ALARA World Congress 2013*. Brisbane.

McLean, J. (2016). How can an organisation learn its way to becoming 'sustaining'? *GCWAL-ALARA World Congress 2016*. Adelaide.

McLean, J. (2017), Embedding sustainability into organisational DNA: A story of complexity. Business School, Doctor of Philosophy thesis, University of Adelaide. Available from: https://digital.library.adelaide.edu.au/dspace/bitstream/2440/112859/1/01front.pdf (Accessed 8 August 2022).

McLean, J. (2020a). *Big Little Shifts: a complexity practitioners guide to complexity for organisational change and adaptation*. Adelaide: The Partnership Pty.

McLean, J. (2020b), *Who are coaches being called to be now?* ICF Germany Conference, [online] 12 November.

McLean, J. and Koth, B. (2016). The conscious workplace: beyond mindfulness and greening in SMEs. *5th Australian positive psychology and wellbeing conference*. Adelaide.

McLean, J. and Wells, S. (2010). Flourishing at the edge of chaos: leading purposeful change and loving it. *Journal of spiritual leadership and management. 4(1)*, 53–61. Available from http://www.slam.org.au/wp-content/uploads/2013/06/JSLaMvol4_2010_McLean.pdf (Accessed 8 August 2022).

Meadows, D. (1994). Envisioning a sustainable world, paper presented at International Society for Ecological Economics, San José, Costa Rica.

Meadows, D. (1998). *Indicators and information systems for sustainable development: A report to the Ballaton group*. TS Institute.

Meadows, D. (2008). *Thinking in systems*. White River Junction, VT: Chelsea Green Publishing.

Meadows, D., Randers, J., and Meadows, D. (2004). *Limits to growth: The 30-year update*, 1st ed. White River Junction, VT: Chelsea Green Publishing.

Meadows, D, Meadows, D, Randers, J., and Behrens III, W.W. (1972). *The limits to growth*. London: Earth Island Ltd.

Merriam-Webster. (2022). Legal definition of 'wilful blindness'. Available at: www.merriam-webster.com/legal/willful%20blindness (Accessed 8 August 2022).

Mikati, I., Benson, A.F., Luben, T.J., Sacks, J.D., and Richmond-Bryant, J. (2018). Disparities in distribution of particulate matter emission sources by race and poverty status. *American Journal of Public Health*. [online] April 2018. Available at: https://ajph.aphapublications.org/doi/10.2105/AJPH.2017.304297 (Accessed 8 August 2022).

Miller, A. (2022). Climate Earth blog. Available at https://climateearthconsulting.com/climate-earth-blog (Accessed 8 August 2022).

Miller, W.R., Forcehimes, A.A., and Zweben, A. (2019). *Treating addiction: A guide for professionals*. Guilford Publications.

Mills, R. (2021). COP26: Prince Charles to tell the world leaders they need to be on 'a warlike footing' to tackle climate crisis, SkyNews, [online] 1 November. Available at: https://news.sky.com/story/cop26-prince-charles-tells-g20-world-leaders-the-future-of-humanity-and-nature-herself-are-at-stake-12456102 (Accessed 8 August 2022).

Millward-Hopkins, J., Steinberger, J.K., Rao, N.D., and Oswald, Y. (2020). Providing decent living with minimum energy: A global scenario. *Global Environmental Change*, 65. [online] November 2020, 102168. Available at: www.sciencedirect.com/science/article/pii/S0959378020307512.

Miranda, E.R. and Braund, E. (2015). *Music with unconventional computing: granular synthesis with the biological computing substrate physarum polycephalum*. Proceedings of 11th Computer Music Multidisciplinary Research: Music, Mind, and Embodiment. Plymouth University.

Montgomery D.R. and Bikle, A. (2016). *The hidden half of nature: The microbial roots of life and health*. New York, NY: W.W. Norton & Company.

Naqoyqatsi (2002). Motion Picture. Institute For Regional Education. Qatsi Productions. USA. Miramax.

NASA (2022a). Vital signs of the planet. Available at: https://climate.nasa.gov/vital-signs/ (Accessed 8 August 2022).

NASA (2022b). Climate change: How do we know? Available at: https://climate.nasa.gov/evidence (Accessed 8 August 2022).

NASA (2022c). Facts – carbon dioxide. Available at: https://climate.nasa.gov/vital-signs/carbon-dioxide (Accessed 8 August 2022).

National Geographic Society (2022). Anthropocene. Available at: www.nationalgeographic.org/encyclopedia/anthropocene (Accessed 8 August 2022).

National Oceanic and Atmospheric Administration (2022). What is ocean acidification? Available at: www.pmel.noaa.gov/co2/story/What+is+Ocean+Acidification%3F (Accessed 8 August 2022).

Nazrul Islam, S. and Winkel, J. (2017). Climate change and social inequality. DESA Working Paper No. 152 ST/ESA/2017/DWP/152. Available at: https://www.un.org/esa/desa/papers/2017/wp152_2017.pdf (Accessed 8 August 2022).

Nelson, Portia (1993). *There's a hole in my sidewalk*. Hillsboro, or: Beyond Words Publishing. Reprinted with the permission of Beyond Words/Atria Books, a division of Simon & Schuster.

Obama, B. (2015). *Remarks by the President at the GLACIER Conference - Anchorage, AK.* [online] 31 August. Available at: https://obamawhitehouse.archives.gov/the-press-of fice/2015/09/01/remarks-president-glacier-conference-anchorage-ak (Accessed 8 August 2022).

Olanday, D. and Rigby, J. (2020). Inside the world's longest and strictest coronavirus lock-down in the Philippines. [Online] *The Telegraph*, 11 July. Available at: www.telegraph.co.uk/global-health/science-and-disease/inside-worlds-longest-strictest-coronavir us-lockdown-philippines (Accessed 8 August 2022).

One Earth (2020). *Video.* [online] 20 August. Romain Pennes. France. Available at: https://youtu.be/QQYgCxu988s (Accessed 8 August 2022).

Oxfam (2021). Press release: Carbon emissions of richest 1% set to be 30 times the 1.5 limit in 2030. [online] 5 November. Available at: www.oxfam.org/en/press-releases/carbon-em issions-richest-1-set-be-30-times-15degc-limit-2030 (Accessed 8 August 2022).

Palmer, S. (2015). Can ecopsychology research inform coaching and positive psychology practice? *International Society for Coaching Psychology, 8(1)*: 11–15.

Pappas, S. (2020). Human-made stuff now outweighs all life on earth, *Scientific American.* [online] 9 December. Available at: www.scientificamerican.com/article/human-made-s tuff-now-outweighs-all-life-on-earth (Accessed 8 August 2022).

Parente, M.C. and Cardoso, C. (2020). This is not a manifesto, this is not the truth. Video. [online] 17 December. Available at: www.youtube.com/watch?v=v_DmryPexno (Acces-sed 8 August 2022).

Parlett, M. (2015). *Future sense: Five explorations of whole intelligence for a world that is waking up.* Leicestershire: Matador.

Perkins, J. (2018). *The new confessions of an American hitman: How America really took over the world.* London: Ebury Press.

Phukan, R. (2021). *Climate justice on the ground: building resilience and strength*, presenta-tion for Climate RealityProject Training, 17–24 October 2021. Available at: www.clima terealityproject.org (Accessed 8 August 2022).

Polman, P. and Winston, A. (2021). *Net positive: How courageous companies thrive by giving more than they take.* Cambridge, MA: Harvard Business Review Press.

Porges, S. (2017). *The pocket guide to the polyvagal theory.* New York, NY: W.W. Norton & Co.

Powaqqatsi (1988). Motion Picture. Institute For Regional Education. USA. Qatsi Pro-ductions. USA. Golan Globus. USA. Cannon Films USA. NorthSouth Productions.

Powers, R. (2018). *The Overstory.* London: Penguin.

Prentice, K. (2020). *Nature's Way: Designing the life you want through the lens of nature and the five seasons.*

Pritchard, A., Richardson, M., Sheffield, D., and McEwan, K. (2020). The relationship between nature connectedness and eudaimonic well-being: A meta-analysis. *Journal of happiness studies.* Vol. *21*, 1145–1167.

Qatsi Trilogy (1982). Motion Picture. Godfrey Reggio. Institute For Regional Education. New Mexico, USA. Qatsi Productions.

Rajan, A. and Lane, D.A. (2000). *Employability: bridging the gap between rhetoric and reality.* Tonbridge: Create.

Ravindra, R. (2014). *The Pilgrim Soul: A path to the sacred.* Wheaton, IL: Quest Books.

Raworth, K. (2017). *Doughnut economics: seven ways to think like a 21st century economist.* London: Random House Books.

Reason, P. and Gillespie, S. (2019) *On presence: essays - drawings.* Dartington: The Letter Press.

Reason P. and Gillespie, S. (2021). *On sentience: essays - drawings.* Dartington: The Letter Press.

Reason, P. and Hawkins, P. (1988). Storytelling as Inquiry. in Reason, P. (ed). *Human Inquiry in Action.* London: Sage Publications.

Reed, B. (2007). Shifting from 'sustainability' to regeneration. *Building Research and Information, 35(6)*: 674–680, doi:10.1080/09613210701475753.

Ren21 (2021). *Renewables 2021 global status report.* Available at: www.ren21.net/wp-con tent/uploads/2019/05/GSR2021_Full_Report.pdf (Accessed 8 August 2022).

Renshaw, B. (2018). *Purpose: The extraordinary benefits of focusing on what matters most.* London: LID Publishing.

Report of the United Nations Conference on the Environment (1972). United Nations.

Richardson, M. et al. (2020). Applying the pathways to nature connectedness at a societal scale: a leverage points perspective. *Ecosystems and people, 16(1)*: 387–401. Available at: www.tandfonline.com/doi/full/10.1080/26395916.2020.1844296 (Accessed 8 August 2022).

Ritchie, H. (2020). *The carbon footprint of foods: are differences explained by the impact of methane? Our world in data.* [online] 10 March. Available at: https://ourworldindata.org/ carbon-footprint-food-methane (Accessed 8 August 2022).

Rilke, R.M. (1929/2012). *Letters to a young poet.* Harmondsworth: Penguin.

Roberts, D. (2020). Social tipping points are the only hope for the climate - A new paper explores how to trigger them. *Vox*, 29 January 2020. Available at: www.vox.com/ener gy-and-environment/2020/1/29/21083250/climate-change-social-tipping-points (Accessed 8 August 2022).

Rooke, D. and Torbert, W.R. (2005). Seven transformations of leadership. *Harvard Business Review, 83(4)*: 66–76.

Ryde, J. (2005). White racial identity and intersubjectivity in psychotherapy. Doctoral Thesis, University of Bath.

Ryde, J. (2009). *Being white in the helping professions.* London: Jessica Kingsley Publishers.

Ryde, J. (2019). *White privilege unmasked.* London: Jessica Kingsley Publishers.

Ryde, J., Seto, L., and Goldvarg, D. (2019). Diversity and inclusion in supervision. In Turner, E. and Palmer, S. (Eds). *The heart of coaching supervision –working with reflection and self-care.* Abingdon: Routledge.

Sahtouris, E. (2005). The biology of business: new laws of nature reveal a better way for business. *World Business Academy, 19(4)*. Available at: https://worldbusiness.org/wp-con tent/uploads/2013/06/pr092205a.pdf (Accessed 8 August 2022).

Salmon, E. (2000). Kincentric ecology: indigenous perceptions of the human–nature relationship. *Ecological Applications, 10(5)*: 1327–1332.

Samsara (2011). Motion picture. Ron Fricke and Mark Magidson Films. USA.

Samuels, R and Prasad, D.K. (1994). *Global warming and the built environment.* London: Spon.

Sanchez, A. (2017). *The four sacred gifts - indigenous wisdom for modern times.* New York, NY: Enliven Books.

Scharmer, C.O. (2009). *Theory U.* San Francisco, CA: Berrett-Koehler Publishers .

Scharmer, O. and Kaufer, K. (2013). *Leading from the emerging future – from ego-system to eco-system economies.* San Francisco, CA: Berrett-Koehler Publishers.

Scharmer, O. (2016). *Theory U –leading from the future as it emerges.* 2nd ed. Oakland, CA: Berrett-Koehler Publishers.

Scotton, N. (2021). One life, one world, one humanity, one future: making something good happen. Available at: https://neilswheel.org (Accessed 8 August 2022).

Scotton, N. (2022). Neil's Wheel. Available at: https://neilswheel.org (Accessed 8 August 2022).

Seed, J., Macy, J., Fleming, P., and Naess, A. (2007). *Thinking like a mountain –towards a council of all beings.* Gabriola Island, BC: New Catalyst Books.

Senge, P. (2022). What the vision does. [online] Available at: www.awakin.org/v2/read/view.php?tid=669 (Accessed 8 August 2022).

Shafak, E. (2021). *The island of missing trees: a novel.* London: Bloomsbury.

Sharpe, B., Hodgson, A., Leicestor, G., Lyon, A., and Fazey, I. (2016). Three horizons: A pathways practice for transformation. *Ecology and Society, 21(47).*

Sherif, M., Harvey, O.J., White, B.J., Hood, W.R., and Sherif, C.W. (1961). *The robbers cave experiment: intergroup conflict and cooperation.* Institute of Group Relations, University of Oklahoma; reprinted by Wesleyan University Press, 1988.

Snorf, K. and Baye, J. (2010). Personal change, climate change, and human development. *Journal of Integral Theory & Practice, 5(1),* 61–87.

Stockholm Resilience Centre (2022). *Planetary boundaries.* Available at: www.stockholm resilience.org/research/planetary-boundaries.html (Accessed 8 August 2022).

Stout-Rostron, S. (2019). *Transformational coaching to lead culturally diverse teams.* London: Routledge.

Tai, M.C. and Lin, C.S. (2001). Developing a culturally relevant bioethics for Asian people. *Journal of Medical Ethics, 27,* 51–54.

Tajfel, H. and Turner, J.C. (1986). The social identity theory of intergroup behavior. In Worchel, S. and Austin, W.G. (Eds), *Psychology of intergroup relations,* 7–24.

The Philippine Daily Inquirer (2019). The Philippines will be underwater in 30 years. [online] The Straits Times, 5 December. Available at: www.straitstimes.com/asia/the-philippines-will-be-underwater-in-30-years-inquirer (Accessed 8 August 2022).

Thomas, L. (2020). Why every environmentalist should be anti-racist. *Vogue.* [online] 8 June. Available at: www.vogue.com/article/why-every-environmentalist-should-be-anti-ra cist (Accessed 8 August 2022).

Tomasdottir, H. (2021). Transforming our impact. Talk given at CCA Transforming Our Impact event. [online] 9 September. Available at: www.climatecoachingalliance.org/event/meeting-or-networking-event-transforming-our-impact/ (Accessed 8 August 2022).

Turner, E. and Palmer, S. (2019) (Eds). *The heart of coaching supervision – working with reflection and self-care.* Abingdon: Routledge.

Turner, E. and Clutterbuck, D. (2019). *All in the small print. A brief study of contracting issues in coaching and supervision.* Presentation to the 8th international coaching supervision conference, Oxford, 11 May.

Turner, E. and Hawkins, P. (2016). Multi-stakeholder contracting in executive/business coaching: an analysis of practice and recommendations for getting maximum value. *International Journal for Evidence-Based Coaching and Mentoring, 14(2):* 48–65. Available from: https://radar.brookes.ac.uk/radar/items/299b7998-8bae-4cd2-b66c-5bcc01bdff59/1 (Accessed 8 August 2022).

Turner, E. and Hawkins, P. (2019). Mastering contracting. In J. Passmore, Underhill, B., and Goldsmith, M. (Eds), *Mastering executive coaching.* Abingdon: Routledge.

UN Environmental Programme. (2018). News article: How Islam can represent a model for environmental stewardship. [online] 21 June. Available at: www.unep.org/news-and-stor ies/story/how-islam-can-represent-model-environmental-stewardship (Accessed 8 August 2022).

UNICEF (2022). Child poverty. [online]. Available at: www.unicef.org/social-policy/child-p overty (Accessed 8 August 2022).

United Nations (2022). The climate crisis –a race we can win. [online] Available at: www.un. org/en/un75/climate-crisis-race-we-can-win (Accessed 8 August 2022).

United Nations (2019). Blog on UN Report: Nature's dangerous decline 'unprecedented'; species extinction rates 'accelerating'. [online] 6 May. Available at: www.un.org/sustaina bledevelopment/blog/2019/05/nature-decline-unprecedented-report (Accessed on 8 August 2022).

United Nations (2021). Secretary-General calls latest IPCC Climate Report Code red for humanity', stressing 'irrefutable' evidence of human influence. [online]. Statement SG/ SM/20847. 9 August. Available at: www.un.org/press/en/2021/sgsm20847.doc.htm (Accessed 8 August 2022).

United Nations (2022). The 17 Goals. Department of Economic and Social Affairs Sustainability Development Goals. Available at: https://sdgs.un.org/goals (Accessed 8 August 2022).

Van Nieuwerburgh, C. and Allaho, R. (2017). *Coaching in Islamic culture.* London: Karnac.

Varela F.J., Maturana H.R., and Uribe, R. (1974). Autopoiesis: The organization of living systems, its characterization and a model. *Biosystems, 5(4)*: 187–196.

Vaughan-Lee, L. (2013) (Ed.). *Spiritual ecology: the cry of the earth.* Point Reyes Station, CA: Golden Sufi Centre.

Višević, D. (2021). Comment on Twitter, 10:53 PM, 26 August 2021. Available at: https:// twitter.com/visevic/status/1431012057143463937?s=20 (Accessed 8 August 2022).

Walter, G. (2022). *Mother Earth's dilemma.* An original poem written for this book.

Ware, B. (2011). *The top five regrets of the dying.* Bloomington, IN: Balboa Press.

WEAll (2022). The Wellbeing Economy Alliance. Available at: https://weall.org/about (Accessed on 8 August 2022).

WeGo ITN (2022). Available at: www.wegoitn.org (Accessed 8 August 2022).

Weinstein, N., Przybylski, A.K., and Ryan, R.M. (2009). Can nature make us more caring? Effects of immersion in nature on intrinsic aspirations and generosity. *Personality and Social Psychology Bulletin, 35(10)*: 1315–1329.

Weintrobe, S. (2012) (Ed.). *Engaging with climate change: psychoanalytic and inter-disciplinary perspectives.* London: Routledge.

Weiskopt, R. and Willmott, H. (2013). Ethics as critical practice: the "Pentagon papers", Deciding responsibly, truth-telling, and the unsettling of organizational morality., *Organization studies, 34(4)*: 469–493. https://doi.org/10.1177/0170840612470256.

Wells, S. and McLean, J. (2013). One way forward to beat the Newtonian habit with a complexity perspective on organisational change, *Systems, 1*: 66–84.

Wells, S. and McLean, J. (2020). Organizational change as emergence: a living systems perspective. In: Metcalf, G., Kijima, K., and Deguchi, H. (Eds) *Handbook of Systems Sciences.* Springer, Singapore. https://doi.org/10.1007/978-981-13-0370-8_16-1.

Wheatley, M.J. (1999). *Leadership and the new science: Discovering order in a chaotic world,* 2nd ed. San Francisco, CA: Berrett-Koehler Publishers.

Whitmore, J. (2017). *Coaching for performance.* 4th ed. London: Nicholas Brealey Publishing.

Whitburn, J., Linklater, W., and Abrahamse, W. (2020). Meta-analysis of human connection to nature and pro environmental behavior. *Conservation Biology, 34(1):* 180–193.

Widdowson, L., Rochester, L., Barbour, P.J., and Hullinger, A.M. (2020). Bridging the team coaching competency gap: A review of the literature. *International Journal of Evidence-based Coaching and Mentoring, 18(2):* 35–50.

Wilber, K. (2006). *Integral spirituality.* Boston, MA: Shambhala Publications.

Wilber, K. (2007). *The integral vision: a very short introduction to the revolutionary integral approach to life, God, the universe, and everything.* Boulder, CO: Shambhala Publications.

Wilkins, C. (2021). *Tips and tricks for talking with politicians,* CCA Political Pod talk, [online] 7 January.

Wilkins, P. (2000). Unconditional positive regard reconsidered. *British Journal of Guidance & Counselling, 28(1):* 23–36. doi:10.1080/030698800109592.

Willimas, F. (2017). *The nature fix why nature makes us happier, healthier, and more creative.* New York, NY: W.W. Norton & Co.

Williams, A. and Whybrow, A. (2013). *The 31 practices: Release the power of your organization values every day.* London: Lid Publishing.

Williams, M. and Penman, D. (2011). *Mindfulness.* London: Piatkus.

Willis, J. and Todorov, A. (2006). First impressions: Making up your mind after a 100-ms exposure to a face. *Psychological Science, 17(7):* 592–598.

Willmott, H. (2021). Sustainability is improving but is it enough. *Focus, January 2021:* 44–45.

Wirzba, N. (2021). *This sacred life: humanity's place in a wounded world.* Cambridge: Cambridge University Press.

Witzany, G. (2012). Key levels of communication in plants. In Witzang, G. and Baluska, F. *Biocommunication in Plants.* Berlin: Springer-Verlag.

Wohlleben, P. (2017). *The hidden life of trees.* London: William Collins.

World Health Organization (2022). Air pollution. [online] Available at: www.who.int/health-topics/air-pollution#tab=tab_1 (Accessed 8 August 2022).

Wrangham, R.W. and Peterson, D. (1996). *Demonic males: apes and the origins of human violence.* Houghton Mifflin.

WWF (2022). Threats: soil erosion and degradation: overview. [online]. Available at: hwww.worldwildlife.org/threats/soil-erosion-and-degradation (Accessed 8 August 2022).

Xu, C., Kohler, T.A., Svenning, J.-C., and Scheffer, M. (2020). Future of the human climate niche. *PNAS,* 26 May, *117(21):* 11350–11355. Available at: www.pnas.org/content/117/21/11350 (Accessed 8 August 2022).

Yanrong Chang, Y. (2008). Cultural "faces" of interpersonal communication in the U.S. and China. *Intercultural Communication Studies, XVII:* 1.

Yunkaporta, T. (2019). Integral theory thoughts from land. LinkedIn article [online], 19 November. Available at: www.linkedin.com/pulse/integral-theory-thoughts-from-land-tyson-yunkaporta (Accessed 8 August 2022).

Index

Note: Locators in *italic* and **bold** refer to figures and tables, respectively.

13th fairy 179, 186, 246
2030 Agenda for Sustainable Development 267

Abrams, David 191
Acknowledgment of Country (welcome protocol) 198
Active Hope approach 92, 102, 157–160, 262
adaptive leadership 226–229
Adoo-Kissi-Debrah, Ella 100
Africa Adaptation Programme (AAP) 201
Al-Jayyousi, Odeh 218
Albrecht, Glenn 95
Allaho, Raja'a 217
Allen, Kathleen 39, 175
Amidon, Elias 31
Andrews, Nadine 223
Angelou, Maya 169
anger 101–103, 106
Anthropocene 11, 85–86, 264
Anthropocene Epoch, The 264
anthropocentrism 181
Arctic sea-ice loss 63
Arctic warming 63, 266
Arnold, Alexandra 278–280
Asia Pacific Alliance of Coaches 193, 211, 212
Aspey, Linda 28, 145, 213
Association for Coaching (AC) 193, 211, 212
Association of Coaching Supervisors (AoCS) 193, 211, 212
Association for Professional Executive Coaching and Supervision (APECS) 193, 211, 212, 214, 215

Atlantic Meridional Overturning Circulation (AMOC) 63
Atomistic Newtonian thinking 115–116
Attenborough, David 77, 152
Awakening in coaching profession: agenda-less conversations 50; DiGirolamo's story 41–42; exercises 37, 40, 46, 49; Hawken's story 40–41; Hawkins's story 46–48; Joanes's story 41–42; Lane's story 35–37; McLean's story 43–46; Turner's story 48; Whybrow's story 37–40; *see also* Eco-Awakening

Baafi, Abena 201
Bacon, Francis 267
Bailey, Jeanine 196–200
Ballard, David 31
Barbes, Dominique 29–30
Bateson, Gregory 6, 11, 30, 264, 267
Bateson, Nora 91
B Corp 240
Bell, Meredith 278–280
Bendell, Jem 100
Berry, Thomas 16, 191
Beyond the darkness, a new world comes (poem by Maryl) 245
Bhateja, Jaya 270–272
biomimicry 196
biophilic/biophilia 125–126, 195–196; 'Biophilia Plan' 196; biophilic activism 125–127
Blakeley-Glover, Jaime 172, 184–185
blame 101–103, 106
Borst, Anamika 121
boundary management 140

Bradshaw, Corey 57
Braiding Sweetgrass (Kimmerer) 91
Braund, Edward 70
Brause, Jan 280
Breaking Boundaries (documentary) 77
Brest, Anne-Marie 278–280
British Broadcasting Corporation
    (BBC) 25
Brock, Vikki G. 42
Buhner, Stephen Harrod 71
butterfly effect 261

Capra, Fritjof 97, 265
capitalism 11, 12, 40–48, 57; see also
    extractive capitalism; see also Awakening
    in the Coaching Profession; see also
    degrowth of economy
carbon cutting 234
carbon dioxide (CO2) emission 55, 63
carbon emissions reduction 55
Carbon Footprint 209
Carson, Rachel 30, 64
Chandler, Fiona 273–274
change: immunity to 164; inner frontier of
    159; mindset 195; see also systemic
    change; see also social change
Chang, Yanrong 216
Charles III, King 65
Chief Seattle 33, 248
Chief Storyteller, Sweden 165
Christianity 251–252
clear view, A (poem by Borst) 120–121
climate-aware coaches 222
Climate Action Team (CAT) 277
climate change 56, 152–153, 162–163, 164,
    192, 223–224, 251–252, 266; exterior
    adaptation to 164; human side of 163;
    Rob and Zoe's conversation on 164–168;
    Snorf's views on 162–164
Climate Coaching Action Day 19, 269
Climate Coaching Alliance (CCA) 1,
    19, 74, 172, 213, 221, 269–270; Asia
    270–272; Canada 272–273; Circle of
    Interest for 221–222; communities 270;
    French-speaking 273–274; Japan
    274–276; Oceania 276–277; Spanish
    speaking 277–278; USA 278–280;
    Yorkshire 280
climate conscious coach domains 139,
    207, 208, 225; business footprint 209;
    communities 209; ethics 210; own life
    209; political engagement 209; practice

209–210; reflection/supervision 210; self
    207, 208; supporting adaptation 210
climate crisis 55, 56, 158, 196, 202–203,
    236, 266
climate emergency 18, 19
climate justice 192, 196
Climate Psychology Alliance (CPA) 213
climate violence 82–83
Clutterbuck, David 140
co-abiding principle 265
Coaches and Mentors of South Africa
    (COMENSA) 193, 211, 212, 215
coaches/coaching 35, 36, 41, 148, 150–151,
    203, 210, 224; adoption of coaching
    skills in societies 43; agenda-less
    conversations 50; applying regenerative
    framework to 39; as client-centred
    facilitation of change 42; climate-aware
    coaches/coaching 222, 224; coaching
    bodies/trainers, exercise to 168; concern
    about state of environment 13–15;
    eco-centric 47; ecological purpose of
    220; Ershad approach to 217; exercise to
    engage imagination 168; gift based 157;
    habits 149–150; investment in resources
    49; modern executive 47; psychology 41;
    purpose of 12; role in changing mindsets
    223; roots of regenerative practice 39–40;
    shifting frame for 38–39; skills from
    sociological perspective 42–43, 42;
    stepping up global challenges 12–13; and
    supervision for social change 81;
    supervision role in 171–174; as way of
    working with people 263
Coaches Rising 2020 summit 13, 189
Coaching and Environmental Health 36
"coaching culture", idea of 43
coaching relationship 137; contracting and
    139; fundamental questions of 139–141
Coates, Robin 31
'code red for humanity' 147
codes of ethics 41, 194 see also Global
    Code of Ethics
Co-evolution 117, 118, 264
Cohen, Ronald 12, 38
Cohen, Zoe 28, 77, 89, 139, 207, 208, 213;
    Nine domains of a climate conscious
    coach 207–211; about climate crisis
    56–57, 58; conversation with Rob on
    climate change and Transition Town
    Network 164–168; about eco-anxiety 99;
    about human-made mass vs. living

biomass 64–65, *64;* about impact of climate and ecological destruction 60; about managed degrowth of economy 65–66; about quantitative planetary boundaries of Earth 58–59, *59;* about social tipping points 66; story of Eco-Awakening 26–27
collaborative inquiry 2, 149–150, 189 *see also* limiting mindsets
Collett, Diana 221–224
Collins, Randall. 42
communities 209; CCA 270–280; Hoop of Life 146
Compassion Cultivation Training (CCT) 107
complexity theory 63, 112–113, 118, 129, 223, 228
Confessions of an Economic Hit Man 46
conformonomics 147–148
consciousness of natural non-human living systems 71
contracting, containing, and connecting framework139–141; 3Cs framework 153–154
conversational process 217; discovery phase 217; effort phase 218; intention phase 217–218; pathways phase 218
Cook, Tim 238
COP26, 73, 216, 222–223
'Corporate Social Responsibility', chimera 21
Council of All Beings, The 86–89; elements to holding 89; stages of 87–89
courage 11, 45, 104, 169 *see also* eco-engaged coaching
Covid-19 pandemic 13, 58, 67, 113, 230, 252, 255, 270–1, 276,
Cox, Charly 19
crisis of imagination, 227 *see also* imagination

David, Susan 140
DDT 27
deep adaptation 100
deep ecology 75, 264–265
Deep Time Walk (DTW) 73, 112; participants' views of 74–77; Lécuyer's views of 73–74; reflection 77–78; timeframe *76*
DeGroot, Michelle 122–123
degrowth of economy 65–66, 265
de la Sablonnière, R. 42

Dellinger, Drew 7–9, 89, 131–133, 261
denial 106; moving through denial 97, 98–99; narratives of 97–98
denialism 99
'developed world' 77–78, 247
DiGirolamo, Joel 41–42
digistraction 147
disavowal 98, 101, 103
diversity 216
Doumbia, Mayme 278–280
Downey, Paula 21–23
Drake, David 20, 23–24, 278–280

Earth: average surface temperature of 55; listening to 69; percentage of mammals in *65;* quantitative planetary boundaries of 58–59, *59;* rights 196
Earth Overshoot Day 97
Earth Stations: A Resource Guide for Planning a Deep Time Walk 74
Eco-Active coaching 6–7, 41, 204, 205, 226; audit 232–233; beneficial social and ecological value assessment **242**; climate conscious coach domains 207–210, *208;* coach training 220–221; Collett's experiences of political activism 221–224; complexity of values 229–232; conversational process 217–218; ecological ethics 221; ecological purpose of 220; exercise 219–221, 224–225; Global Code of Ethics 214–217; Hawken's practical questions 242–244, *243;* Joint Global Statement on Climate Change 211–213; McLean's experience on adaptive leadership 226–229; Miller exploring Earth-centred values 234–236; need of resources and suppliers 239–241; questionnaire 238; SDGs, discussions about 236–238; SMART, discussions about 239; vision and values as indicators of success 232
eco-anxiety 99, 100
Eco-Awakening 4–5; Barbes's story 29–30; Cohen's story 26–27; Downey's story 21–23; Drake's story 23–24; exercise 16–17, 21, 33; Gillespie's story 17–18; Hall's story 18–19; Hawkins's story 30–33; McLean's story 19; Turner's story 24–26; Whybrow's story 27–29; *see also* Awakening in coaching profession
Eco-Aware coaching 5–6, 91, 111; Degroot's experience eco-awareness

122–123; Ecological Awareness Cycle
96–106; emotional, cognitive and
spiritual awareness development 90–91;
exercise 116–119; Exploring Difficulty
practice 106; Hawkins' emotional
reactions and responses 93–96;
Uchendu's exploration of power and
privilege in Africa 102–103; Hutchins'
experience of Nature Immersions for
leaders and coaches 127–129; inner
practices 106; Jayne's experience of
Earth Connection 124–125; learning to
think eco-systemically 111–113; Loving
Kindness Meditation 107–108; reflection
113–116, 130–131; Sanchez's experience
of eco-spiritual awareness 121–122;
Seto's experience of eco-awareness
123–124; sustainability 118, Tedoldi's
coaching for biophilic activism 125–127;
working through emotional reactions
and responses 92–93
ecocide 84–85
Eco-Engaged coach development and
training 187; awareness of joint
professional body statement **193**; Bailey's
experiences in 196–200; current training
191; future training **193**; joint
professional statement with students **194**;
MCI 201–202; reflection 187–188;
responses from coaches 202–204;
Tedoldi's experiences in 195–196;
Turner's experiences in 188–194; *see also*
Eco-Engaged supervision and training
Eco-Engaged coaching 6, 153, 168–169;
Howard-Vyse's Active Hope approach
157–160; climate and ecology integrated
in practice 138–151; contracting process
in 138–139; exercises 168–169; Gift
Based Coaching experiment 157;
Gorham's 3 Cs framework 153–155;
Great Turning concept 157; Hopkins
and Cohen's conversation on climate
change 164–168; Monro coaching
experiences 155–158; questions and
thoughts in 135–138; Shetty's coaching
experiences 152–153; Snorf's views on
climate change 162–164; Wood's act of
observing 160–161; Yunkaporta's
experiences 162
Eco-Engaged supervision and training 6,
170; changing supervisory questions **176**;
experiments in 175; nature metaphors
178–180; Neil's Wheel 176–178, **177**; role
in moving from ego to eco coaching
171–174; Seven-eyed model 180–181,
180; *Tortuga Voice, The* 181, 182–186;
see also Eco-Engaged coach
development and training
eco-grief 100
Eco-Informed coaching 5, 51; Charmaine
about coach for social change 80–82;
climate change impact in Philippines
67–68; collective building of climate
issues 54–56; communication and
collaboration in eco-systems 69–71;
consciousness of natural non-human
living systems 71; *Council of All Beings,
The* 86–89; DTW 73–77; Earth systems'
tipping points 63; exercise 54, 57;
human-made mass *vs.* living biomass
64–65, *64*; knowing within ourselves 72;
Mehta's views about ecocide 84–85; need
for managed degrowth of economy
65–66; quantitative planetary boundaries
of Earth 58–59, *59*; responsibility for
excess emissions 61–62, *61, 62*; scientific
modelling about climate crisis 56–57,
62–63; Sikri's views about climate
violence 82–83; social tipping points 66;
Suppiah's views about Anthropocene
85–86; Symons addressing truth about
social inequality 78–80
Ecological Awareness Cycle 96, *96*;
addressing pain, grief and trauma
99–101; anger and blame 101–103;
dealing with guilt and shame 103;
moving through denial 97–99; taking
responsibility 103–106
ecological crisis 6, 55, 202–203, 236
ecological ethics 221
ecology 203, 265; centrality of 136–137;
deep 75, 264–265; integrated in practice
138–151; kincentric 267; social 265;
spiritual 31
Eco-Phase Cycle 4, *4*, 7, 26, 212, 246;
Eco-Active coaching 6–7;
Eco-Awakening 4–5; Eco-Aware
coaching 5–6; Eco-Engaged coaching 6;
Eco-Engaged supervision and training 6;
Eco-Informed coaching 5
eco-psychology 195–196
eco-somatic coaching 195
eco-spiritual awareness 120 *see also*
Eco-Aware coaching

eco-systemic awareness 91, 111, 117, 128, 156, 265
eco-systemic literacy 170, 265–266
ego-coaching 253–254 *see also* coaching: modern executive
egoism 254
Eight Core Competencies Framework 195
Eight Natural Archetypes 196
Einzig, Hetty 38
Elder, Mohican 249
Elhacham, Emily *64*
Elkington, John 141
emergence 118, 130, 172; of Asia 216; of awareness of climate and ecological crisis 135; of living system 270
emotional intelligence 195
emotions 234; *see also* ecological awareness cycle 96–105
Eng, Christopher 278–280
England Economic Heathlands Draft Transport Strategy 37
environmental revolution 226
Ershad approach to coaching 215, 217–218
ethics 210, 215; African concepts of 217; in coaching 216; ecological 221; ethical investments 241; of interconnectedness 267; spiritual model influencing 217; see also Global Code of Ethics; *see also* social change ; *see also* violence
European Association for Supervision and Coaching (EASC) 193, 211, 212
European Mentoring and Coaching Council (EMCC) 193, 211, 212, 215
Exploring Difficulty practice 106
Extinction Rebellion (XR) 18–19, 27
extractive capitalism 57; *see also* capitalism

face management 216
fertile soil 55
First Nations – 38, 122, 235
Flanagan, Lilith Joanna 69–71
Fleming, Pat 87
Flowers, Betty 31
Forde, Cindy 104
Forest Coaching®, 195, 196
fossil fuel use 55
Fourier, Joseph 266
Freeman, Meredith 69
Fuller, Buckminster 141
future-back inquiry 150
future forward inquiry 150

Future of Coaching Collaboration (FCC) 214

Galilei, Galileo 267
General Systems Theory project 268
generative listening 147
Gen Z 86, 105
Gestalt Organisation and System Development approach 201
Gift Based Coaching experiment 157
Gillespie, Ed 44, 53
Gillespie, Sally 11, 17–18, 99, 147
global change 91
Global Code of Ethics 12, 79, 212, 214–217
Global North 98, 109, 265, 266
global sea level, rise of 55
Global South 109, 266
Global Supervisors' Network 213
global warming 28 *see also* climate change
Goodall, Jane 205
Goodchild, Melanie 248
Gorham, Catherine 153–155
Graves, Clare 128
Gray, Heather-Jane 276–277
Great Turning concept 11, 157
green economic growth 65
greenhouse effect 266
Greenland melting 63
greenwashing 53, 266–267
grief 99–101, 106
Gross Domestic Product (GDP) 65
guilt 103, 106
Guterres, António 56, 147

Hall, Liz 18–19
Hamalainen, R. P. 229
Harding, Stephan 73, 74
Hawken, Paul 242; Awakening in Coaching profession 40–41; practical questions 242–244, *243*
Hawkins, Peter 1, 2, 4, 25, 28, 29, 38, 113, 118, 129–130, 147, 172, 180, 188, 194, 199, 221, 263; addressing pain, grief and trauma 99; about anger and blame 101–102; Awakening in Coaching profession 46–48; about centrality of ecology 136–137; about coaching mindsets from training 188–189; about eco-anxiety and eco-grief 100; about eco-centric awareness 150–151; on Eco-Engaged coaching 148–150; Eco-Phase Cycle model 212; about

304 Index

Ecological Awareness Cycle 97; about ego-coaching 253–254; emotional reactions and responses 93–96; exploring theme of indigenous orphans 247–248; about guilt and shame 103; about interdependency of living systems 114–115; key steps in shifting paradigm 115–116; Nasrudin stories 114; about resources and suppliers 239–241; SDG goals 236–238; Seven-eyed model 180–181, 180; SMART 230–231; about spiritual listening 148; story of Eco-Awakening 30–33; about taking responsibility 103–105; views on Opening the Seven Levels 246–247; also see systemic coaching
Heffernan, Margaret 25, 50, 82, 97, 188
Heifetz, Ronald 226
Henderson, Gosia 172, 173, 182, 185
Heraclitus 268
hieroglyphic stairway (poem by Dellinger)
Hillman, James 47, 226
Hillman, Jim 47, 94, 226
Hogan, Clover 104
holarchy 112–113, 112
Holding Space: A Guide to Organising and Facilitating a Deep Time Walk 74
holons 112
Homo sapiens 11, 85, 264; propensity 41–42; survival of 41
Hoop of Life community 146
Hopkins, Rob 164–168
Horn, Jen 67–68
Howard-Vyse, Alice 157–160
Huet, Muriel 74
human climate niche 62
humaneness 217
humanity 1
Hutchins, Giles 127–129, 130
hymn to the sacred body of the universe (poem by Dellinger) 131–133

imagination 164–167, 227, 246; see also crisis of imagination
'Immunity to Change' 164
indigenous coach training 196 – 200
indigenous wisdom – traditions, connection, coaching and learning from 80, 82, 91, 121–124, 146, 196–200, 206, 227, 229, 247, 248–49, 267 definition, 249
Industrial Revolution 55

injustice, racial 79
injustice, social 4, 68, 79, 80–81, 243, 250, 267
Institute of Coaching (IOC) 193, 211, 212
Integrative Development 24
Integrative Psychotherapy (Hawkins and Ryde) 93, 94
interbeing 6, 18, 40, 91, 111, 157, 175, 247, 255, 267
Intergovernmental Panel on Climate Change (IPCC) 147
integral theory, 128, 163, 216 ; see also Snorf, Kevin; see also Yunkaporta, Tyson, 162 ; see also Wilber, Ken
International Association of Coaching 193, 211
International Coaching Federation (ICF) 19, 41,190, 191, 195, 197; 211, 212, Code of Ethics 215, 221; shift in requirements 190; 255, see also Eight Core Coaching Competencies
International Coaching Federation Australia (ICFA) 19, 20
International Criminal Court (ICC) 84
International Society for Coaching Psychology 193, 211, 212
Ireland 240–241
Ison, Ray 268

Jaworski, Jo 31
Jayne, Tabitha 124–125
Joanes, Tina 41–42
Johnson, Chris 12
Johnstone, Chris 3, 97, 158–160, 262
Joint Global Statement on Climate Change 12, 41, 211–213, 214

Kahane, Adam 31
Kalyan, Srivi 178
Kanelidou, Katerina 172, 173, 177, 185
karakias protocols 197–200
Kashtan, Miki 78
Kaufer, Katrin 13
Kegan, Bob 106
Kegan, Robert 164
Khan, Inayat 31
Khan, Reshma 252–253
Khomsi, Kenza 172, 182, 183
Kiehl, Jeff 99, 147
Kimmerer, Robin Wall 72, 91, 248
Kincentric Ecology (Salmon) 80
kincentric ecology 174, 267

King Jr, Martin Luther 111
King, Martin Luther 267
Klein, Naomi 28, 164
Kline, Nancy 147, 148
Kolbert, Elizabeth 55
Korten, David 157
Krznaric, Roman 72, 205
Kumar, Satish 2–3

Laloux, Frederic 128
Lane, David 35–37, 215
Lao Tzu 128
Lead Climate Negotiator for Ghana 201
learning integration 245; Hawkins' about
    ego-coaching 253–254; Hawkins' views
    on *Opening the Seven Levels* 246–247;
    Khan's coaching experiences of
    eco-systems 252–253; McLean's pathway
    from ego to eco coaching 254–255;
    Opio's thoughts about climate change
    251–252; Whybrow's thoughts on
    indigenous wisdom traditions 248–250
Lécuyer, Roselyne 73–74, 273–274
Lenton, Timothy 63
Lewis, Jeremy J. 280
'Liege-Foodbelt, The' project 165
limiting mindsets 110, 187 – 190; *see also*
    Eco-Aware – Shifting our Thinking
Limits to Growth report 62–63
listening 147; to earth 69; generative 147;
    spiritual 148; *see also* Eco-Informed
    coaching
living organism 3
living systems: of Earth 10–11;
    interdependency of 114–115; principles
    172–173, 175; sustainability of 117–118
Logic of Life 128
Loving Kindness Meditation (LKM)
    107–108
Luisi, Pier Luigi 97

Macy, Joanna 34, 54, 87, 93, 97, 157,
    159–160, 191, 262
Manos, Jarid 53, 91, 174
Māori karakia 197–198
Margulis, Lyn 264
Maruska, Don 278–280
Maryl, Jenny 245
Mathews, Freya 151, 264
McLean, Josie 1, 2, 19, 111, 199, 226, 228,
    254, 260, 263; Awakening in Coaching
    profession 43–46; about climate crisis

56–57; about CCA 213–214; concern
    about climate change 115; about
    holarchy 112–114; about impact of
    climate and ecological destruction 60;
    indicators of success 231–232; about
    moving through denial 97; about need
    for managed degrowth of economy 66;
    pathway from ego to eco coaching
    254–255; about quantitative planetary
    boundaries of Earth 58–59, *59;* research
    experience on adaptive leadership
    226–229; story of Eco-Awakening 19;
    about strange attractors concept 63–64;
    about sustainability of living system
    117–119; values embedded in story
    228–229; views of DTW 74–77; vision
    and values audit 232–234
McMordie, Mark 19, 96, 106, 107–108
McPhee, John 72
Meadows, Donella 226, 227, 232
Mean Annual Temperature 62
Megginson, David 140
Mehta, Jojo 84–85
mentoring 41
Mentoring and Coaching Initiative (MCI)
    201–202
meta-awareness 106
Metzler, Phil 278–280
Millennials 86, 105
Millennium Development Goals 267
Miller, Andy 234–236, 278–280
Millward-Hopkins, Joel 65
mindfulness 18–19, 39, 106–108, 154,
    195–196, 198, 247–248; *see also* Loving
    Kindness Meditation
Mindfulness Based Stress Reduction
    (MBSR) 106
Mindful Self Compassion 107
Miranda, Eduardo, R. 70
modern executive coaching 47
Monro, Heather 155–158
Montañez, Amanda *64*
Morosi, Andra 273–274
Mother Earth's Dilemma poem (by Walter)
    51–52
Mrenica, Janet 172, 173–174, 182–184,
    272–273, 278–280
Murdoch, Edna 191

Naess, Arne 73, 264
NASA 55, 78
National Geographc Society 264

nature 14, 31, 46, 86, 280; Coaching 195, 196; metaphors 178; nurtures 153; as place and source of physical and mental healing 94
negation 99
Neil's Wheel 141, *142*, 156, 176–178, *177;* beta-testers of 145; examples of 143, *144*; Five Freedoms and Four Mantras 143; neutral in politics and persuasion 142–143; online tool for coaching community 141; 'Wheel of Life' 142
Nelson, Portia 109
Nelson, Winfred 201
neo-cortex cognitive awakening 31
neoliberalism 78
Newtonian Atomistic thinking 6, 90, 254, 267
Newton, Isaac 267
Nine Domains of a Climate Conscious Coach 207–11, *208*
Noiles, Lennie 278–280
non-renewable geo-resource 55
North–South divide 266
Nxumalo, Musa 172, 173, 185

Obama, Barack 153
Okyere-Nyako, Akua Amoa 201
Omertà of Consultancy 44
online eco-somatic training 195
ontopoetics 151
Opening the Seven Levels 246–247
Opio, Denis 251–252
organic pollutants, man-made persistent 59–60
organisational purpose, idea of 37
Orr, David 265
Oshry, Barry 31
Oswald, Yannick 65
'outside-in' inquiry 150
Oxfam 61

Pachamama Alliance 28
pain 99–101
Palmer, Stephen 12, 28, 191
Parenting in a Changing Climate (Bechard) 207
Paris Agreement 60
pathological violence 53
Perkins, John 46
personal integrity 40
Petero, Rachel 197, 199, 200

poem(s): clear view, A by Borst 120–121; hymn to the sacred body of the universe by Dellinger 131–133; Mother Earth's Dilemma by Walter 51–52; Beyond the darkness, a new world comes by Maryl 245; Wheel, The by Scotton 134–135
political engagement 209, 222, 224
Polman, Paul 205, 207, 238, 239
Porges, Stephen 107
Potsdam Institute for Climate Impact Research (PIK) 77
priming 235
productive land 55
profound listening 68, 69
Promise That Changes Everything, The – I Won't Interrupt You (Kline) 147

Qatsi film trilogy 235
questions/questioning 149; changing supervisory *176*; of coaching relationship 139–141; in Eco-Engaged coaching 135–138; Hawken's practical questions 242–244, *243*; 'what if' question 164–168
Quigley, Marie 199

Rao, Narasimha D. 65
Ravindra, Ravi 148, 254
Raworth, Kate 38
Reason, Peter 31
reductionist approach 236, 267, 268
Reed, Bill 38
regeneration 128, 196, 226; regenerative coaching 39–40, 241; regenerative practice, roots of 39–40 ; *see also* Alison's story 27–28
Regeneration, Age of 128
Regenerative Leadership 127–128
Regenerative Leadership Consciousness 128
Renewal Associates 32
research 17, 20, 22, 61, 65, 68, 69–71, 77–78, 106, 107, 124, 139, 164, 165, 266, climate crisis and coach training 190–194; Hawkins, Turner and Roell 13–15, 135–136; Joint Statement research 212; McLean research 227–208; Roche research 80–81
resourcing ourselves 256–263
Revans, Reg 118
Reweaving Our Human Fabric (Kashtan) 78
Rilke, R. M. 165

Roberts, David 66
Roche, Charmaine 80–82
Rockström, Johan 77, 78
Roell, Christiaan, 13, 135
Ross, Kubler 93
Rumi, Mevlana 16, 28
Ryde, Judy 93, 103, 110

Saarinen, Esa. 229
Sahtouris, Elisabet 111, 255
Salmon, Enrique 80, 267
Sanchez, Anita 248–250; about
    eco-spiritual awareness
    121–122;'Andrea', 146, experiences
    of connection with ecology 145–146;
    life-giving connection to others 262
Scharmer, Otto 13, 31, 92
Schwenk, Gil 180
Scotton, Neil 19, 134; examples of
    Neil's Wheel 144; Wheel, The
    134–135, 142
SDGs – see Sustainable Development
    Goals
Seed, John 54
self-awareness 208
Self Compassion Break 107
Semedo, Maria-Helena 55
Senge, Peter 31, 228
Separation, Age of 128
Separation, Story of 156, 157
Seto, Lily 122–124
Seven-eyed model 180–181, 180
shadow work 136, 155 – 157, 160–161
shame 103, 106
Shetty, Rashmi 152–153, 270–272
Sikri, Kanishka 82–83
Silent Spring (Carson) 64
slavery, legacy of 79
slime mould 70–71
Snorf, Kevin 162–164
social change, coaching and supervision for
    80–83, 81
social ecology 265
social inequality 78–80
Solastalgia 95
somatic coaching 195
Sovereign Wealth Fund 241
species and biodiversity, loss of 55
Specific, Measurable, Achievable, Realistic
    and Timely (SMART) 230–231, 239
Spinoza, Baruch 267
spiritual ecology 31

spiritual listening 148
stages of adult development, 106, 163 see
    also shadow work
Stansfield, Nigel 12
Staton, Tamara 278–280
Steinberger, Julia K. 65
Stop Ecocide Foundation 85
Stop Ecocide International 84–85
Stout-Rostron, Sunny 217, 218
strange attractors concept 63
supervision 41, 49, 81, 154, 170–186, 259
Suppiah, Sam 85–86
Sussfeld, Jerome 74
sustainability 196, 220, 226
sustainable development 36
Sustainable Development Goals (SDGs)
    79, 236–238, 267
Symons, Rita 211; addressing truth about
    social inequality 78–80; about Joint
    Global Statement on Climate Change
    211–213
systemic awareness 30, 90, 111, 265, 268,
systemic change 31, 159, 270, 280
Systemic Coaching (Hawkins and Turner)
    12, 13, 25, 47, 48, 93, 97, 139, 174, 179,
    188, 208, 268
systemic supervision frameworks 172
Systemic Team Coaching 32
systems thinking 6, 31, 44, 111, 196,
    228, 268

Teal-Evolutionary 128
Tedoldi, Diana 125–126, 192,
    194–196
theory U 92, 93, 106, 135, 147 see also
    Scharmer, Otto
Thich Nhat Hanh 18, 91, 111, 267
This changes everything (Klein) 28
Thunberg, Greta 5, 18, 53
tipping points 57, 63, 66, 78, 115, see also
    Eco-informed coaching
Tomasdottir, Halle 223, 224
Torbert, Bill 31
Tortuga Voice, The 181, 182–186
training 3, 25, 32, 49, 69, 81, 97, 107,
    124, 167, 168, 170, 183, 187–204, 215,
    220–21
Transition Town Network, 164
trauma 99–101
triple bottom line approach 142
Turner, Eve 1, 2, 19, 109, 182, 199, 261,
    262–263; Awakening in Coaching

profession 48; about CCA 213–214;
about *Council of All Beings, The* 86–89;
about contracting 139–141; about
Eco-Active coaching 206; about
Eco-Engaged coach development and
training 188–194; about listening
147–148; on nature metaphors 178–180;
story of Eco-Awakening 24–26;
supervision role in eco coaching
172–184, ; about *Tortuga Voice,
The* 182; seven-generation perspective
206–207; Joint Global Statement on
climate change, 214
Turner, Tammy 199
Typhoon Haiyan 67
Typhoon Odette *see* Typhoon Rai
Typhoon Rai 67

Ubuntu 91, 215, 217
Uchendu, Jennifer 102–103
Ulrich, Stephan 172, 182, 184–186
UN Conference of Environment 36
UN Inter-Governmental Panel on Climate
Change (IPCC) 56
Usui, Aya 274

Valdivielso-Martínez, Elsa 277–278
values: audit 232–244, **233**; beneficial social
and ecological value assessment **242**;
complexity of 229–232; as indicators
of success 232; reprioritizing 230; social
and ecological assessment 241–3,
Earth-centered 234–235
Van Collier, Salome 194
Van Nieuwerburgh, C. 217
Vaughan-Lee, Llewellyn 33
violence as entry point to climate action
82–83; *see also* Eco-Informed
Caoching

vision: audit 232–244, **233**; as cradle
for experiments *231;* as indicators of
success 232
Viswanathan, Vaishnavi 172, 179, 182

Walter, Gillian 51–52
Ware, Bronnie 233
war-time style mobilisation 65
WeGo 222
Weintrobe, S. 99
Well-being of Future Generations Act 223
Wells, Sam 227
whakapapa 200
'what if' question 164–168
'Wheel of Life' 142
Wheel, The (poem by Scotton) 134
White Western Privilege 187, 268
Whitmore, Sir John 11, 19, 20, 35, 38
Whybrow, Alison 1, 2, 19, 186, 199, 213,
241, 262; Awakening in Coaching
profession 37–40; narratives of denial
97–98; and Neil's Wheel 176–177, *177;*
about regeneration 242–243; story of
Eco-Awakening 27–29; supervision role
in eco coaching 172–185, 175; thoughts
on indigenous wisdom traditions
248–250
Wilber, Ken 128, 163
Wilful Blindness (Heffernan) 25, 82, 97, 188
Winston, Andrew 207
With the Earth in Mind tool 145
Witzany, Günther 71
Woodford, Robert 74, 77
Wood, John 160–161
Woodland Trust 32
World Health Organisation (WHO) 100

'younger brothers of Creation' 71–72, 248
Yunkaporta, Tyson 162